Focus on Computer Graphics

Tutorials and Perspectives in Computer Graphics

Edited by W.T. Hewitt, R. Gnatz, and W. Hansmann

P. Brunet F. W. Jansen (Eds.)

Photorealistic Rendering in Computer Graphics

Proceedings of the Second Eurographics Workshop on Rendering

With 175 Figures, 24 in Colour

Springer-Verlag
Berlin Heidelberg GmbH

Focus on Computer Graphics

Edited by W. T. Hewitt, R. Gnatz, and W. Hansmann
for EUROGRAPHICS –
The European Association for Computer Graphics
P. O. Box 16, CH-1288 Aire-la-Ville, Switzerland

Volume Editors

P. Brunet

Dpt. Llenguatges i Sistemes Informàtics, U.P.C
Avda. Diagonal, 647 planta 8
E-08028 Barcelona, Spain

F. W. Jansen

TU Delft, TWI
Julianalaan 132
2628 BL Delft, The Netherlands

ISBN 978-3-642-63416-1 ISBN 978-3-642-57963-9 (eBook)
DOI 10.1007/978-3-642-57963-9

CIP data applied for

Cover: Konzept & Design Künkel, Lopka GmbH, Ilvesheim, FRG
Typesetting: Camera ready copy by authors/editors
45/3140 – 5 4 3 2 1 0 – Printed on acid-free paper

Preface

Over the last decade enormous advances have been made in photo-realistic rendering of objects, interiors, buildings and general scenes, both in the accurate modeling of the local and global reflection of light, as well as in the complexity of scenes that can be displayed. In addition, faster graphics hardware and parallel processing have reduced picture generation times that once were counted in hours to minutes and seconds. Photo-realistic rendering has now become an important simulation tool in lighting design, interior design and architecture.

Many of the recent advances are reflected in the twenty-five papers collected in this book. Original versions of the papers were presented at the second Eurographics Workshop on Rendering, held in May 1991 at the University of Catalunya in Barcelona, Spain. The workshop brought together 47 researchers from all the leading research groups in the US and Europe. The papers present new contributions both to improve the rendering quality as well as to improve the efficiency and speed of the rendering process.

A major topic in photorealistic rendering concerns the local and global reflection of light. Conventional ray tracing techniques calculate the shading of objects by determining only the reflection of light that is directly received from light sources. Indirectly reflected light between objects in a scene cannot be accounted for with ray tracing unless expensive stochastic sampling techniques are applied. Radiosity algorithms take a completely different approach in that the global illumination in a scene is calculated by determining the light exchange between objects in a scene. This also accounts for the indirect diffuse reflection. However, radiosity algorithms do not account for specular reflections and shadow accuracy is in general low. Combinations of ray tracing and radiosity, also known as two-pass methods, provide a solution that offers the best of both worlds. Recently a large number of new meshing and stochastic sampling techniques have been developed to further enhance these two-pass methods. The papers collected in this volume give a good introduction to these new developments.

The first paper, 'The State of the Art in Physically-based Rendering and its Impact on Future Applications' by *Sillion,* discusses some of the recent developments, i.e., incremental radiosity calculations for objects moving in a scene, improved general reflectance distributions for bidirectional diffuse and specular reflection, and hierarchical algorithms to handle complex scenes.

The paper 'Adaptive Shadow Testing for Ray Tracing', by *Ward,* is the first of a number of papers on stochastic sampling techniques. A method is presented that samples the light of the most important light sources and estimates the received energy of the other light sources by applying the same probability factor found for the important light sources to these light sources also. In 'An Adaptive Sampling Technique for Multidimensional Integration by Ray-Tracing' by *Schlick*, an adaptive stochastic ray tracing method is presented that is based on an N-rooks sampling technique that distributes samples pseudo-randomly in image space while still allowing easy reconstruction of pixel values. *Lange,* in her paper 'The Simulation of Radiant Light Transfer with Stochastic Ray-Tracing', presents techniques and weighting factors for diffuse and specular reflection for use in Kajiya's path tracing algorithm. *Kirk and Arvo* discuss sampling strategies in 'Unbiased Variance Reduction for Global Illumination' and present a method that uses conic shafts oriented towards the light sources to localize those directions where the sampling density should be increased. In 'Direct Lighting Calculation by Monte Carlo Integration', *Shirley and Wang* propose a probability function for sampling light from multiple light sources simultaneously. In general, sampling efficiency can be increased by applying stochastic (importance) sampling or deterministic (structural, directional) sampling methods that direct the sampling effort to those areas where important contributions may be expected (i.e., the light sources). This can either be done by applying knowledge about the environment (position of light sources, etc.) or by adapting the sampling directions during the sampling process. *Drettakis and Fiume* in 'Structure-Directed Sampling, Reconstruction and Data Representation for Global Illumination' explore a structural approach. In the same style, *Kok and Jansen* in 'Source Selection for the Direct Lighting Computation in Global Illumination', use the information from a radiosity preprocessing to identify the most important light sources, to which in a second pass shadow rays are cast to improve the shadow accuracy of the patch radiosities.

Pure radiosity methods calculate the diffusely reflected light with a projective (hemi-cube or hemi-sphere) method or with ray tracing. Improvements are sought in a better discretization of the hemi-sphere and in new adaptive patch subdivision techniques to represent more accurately shading gradients. Also visibility preprocessing and spatial subdivision techniques are of interest. In 'The Constant Radiosity Assumption Syndrome', *Tampieri and Lischinski* propose to take into account the distribution of light over a shooting patch. In 'Some Experiments on Adaptive Subdivision in Progressive Radiosity', *Vedel and Puech* compare quadratic and cubic interpolation techniques to represent more accurately the radiosity shading over a patch. In 'Radiosity in Computer Graphics: A Proposed Alternative to the Hemi-cube Algorithm', *Gatenby and Hewitt* explore a triangular hemi-sphere discretization. In 'Physics-Based Lighting Models: Implementation Issues', *Tellier and Bouatouch* report on an efficiency improving method that uses a visibility graph. *Haines and Wallace* in 'Shaft Culling for Efficient Ray-Cast Radiosity' describe a shaft culling algorithm that reduces the number of ray object intersections for a bundle of rays shot from one patch to another by building a shaft volume between the two patches and culling those objects that do not intersect the shaft volume.

Multi-processing is an interesting option for radiosity and ray tracing. Rendering times can be reduced and large model data bases can be distributed over multiple processors. Model distribution can be based on spatial subdivision techniques or on load distribution considerations. Load distribution in its turn can be done by distributing the rays over the processors and passing model data between the processors, or by communicating rays between processors. With a progressive radiosity algorithm, additional information has to be broadcast to determine the patch with the highest unshot energy. In 'Progressive Refinement Radiosity on a Transputer Network', *Feda and Purgathofer* describe a multi-processor architecture with one master processor and two subsystems for radiosity preprocessing and rendering; the rendering improves continuously as the radiosity solution progresses. *Chalmers and Paddon,* in their paper 'Parallel Processing of Progressive Refinement Radiosity Methods', compare different network topologies, such as ring, hypercube, torus and a minimum path configuration; the latter seems to perform the best. In 'Two Parallel Approaches for a Progressive Radiosity', *Guitton et al.* compare a stochastic and a deterministic (each ray directed to a specific patch) shooting method. In 'An Extended Radiosity Using Parallel Ray-Traced Specular Transfers, *Jessel et al.* exploit a voxel space partitioning to distribute the model data over the processors; while *İşler et al.*, in 'Subdivision of 3D Space Based on the Graph Partitioning for Parallel Ray Tracing', convert the load distribution problem into a graph partitioning problem that is solved by a heuristic algorithm. Quadratic performances can be avoided by hierarchically subdividing the scene into cells that are separated by virtual walls. A global solution is then derived after several iterations from the local solutions. Experimental implementations of this method are described in 'Divide and Conquer Radiosity', by *van Liere,* and in 'On the Division of Environments by Virtual Walls for Radiosity Computation' by *Arnaldi et al.*

The last five papers deal with color and texture and describe applications of photorealistic rendering. In 'Texture Shaping: A Method for Modeling Arbitrarily-Shaped Volume Objects in Texture Space', *Sakas and Kernke* explore a modeling technique for texture-shaped objects that combines object shape and three-dimensional texture in the frequency domain and converts the result back to space domain. *Tastl and Purgathofer* report in 'Color Spaces and Human Color Perception' on experiments on what for human perception are uniformly distributed color spaces. *Anderson and Grant* present in 'Visulux: A Radiosity Based Lighting Design Tool' an application of radiosity for lighting design in buildings. *Brivio et al.*, in 'Synthesis of Multispectral Images of Natural Landscape', visualize natural terrain models obtained from satellite data. *Clavé and Gross* present results of city street lighting simulation in 'A Rendering Pipeline for Street Lighting Simulation'.

This volume brings together a large number of interesting papers on all the main issues of photorealistic rendering and gives a full picture of the current state of the art in rendering research. It will be of interest for researchers and students in computer graphics, and as well for designers who apply rendering techniques for realistic simulation in lighting design, interior design and architecture.

We would like to thank all the authors for providing carefully revised and elaborated versions of their presentations for this volume. We want to express our special thanks to Xavier Pueyo and his staff of the Universitat Politècnica de Catalunya for organizing the workshop and making it such a special event. We thank Arjan Kok for his enormous help with the editing and reformatting of the book.

Barcelona, Delft, February 1993 *Pere Brunet, Frederik W. Jansen*

Table of Contents

The State of the Art in Physically-based Rendering and its Impact on Future Applications

François Sillion

Abstract

The simulation of global illumination has become a major focus in computer graphics, as physically based and accurate computation techniques are developed. We review some of the most recent advances, such as incremental radiosity computations for changing environments, and the simulation of arbitrary reflectance distributions. A discussion of persistent problems in physically-based illumination simulation follows, with an emphasis on the design issues faced by application developers.

1 Possible applications for physically accurate simulations

In recent years, many new applications of sophisticated computer graphics have emerged, as new algorithms became available to perform physically meaningful simulations. Early computer graphics algorithms treated light and illumination as a necessary but otherwise unimportant step, as they focused on hidden surface elimination, or surface representation issues.

With the introduction of the radiosity method in 1984 [GTGB84], the problem of computing illumination in a scene began to be viewed as the simulation of a physical problem : that of light transfer and energy balance. Radiosity methods have quickly attracted a lot of attention, and have matured surprisingly quickly, to become a reference for high-quality rendering.

Why is radiosity needed ? or more generally, why is a faithful simulation of the physical behavior of light needed ? These questions were frequently asked a few years ago, when the most remarkable feature of radiosity programs was their running time. The introduction of radiosity occured at a time when robust and simple shading algorithms were being implemented in hardware, and when "high quality rendering" was defined as *ray tracing* with shadows and specular reflections. In contrast, the early radiosity algorithms were slow, not very robust in that they produced visible artifacts, and limited to diffuse reflectors. Their usefulness was therefore questionable.

In this context, the primary effect of these early developments on the field of computer graphics has been to trigger a growing awareness of the need for accuracy in rendering. Aside from the entertainment applications, most of the professional users of computer graphics actually need some kind of realism in their pictures :

Industrial design tasks are an obvious example, where designers want to experiment with very well defined lighting cases, and to obtain reproducible and meaningful images. In the auto industry, for example, special light sources are used to reveal any discontinuities in the shape of the surfaces, and they have to be modeled accurately. Some applications explicitly require quantitative accuracy, such as lighting design, when norms have to be met by the simulated light distribution. In architectural design applications, the simulation must also be in quantitative agreement with reality, if decisions are to be based on the resulting images. More generally, studies in human perception show that some visual cues are very important to help understand synthetic pictures, and that shadows constitute a very important aid in the comprehension of a 3D environment from a 2D picture [Wan91]. Similarly, global illumination effects can be expected to help in the understanding of an image.

To be of practical use in any of these fields, the radiosity method had to be extended in a number of directions, such as non-planar geometries, non-diffuse reflectors, or non-static environments. We will present some new developments and see that the frontier is still moving, as more and more attention is devoted to research on the subject.

However, despite all the advances that have been made since the early days, physically based rendering is not yet widely used. An effort is needed to integrate all the *state of the art* research techniques into versatile, usable systems. On the one hand, new illumination techniques and algorithms should be developed with the final user in mind, and the constraints imposed by a global design environment should be considered from the beginning. On the other hand, new graphics applications should be designed with the notion that physically based rendering will be applied somewhere in the process.

2 Recent developments in radiosity

Research on physically based rendering has been gaining momentum in the last few years, and many new ideas will likely be introduced at this workshop. Some innovative extensions to the basic radiosity method are presented below :

2.1 Incremental changes to a solution

Because of the global character of the illumination, radiosity methods were originally limited to static environment : any change in the scene geometry required a complete recalculation of the environment.

Incremental methods have been proposed [GSG90, Che90], where the effect of a change in the environment is computed in a progressive manner. In the case where a new object is added to a scene, a number of *form factors* can potentially be modified : those between any patch and a patch on the new object, and those between any two patches that become occluded by the new object.

Incremental redistribution methods extend the *progressive radiosity* approach by evaluating these form factor changes in an ordered fashion, treating first the "most important" ones. In effect, changes can be applied anytime in the course of a radiosity solution : the energy that was propagated before the change is simply *redistributed* according to the new geometry. For any patch that had "shot" a significant amount of energy in the direction of the new object, that energy is redistributed from the shadowed region towards the surface of the new object, while other portions of the scene are left unchanged (Figure 1).

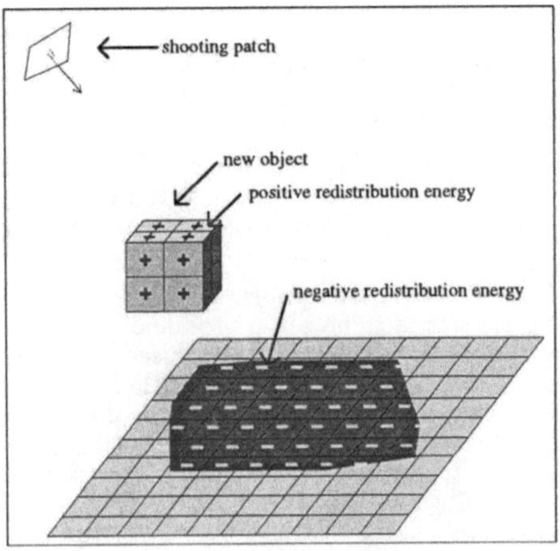

Fig 1. Redistributing positive and negative energy contributions

Because of the global nature of illumination, with multiple reflections accounting for a significant portion of light transfer, it is not sufficient to merely redistribute the energy in the scene between the new object and its shadow. A patch that now falls into the shadow of a new object could very well have reflected some of the energy it was receiving prior to the change. The redistributed energy must therefore be propagated in the scene, which is done using a regular progressive radiosity step.

Incremental methods can be thought of as a mere rewriting of the general energy balance equations, with a finer granularity that allows to isolate the most important terms so they can be computed first and very quickly (Figure 8, on color plates). These techniques represent a significant first step towards the *rendering while modeling* paradigm.

2.2 General reflectance distributions

One can look at the problem of simulating global illumination in terms of a tradeoff between storage space and time. At one end of the spectrum, radiosity methods store illumination information on the surfaces, so that it is available for later reuse. This is affordable for diffuse surfaces since only a scalar radiosity value need be stored. Since the illumination information is present on the surfaces, the solution obtained is *view-independent* and can be used to render images from different viewpoints.

At the other extreme is conventional ray tracing, where illumination is computed on the fly at each intersection of a ray with an object, and is never stored. No storage space is therefore required, at the expense of recomputation time. What makes this cost tolerable is the small number of rays to trace if only ideal specular reflectors are used (Figure 2).

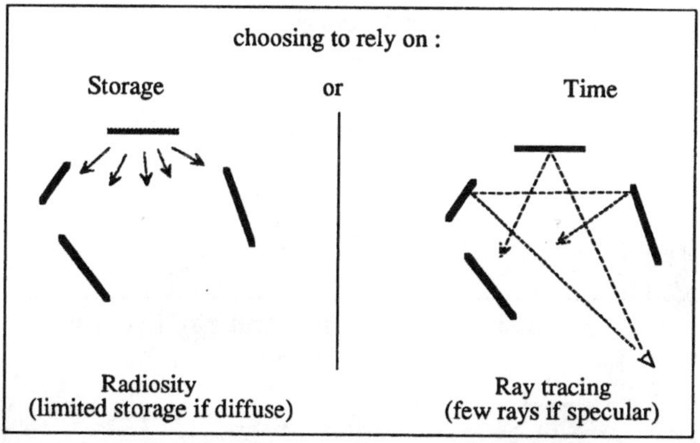

Fig 2. Basic strategies for global illumination simulation

Computer graphics traditionally considers reflection as a mixture of ideal diffuse and ideal specular behaviors, and several extensions of radiosity and ray tracing have been proposed to allow the combination of both types of reflectors [WCG87, SP89, Hec90].

As noted by Ward *et al* [WRC88], the slowly varying diffuse term is too costly to recompute at each pixel, and should be stored instead. *Two-pass* methods combining radiosity and ray tracing do exactly that, by considering the impact of specular reflectors on a view-independent diffuse distribution of light (Figure 3).

However, most real materials cannot be closely approximated by a superposition of ideal diffuse and ideal specular reflectors, but exhibit instead a *directional diffuse* character [HTSG91]. Figure 4 shows a classification of reflectance types in the general case.

A more general algorithm is therefore needed to accurately simulate global illumination. Direct extensions of radiosity and ray tracing have attempted to maintain the underlying choice — storing much more information for the directional radiosity method [ICG86] and tracing many more rays in the case of distribution ray tracing [CPC84] and path tracing [Kaj86] (Figure 5). Despite the continued growth in computing power we have experienced in recent years, these brute force solutions are nowhere near being practical.

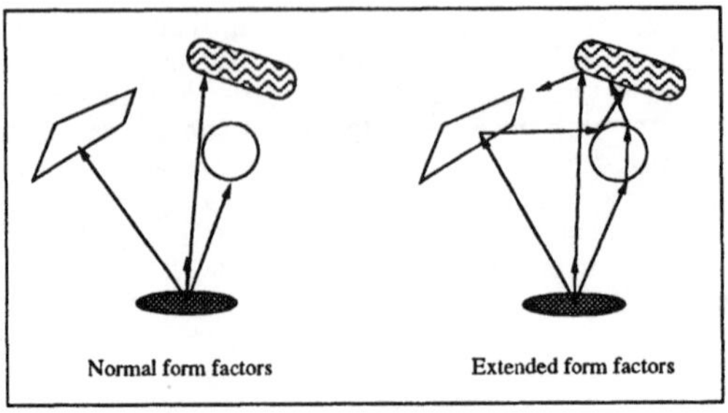

Fig 3. Two-pass methods capture the effect of specular reflection on the diffuse distribution of light in the first pass using *extended form factors*, and render view-dependent specular effects in the second pass

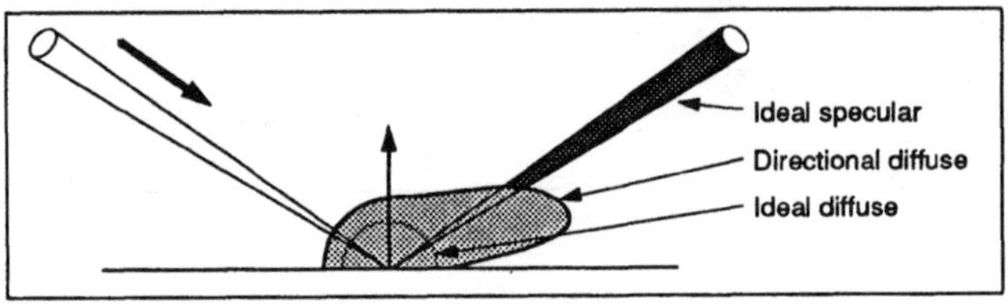

Fig 4. Different components of a general BRDF

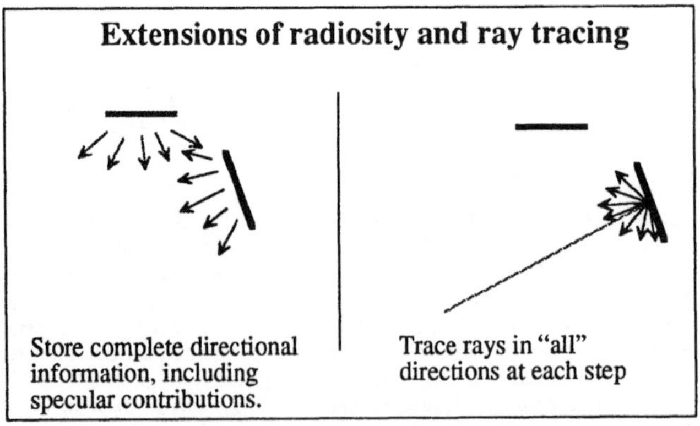

Fig 5. Basic strategies become very expensive in the general case

Storage issues

If one considers the information to be stored across the surfaces of a scene, it takes the form of a directional distribution of intensity leaving the surface. In informal terms, it is easy to store a low-pass filtered version of this distribution, which could be compactly represented, so as to avoid having to sample it with many reflection rays every time the surface is examined. However, very high-frequency, rapidly varying information is more costly to store, and can instead be reevaluated many times, because it is strongly directional and therefore requires very few rays. Applying these simple ideas results in a common-sense definition for a general global illumination simulator [LSS90].

We will present briefly our general system for the accurate simulation of arbitrary reflectance functions [SAWG91]. Using the classification of reflectance types introduced in Figure 4, this system can be described by the following characteristics :

- It is based on radiosity, so that progressive or incremental solution techniques can be used.
- It treats ideal specular reflection separately to eliminate storage of high frequency information.
- It maintains a *directional intensity distribution* on the surfaces, resulting from the smooth *directional diffuse* component of the reflectance function.

As in previous two-pass approaches, the first pass computes and stores the directional intensity distributions, including the effect of ideal specular reflections, which is computed on the fly and not stored. The second, view-dependent pass performs a new evaluation of ideal specular reflection, resulting in a complete simulation of all posible light transfers.

The simple conceptual difference that consists in replacing a scalar radiosity value by an intensity distribution introduces numerous practical problems. One of them is of course the issue of the representation of a directional distribution : previous attempts to discretize the space of directions have shown that aliasing is very severe for all practical resolutions. It is therefore preferable to keep a continuous description of the distributions.

It is easily shown that the intensity distribution resulting from the reflection of incident light by a surface depends solely on the local bidirectional reflectance distribution function (BRDF) of the material. Figure 6 shows how a continuous function — describing the BRDF as a function of outgoing angle, for the given incident angle — must simply be scaled and rotated to obtain the reflected intensity distribution.

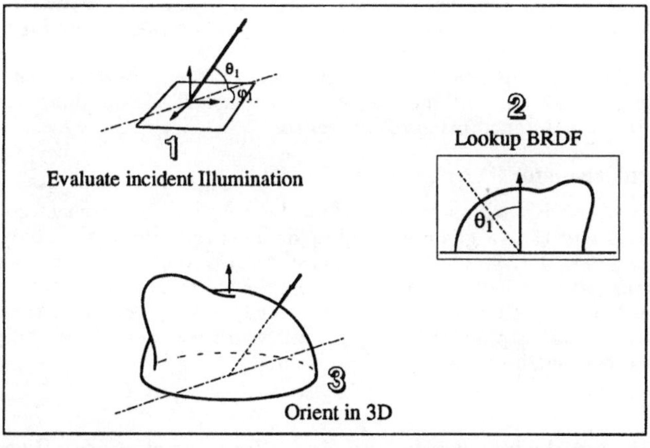

Fig 6. Using a stored representation of the BRDF to obtain the distribution of reflected light

Several options are available to encode the continuous distributions : spline surfaces could be used, either with a fixed mesh of control points, or using hierarchical refinement to approximate more closely the actual BRDFs. However, fixed meshes would result in poor approximation of highly directional BRDFs, while variable meshes would generate an increase in storage space as the solution progresses. This is because different contributions, originating from different direction, must be added together to represent the total intensity reflected at a given point (Figure 7). Harmonic decompositions can be used instead, where more terms are used to represent more directional BRDFs (because of their higher frequency). An implementation based on spherical harmonics has been used to render Figure 9 (on color plates).

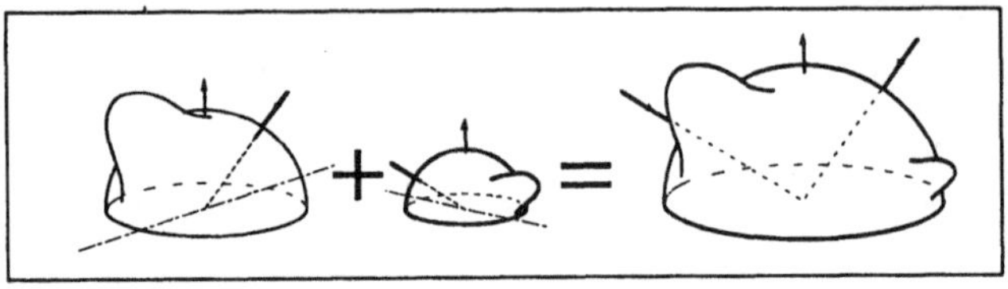

Fig 7. Intensity distributions must be added together as the progressive solution progresses

2.3 Hierarchical algorithms

The complexity of the global illumination problem is such that most solutions presented to date could only deal with fairly simple environments. In effect, because of the global coupling that can potentially occur between any two objects in a scene, the resources needed to simulate the illumination between N objects grow as N^2 with simple radiosity techniques.

However, as the environment becomes more and more complex, the argument can be made that the influence of a particular object or radiating patch will not extend to the entire scene, but will instead be limited to some portion of it. Several researchers have sought to reduce the complexity of the simulation, generally by adopting some kind of hierarchical description, where different levels of detail can be used as needed.

Partitioning the environment

One natural idea that comes up when trying to reduce the complexity of the global illumination problem is to partition the scene in several sub-environments. A global solution can be computed inside each sub-environment, provided that exact boundary conditions are specified. A possible algorithm is then to iteratively refine these boundary conditions as all sub-solutions are progressing [NN89, FFOC90]

The treatment of the boundaries is a very difficult problem, however, as continuous functions must be accurately maintained across the surface of the boundaries, describing at each point a directional distribution of intensities.

Hierarchical light transfer

Another approach starts by acknowledging that the solution will always be computed with a certain error, and that a given level of error is acceptable in the computation as long as it remains under control. The subdivision of large patches can therefore be driven by the energy exchange that actually occurs with other objects : for distant objects, no subdivision is needed as the form factor is necessarily small, but for closer objects a hierarchical subdivision takes place [HS90]. This allows all transfers to be computed with the same accuracy, without unnecessary subdivision.

Levels of detail

Yet another use of hierarchy is in speeding up display for interactive uses. By maintaining different levels of detail in the definition of the scene and its illumination, Airey *et al* [ARFPB90] allow the display of interactive walkthroughs in fairly complex environments : a simpler solution is used to display frames very quickly while the user is moving through the scene, and the complete subdivided description is used to display static images.

3 New problems introduced

We have seen that the general problem of physically accurate rendering is being approached from many different directions. While significant progress has been made in each of the particular areas investigated, almost all of these approaches have also revealed or emphasized some new problems.

In other words, as algorithms are developed to overcome some limitations of the original radiosity approach, some more subtle issues come to light. We review here some of these issues and discuss their influence on the practicality of physically based illumination simulation.

3.1 Controlling the error

All global illumination algorithms represent a compromise, where an approximation to the "true" distribution of illumination is computed. Unfortunately in most cases no measure of the error involved in the approximation is given.

The original radiosity formulation, for example, contained several sources of error : one of them was the assumption that the radiosity of a patch be constant across the patch. Another appeared if the subdivision of the objects into patches did not follow the distribution of illumination. Aliasing artifacts due to the early hemi-cube algorithms were another problem. Significantly, even though the computation of the form factor has been made quite accurate [WEH89, BRW89], the other problems have received little attention to date.

Things became even worse with the introduction of *progressive radiosity* [CCWG88] : during the course of the solution, approximate results are used for display, but no quantitative measure of the accuracy achieved is available. Actually, even if a "converged" solution is available for comparison, it still differs from the correct "continuous" solution, and measures such as "within 1% of the converged solution" are not acceptable because it is not known what this converged solution represents.

With the advent of the incremental techniques described earlier, the results displayed during the simulation are even more difficult to relate to an accurate solution : when light is redistributed in response to a change in the environment, another error is introduced because that redistributed light is not instantly propagated. In effect, multiple changes in the environment result in a situation where different errors add up in an uncontrolled manner. George *et al* [GSG90] suggest to maintain a log to eventually eliminate most of the error when time becomes available, but this does not solve the problem of the immediate display during interactive use.

Furthermore, when negative energy contributions are propagated as well as positive ones [PSV90], the notion of error becomes more complex : it is no longer possible to characterize a solution by the "percentage of energy actually shot", for example.

A comprehensive study of different sources of error in the simulation process is needed, as well as tools to measure and control the resulting global error.

3.2 Discretization issues

All solution methods presented to date rely on some sort of subdivision to encode the variation of illumination across the surfaces. The subdivision can be explicit, for example in the form of a mesh of polygons, or implicit in the form of an illumination texture [Hec90].

Some simple techniques for adaptive refinement of the radiosity mesh have been suggested [CGIB86], but no satisfactory automatic meshing method has been demonstrated. This may have to do with the fact that the concept of error in a lighting solution is poorly understood. Any automatic meshing procedure will have to make decisions based on an appreciation of the local error, and a necessary first step is to define what is acceptable. Many different metrics are available to judge the quality of a simulation, some local and some global, some physical and some perceptual. Each of these metrics can be wrong in some cases. The emphasis should shift towards the production of "better" meshes, as opposed to the goal of having as few polygons as possible. In any case, fully automatic meshers are needed before non-specialists can use radiosity techniques.

Even if the mesh was perfectly adapted to the scene and its illumination, all discrete algorithms will fall short of approaching the continuous solutions. Illumination functions are multi-dimensional, since at each point on a surface there is a hemisphere of outgoing directions. Methods are needed to compute continuous solutions for these functions, without unnecessary assumptions such as uniformity or constant radiosity across a patch.

An accurate method for the evaluation of ideal specular transfers occuring between an emitter and receiving surfaces is also needed. The only algorithms available to date are either limited to planar surfaces or prone to aliasing artifacts. This is because it is very difficult to enumerate the possible paths for specular reflection between two fixed points.

Finally, even though hierarchical methods can drastically reduce the subdivision of large objects, bottom-up hierarchies should be allowed, where nearby objects could be gathered together for the purpose of computing their effect on a distant object. This is necessary in complex scenes consisting of millions of small objects. Top-down hierarchy [HS90] will not result in any savings in that case, since all form factors are likely very small.

3.3 Modeling or rendering ?

Some of the new techniques mentioned in this paper have contributed to blur the traditional distinction between modeling and rendering. The question of where to put this distinction is not merely academic : it is very important in the design of complete, integrated applications. If incremental radiosity is to be used in architectural design, for example, it is inconvenient if not impractical to use separate modeling and rendering programs.

Another related issue is the decision of whether the illumination should be part of the model or not. Arguments can be made for both choices, but the final usage intended for the application should be the key to that decision. Who has never wished to change the lighting slightly in a picture that was computed long ago ? This is practically impossible if the geometrical model and the illumination solution are kept separate. This issue also has deeper implications, in connection with the problem of discretization and meshing : should meshing information be kept in the model, knowing that it depends on the illumination ?

4 Conclusions

A number of recent advances in physically-based rendering have been presented. Radiosity techniques have established themselves as a very powerful illumination simulation techniques, and new applications can be envisioned that will benefit from their physical accuracy. We have seen that most of the limitations of the original radiosity method are being overcome one by one, but that new problems arise which make the integration of all these advances a challenge.

Developers of computer graphics applications should realize that the use of physical models for light reflection and global illumination may require a complete rethinking of the applications, and that specific requirements of these methods have to be incorporated in the design very early on.

Physically based rendering has reached a point where it can make new application areas benefit from computer graphics. Some of the techniques developed for graphics simulation may prove useful in other engineering techniques as well. For example, the formulation of progressive radiosity, which is nothing but a reordering of the steps in an iterative matrix solution technique, originated from the desire to display a picture early in the solution process. However the idea of a continuously improving solution may in fact appeal to many different areas.

References

[ARFPB90] John M. Airey, John R. Rohlf, and Jr. Frederick P. Brooks. Towards image realism with interactive update rates in complex virtual building environments. *Computer Graphics*, March 1990. Proceedings 1990 Symposium on Interactive 3D Graphics in Snowbird.

[BRW89] Daniel R. Baum, Holly E. Rushmeier, and James M. Winget. Improving radiosity solutions through the use of analytically determined form-factors. *Computer Graphics*, 23(3):325–334, July 1989. Proceedings SIGGRAPH '89 in Boston.

[CCWG88] Michael F. Cohen, Shenchang Eric Chen, John R. Wallace, and Donald P. Greenberg. A progressive refinement approach to fast radiosity image generation. *Computer Graphics*, 22(4):75–84, August 1988. Proceedings SIGGRAPH '88 in Atlanta.

[CGIB86] Michael F. Cohen, Donald P. Greenberg, David S. Immel, and Philip J. Brock. An efficient radiosity approach for realistic image synthesis. *IEEE Computer Graphics and Applications*, 6(3):25–35, March 1986.

[Che90] Shenchang Eric Chen. Incremental radiosity: An extension of progressive radiosity to an interactive image synthesis system. *Computer Graphics*, 24(4):135–144, August 1990. Proceedings SIGGRAPH '90 in Dallas.

[CPC84] Robert L. Cook, Thomas Porter, and Loren Carpenter. Distributed ray tracing. *Computer Graphics*, 18:137–147, July 1984. Proceedings SIGGRAPH '84 in Minneapolis.

[FFOC90] Alain Fournier, Eugene Fiume, Marc Ouellette, and Chuan K. Chee. Fiat lux. Technical Report 90-1, University of Toronto, Dynamics Graphics Project, January 1990.

[GSG90] David W. George, François Sillion, and Donald P. Greenberg. Radiosity redistribution for dynamic environments. *IEEE Computer Graphics and Applications*, 10(4), July 1990.

[GTGB84] Cindy M. Goral, Kenneth E. Torrance, Donald P. Greenberg, and Bennett Battaile. Modeling the interaction of light between diffuse surfaces. *Computer Graphics*, 18(3):213–222, July 1984. Proceedings SIGGRAPH '84 in Minneapolis.

[Hec90] Paul S. Heckbert. Adaptive radiosity textures for bidirectional ray tracing. *Computer Graphics*, 24(4):145–154, August 1990. Proceedings SIGGRAPH '90 in Dallas.

[HS90] Pat Hanrahan and David Saltzman. A rapid hierarchical radiosity algorithm for unoccluded environments, June 1990. Proceedings of the Eurographics Workshop on Photosimulation, Realism and Physics in Computer Graphics (Rennes, France).

[HTSG91] XiaoDong He, Kenneth E. Torrance, François Sillion, and Donald P. Greenberg. A comprehensive physical model for light reflection. *Computer Graphics*, 25(4), August 1991. Proceedings SIGGRAPH '91 in Las Vegas.

[ICG86] David S. Immel, Michael F. Cohen, and Donald P. Greenberg. A radiosity method for non-diffuse environments. *Computer Graphics*, 20(4):133–142, August 1986. Proceedings SIGGRAPH '86 in Dallas.

[Kaj86] James T. Kajiya. The rendering equation. *Computer Graphics*, 20(4):143–150, August 1986. Proceedings SIGGRAPH '86 in Dallas.

[LSS90] Bertrand Le Saec and Christophe Schlick. A progressive ray-tracing based radiosity with general reflectance functions, June 1990. Proceedings of the Eurographics Workshop on Photosimulation, Realism and Physics in Computer Graphics (Rennes, France).

[NN89] Lázló Neumann and Attila Neumann. Photosimulation : Interreflection with arbitrary reflectance models and illumination. *Computer Graphics Forum*, 8(1):21–34, 1989.

[PSV90] Claude Puech, François Sillion, and Christophe Vedel. Improving interaction with radiosity-based lighting simulation programs. *Computer Graphics*, March 1990. Proceedings 1990 Symposium on Interactive 3D Graphics in Snowbird.

[SAWG91] François Sillion, James Arvo, Stephen Westin, and Donald P. Greenberg. A global illumination solution for general reflectance distributions. *Computer Graphics*, 25(4), August 1991. Proceedings SIGGRAPH '91 in Las Vegas.

[SP89] François Sillion and Claude Puech. A general two-pass method integrating specular and diffuse reflection. *Computer Graphics*, 23(4), August 1989. Proceedings SIGGRAPH '89 in Boston.

[Wan91] Leonard Wanger. Perceiving spatial relationships in computer generated images. Master's thesis, Cornell University, Ithaca, New York, May 1991.

[WCG87] John R. Wallace, Michael F. Cohen, and Donald P. Greenberg. A two-pass solution to the rendering equation : a synthesis of ray-tracing and radiosity methods. *Computer Graphics*, 21(4):311–320, July 1987. Proceedings SIGGRAPH '87 in Anaheim.

[WEH89] John R. Wallace, Kells A. Elmquist, and Eric A. Haines. A ray tracing algorithm for progressive radiosity. *Computer Graphics*, 23(3):315–324, July 1989. Proceedings SIGGRAPH '89 in Boston.

[WRC88] Gregory J. Ward, Francis M. Rubinstein, and Robert D. Clear. A ray tracing solution for diffuse interreflection. *Computer Graphics*, 22(4):85–92, August 1988. Proceedings SIGGRAPH '88 in Atlanta.

Adaptive Shadow Testing for Ray Tracing

Gregory J. Ward

Abstract

We present a simple technique for improving the efficiency of ray tracing in scenes with many light sources. The sources are sorted according to their potential contribution, and only those sources whose shadows are above a specified threshold are tested. The remainder are added into the result in proportion to a statistical estimate of their visibility. The algorithm requires very little storage, and produces no visible artifacts.

1 Introduction

Since ray tracing was introduced to the computer graphics community by Whitted [Whitted80] it has proven to be a powerful approach to difficult problems in global illumination [Wallace89] [Shirley90]. In particular, the view-dependent nature of ray tracing offers important advantages over view-independent radiosity solutions for non-diffuse (ie. interesting) environments [Immel86] [Kajiya86]. However, there are illumination problems where the approach of a pixel-independent calculation is a liability rather than an asset. Rendering scenes with significant diffuse interreflection is one such problem [Ward88] [Heckbert90], and scenes with many light sources represent another. Unfortunately, these are both common occurrences in the real world, and as such cannot be ignored if rendering realism is the goal.

Visibility testing is the most time-consuming part of a global illumination calculation, and the visibility of light sources is particularly important since they determine the initial lighting distribution. If we could assume that all of the light sources were visible at every point, the calculation would reduce to a few simple operations. Unfortunately, it is almost always necessary to check for occlusions from light sources (ie. perform shadow testing). Spatial subdivision techniques for accelerating ray tracing are essential for general efficiency [Glassner84], but shadow testing still remains the single most time-consuming portion of most ray traced renderings.

Haines introduced a spatial sorting technique to minimize the amount of time required for shadow testing in a polygonal environment with point light sources [Haines86]. The scene is sorted according to the polygons visible from each source, and shadow testing takes advantage of the fact that it does not matter where obstruction occurs between the source and test point in question. (Normally, ray tracing algorithms are optimized to find the *first* intersection -- here we just want to find *some* intersection.) Although this approach offers significant reductions in calculation time, it requires O(N*M) storage, where N is the number of light sources and M is the number of polygons in the scene. These storage costs can get out of hand for scenes with tens of thousands of surfaces and more than a hundred light sources. (It should be noted that curved surfaces can be included using convex polyhedral bounding volumes, though implementation of the algorithm becomes more difficult.)

In this paper we present a very simple approach to light source testing that trades accuracy, rather than storage, for increased speed. The algorithm has been tested on many different models and has been found to be robust and reliable, offering speed increases between 20%

and 80% in most environments. Furthermore, the user can control the accuracy and reliability of the technique, adapting it to suit his or her requirements. If the user specifies an error bound of zero, the algorithm degenerates to the base case with very little overhead, providing straightforward validation and comparison of results.

2 Concept

In scenes with many light sources, only a few will create strong shadows in any one part of the scene. These will generally be the sources with the highest concentration of light in that section due to source brightness, direction and proximity. The remainder of the light sources will contribute, but without causing any significant contrast gradients. Since it is contrast that stimulates the human visual system, lack of contrast translates to low importance for visual studies.

This observation leads to a simple optimization: we can perform shadow testing on the sources with the highest potential contributions first, and quit testing when the remainder of the contributions is below some threshold (see Figure 1). By itself, this approach can guarantee an absolute error below a given threshold. However, it is still not optimal since we do not know what to do with the remainder of untested sources. Do we add them all in, leave them all out, or add in some and leave out others? How we answer this question in effect determines the efficiency of our algorithm. If we do a good job guessing at the visibility of light sources we do *not* test, our results will be very close to those of the full calculation without the associated cost.

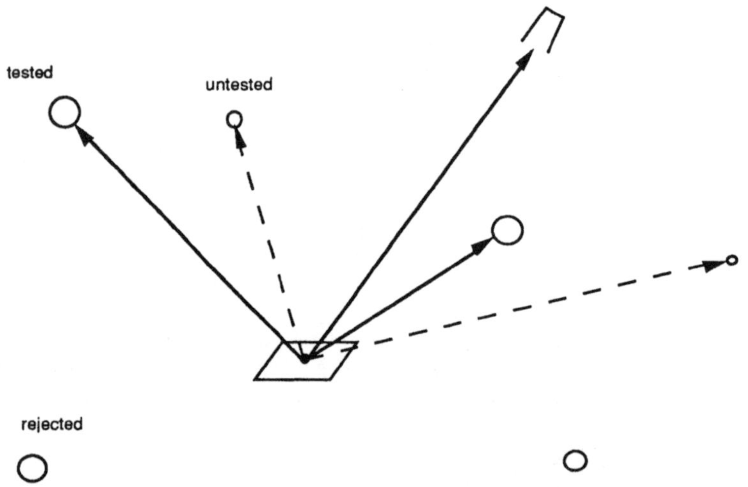

Fig. 1: The brightest sources are tested first, and a fraction of the potential contributions from untested sources are added in as an approximation

We turn to statistics to provide us with a reasonable estimate of visibility for our untested light sources. If a light source is untested, we use the fraction of hits to that source as a multiplier for its contribution. (We maintain these statistics as light sources are tested. Thus, our estimate will be crude at the beginning and improve during the course of the calculation.)

We use the fraction of sources hit for the tested sources at this point as a second multiplier. In terms of random variables:

```
The probability of seeing source i at p is:
    P(Xi,Yp)  =  P(Xi) P(Yp)
where: Xi is a random variable for hits on source i
       Yp is a random variable for hits at point p
```

This assumes that source visibility at our test point is independent of visibility at each source. This is of course an approximation, but it holds well for interior spaces, where obstructions are many near the floor and the ceiling, and few in between.

Note that we do not test sources based on the probability that they are visible. Rather, we use the probabilities for untested sources as a multiplier to approximate their contribution. A weak source deemed unworthy of shadow testing at a particular point is not used or thrown out based on a random variable -- such a procedure would result in visible noise that is completely unnecessary. Using probabilities as coefficients in the way given here yields smooth shading and visually pleasing results without compromising accuracy.

3 Algorithm

We can refine our algorithm by adding a parameter to control the reliability of our source calculation. If we are not terribly concerned about accuracy but want a nice picture, we can stop source testing as soon as the next contribution is below some threshold. This will not guarantee numerical accuracy, but it will at least guarantee that all visually significant shadows will be tested. We introduce a variable called *certainty*, **c**, that can be set continuously between 0 and 1 to control reliability. A value of 0 for **c** means that the potential contribution of any untested source is below the *tolerance*, **t**, thus guaranteeing only contrast, and a value of 1 for **c** means that the sum total of all untested sources is below **t**, thus guaranteeing absolute accuracy.

The algorithm can be written as the following steps (see below for comments):

```
1) Compute potential contributions from all light
    sources in front of this point.
2) Sort the contributions in descending order.            A
3) Compute r(i), the sum of the next N^c contributions
    smaller than i, where N is the number of light
    sources and c is the certainty.
4) Initialize the sum, S, hits, v, and tests, w, to 0.
    Foreach  contribution in our sorted list  do
      if  S t > r(i)  then                                B
        go to step 5
      increment our test counter, w
      increment the test counter for source i, w(i)
      if  source i is visible from this point  then       C
        increment our hit counter, v
        increment the hit counter for source i, v(i)
        add contribution for source i to S
```

```
5) Foreach  untested contribution  do
     multiply contribution by v/w and v(i)/w(i)              D
     add weighted contribution to S
6) Return S
```

Comments:

A This is the only part of the calculation that is extra beyond standard light source testing. The quicker sort algorithm (qsort C library routine) is fast enough in comparison with the rest of the calculation that the extra time is insignificant.

B **S** is the sum of visible contributions tested so far, thus the test **S t > r(i)** checks to see if the remaining N^c contributions are below the threshold. If true, then we have satisfied our visibility testing requirements and can go on to approximate the remaining contributions to **S**.

C This is the actual source visibility test, and this is where the real cost of the direct calculation is incurred. The whole point is to minimize the number of light sources that must be tested here for visibility (see **B** above).

D The ratios, **v/w** and **v(i)/w(i)**, are the estimated probabilities of seeing any source from our test point and seeing source **i** from any point, respectively.

4 Results

The graphs in Figure 3 show the adaptive shadow testing algorithm's performance for the model shown in Figure 2, a conference room with 24 ceiling-mounted fluorescent fixtures with high angle cutoff (resulting from small cube parabolic louvres).

Figure 3a shows the fraction of light sources tested for visibility when the algorithm is applied with different values of **t** and **c**. Note that with a target accuracy (tolerance) of 0, all of the candidate light sources are tested no matter what value is given for certainty.

The overall rendering time is roughly proportional to the fraction of light sources tested. If an average of 50% of the light sources are tested, the rendering takes a little bit more than 50% as long. The "little bit more" is the time required to follow non-shadow rays (eg. eye rays) and compute their shading plus the time required to calculate the potential contributions of untested sources.

Figures 3b and 3c show the average and maximum error corresponding to the different settings, as compared with a fully tested source calculation. In this example, the average error is always kept below the requested tolerance, and even the maximum error (most deviant pixel) is kept within limits for certainties better than 25%.

Fig. 2: Conference room model used to test shadow algorithm

What happens to the calculation as we increase the number of light sources? For the purposes of comparison, we used the same scene and replaced each rectangular source with 8 smaller sources and repeated the tests. The resulting fraction of shadow tests for this modified scene is shown in Figure 4. The most noticeable difference is the overall drop in the fraction of sources tested, which indicates that the algorithm's performance improves as light sources are added to the scene. (It still takes longer of course.) Also, there is a larger spread between the different certainties. It should be noted that further improvement in performance would be apparent if instead of increasing the number of sources in the same locations, the overall dimension of the space were increased. A larger room with the same ceiling height would have proportionally more sources with small potential contributions that would be excluded from visibility testing by our algorithm. In fairness, though, the same cannot be said for multiple floors with all light sources turned on, since the lights in the floors above would be tested even though they cannot possibly contribute to the image. In general, it is better to exclude such non-contributing sources from the calculation.

Figure 5a shows the same conference room with floating furniture -- a more difficult case for the shadow testing algorithm. Figure 5b shows the same scene rendered with adaptive shadow testing using a tolerance of 20% and a certainty of 50%. This resulted in an 80% reduction in shadow testing with an average pixel error of less than 5%. Figure 5c shows the squared error between the two images (times 100). The errors are largest where surfaces are fully illuminated, because the algorithm uses statistics that say those areas probably do not see all of the light sources when in fact they do. However, these errors are not very visible because the eye is not good at making absolute comparisons.

Figure 6 shows the fraction of shadow tests for 192 high-cutoff sources in the scene with floating furniture. Compared to Figure 4, there is a greater cost associated with the certainty parameter. Figure 7a shows the same test with diffuse (cosine emitting) sources. Here we see the "pretty bad case" performance of our algorithm (as opposed to "worst case"). The certainty

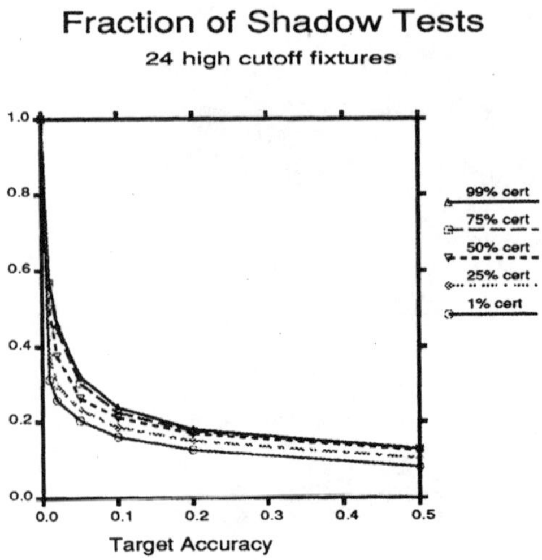

Fig. 3a: Reduction in shadow tests for changes in parameters

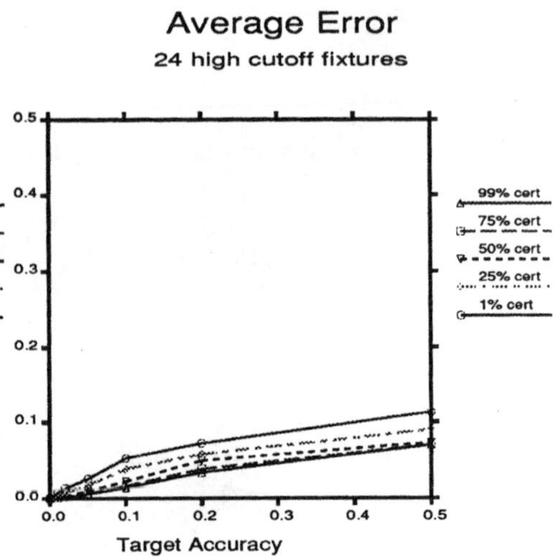

Fig. 3b: The average error here is much less than t

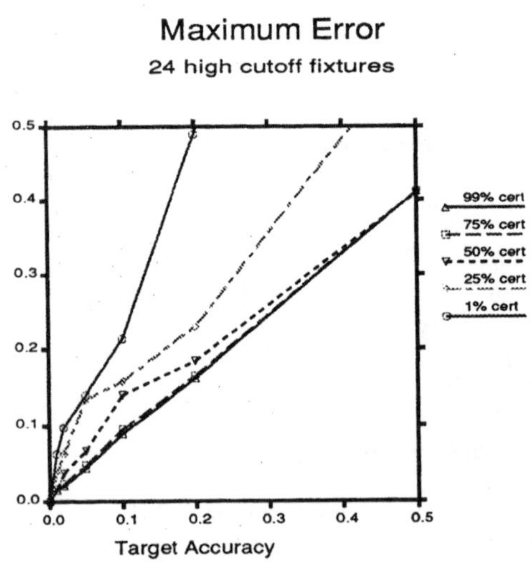

Fig. 3c: Most deviant pixel as a function of parameters

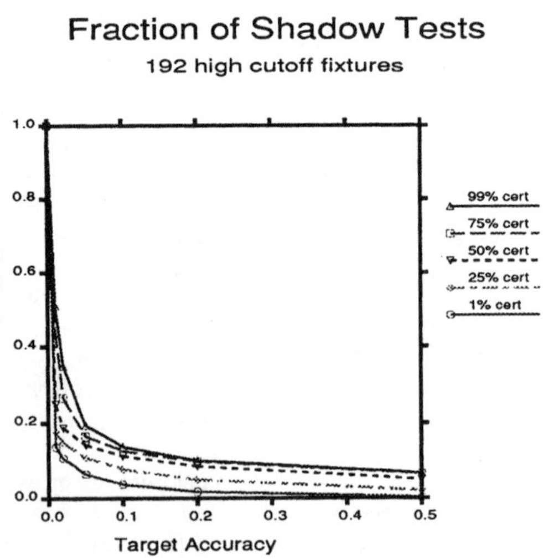

Fig. 4: Test repeated with 8 times as many light sources

Fig. 5a, 5b and 5c: The fully tested shadows in the top image do not look very different from the middle image where only 20% were tested. The bottom image shows the squared difference between the two renderings

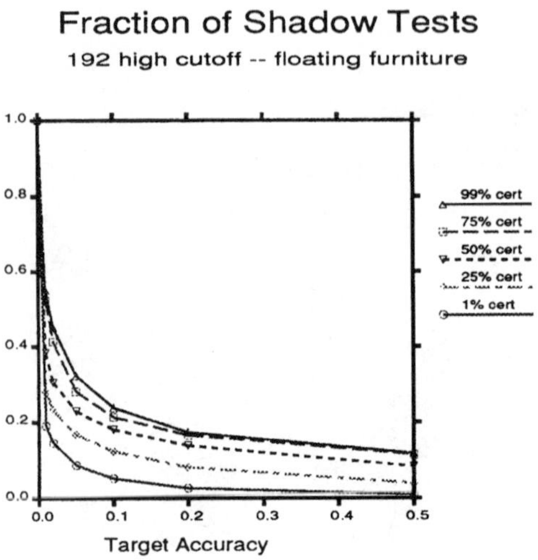

Fig. 6: Floating furniture case has more complex shadows

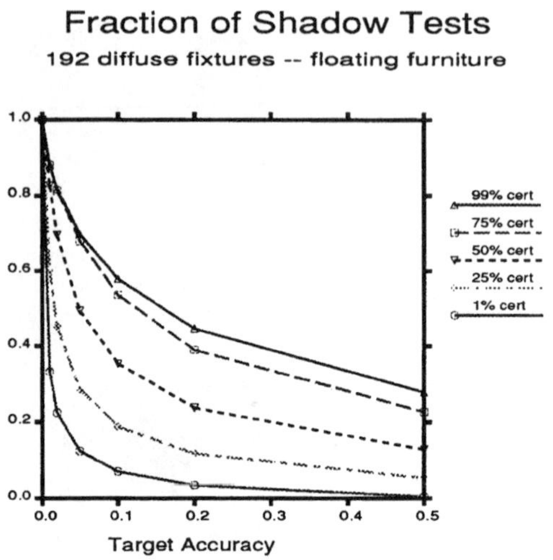

Fig. 7a: The same scene with diffuse light source

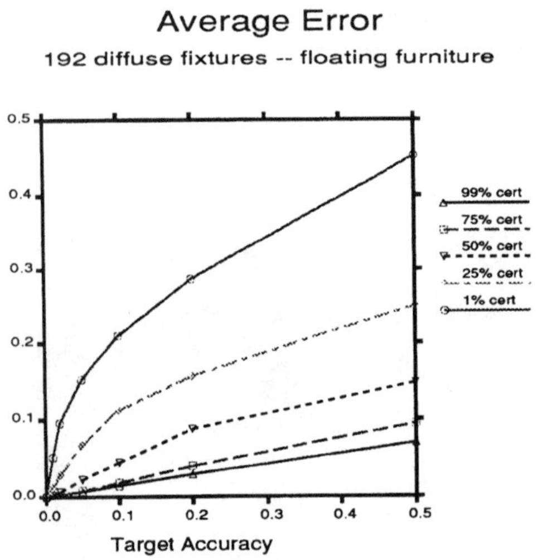

Fig. 7b: Average error for diffuse-floating case

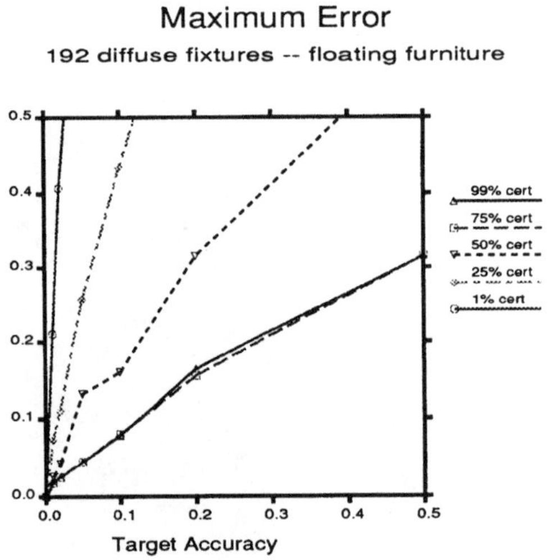

Fig. 7c: The most deviant pixel for diffuse-floating case

plays an even greater role here since diffuse light sources can contribute significant amounts in shadowed areas even when they are on the opposite side of the room.

Figure 7b and 7c show the average and maximum error for diffuse sources and floating furniture, and this is where we see the true performance of our statistical visibility approximation. For certainties above 25%, the average error is maintained within the specified tolerance. This means that even though visibility is not being tested sufficiently, the approximation is coming up with reasonable guesses for the untested sources most of the time. The maximum error, however, falls out of range for 50% certainty, and becomes quite large for 1% certainty. This points out the importance of setting the certainty value with care. If the specified tolerance is truly required, a high certainty should be given. If nice-looking shadows are all that matter, a low certainty will maintain proper contrast boundaries while minimizing tests for shadows that would be invisible to the eye.

5 Conclusion

We have presented a simple optimization for the calculation of the direct light component in global illumination computations using ray tracing. The most important sources for a point are tested first, and less important sources are only tested if their visibility is deemed important to the calculation. If none of the sources is visible, all of the sources that could contribute are tested for visibility. Thus it is fruitless to apply this technique inside of walls and under carpeting.

An important feature of this algorithm is that it *avoids* stochastic sampling. In computer graphics, smooth shading is valued as highly as correct results, and it is best to avoid random noise if the raw artifacts are not visible. By choosing contrast as the primary criteria, our calculation manages to avoid offending the eye in its speed-for-accuracy tradeoff.

Although contrast is a good visual criterion, it does not determine absolute accuracy in a lighting simulation. Thus, a certainty parameter is provided to indicate the relative importance of accuracy. Absolute accuracy can be guaranteed by the algorithm at a slightly higher cost. However, the statistical estimate of source visibility works well enough that such a guarantee is not really necessary.

Another advantage of the source visibility approximation is that it does not rely on point sampling coherence. The calculations can take place at random locations throughout the scene with no loss in speed. Many light source optimizations rely on the next point being close to the last one, as might hold in a simple scanline traversal of an image. Unfortunately, this is not usually the case in a global illumination calculation, where rays are followed all over the scene in no particular order. By maintaining global statistics on source visibility, our algorithm makes use of information that is available and applicable at any point in the scene. Initial tests will have less information to go on, but the statistics build quickly and coherently in the places where they are needed so that artifacts are not visible.

The overhead costs of the algorithm are minimal. The storage overhead is a few additional words per light source for keeping track of test and hit counts. The only additional computation required is the sorting of the potential light source contributions before each test sequence, and we have found this step to be inconsequential in all of the test cases we have studied.

A chief feature of this adaptive light source testing algorithm is its simplicity. In less than a page of C code, a procedure that provides up to a 70% reduction in calculation time can be written. Furthermore, the approach is orthogonal to most other global illumination techniques, and can be added to existing direct light calculations and optimizations.

One optimization that works very well with this algorithm is to relax the tolerance for spawned rays that contribute less to the final pixel value. Other variations such as stochastic sampling of area sources for accurate penumbras work very well also.

Acknowledgements

Thanks goes to Anat Grynberg, who created the conference room model, and to Charles Erhlich, whose abuse of my rendering software compelled me to develop a better direct calculation in the first place. Thanks also to Peter Shirley and Erik Jansen for their helpful feedback. This work was supported by the Assistant Secretary for Conservation and Renewable Energy, Office of Building Energy Research and Development, Buildings Equipment Division of the U.S. Department of Energy under Contract No. DE-AC03-76SF00098. Current research in this area is being sponsored by the LESO Solar Energy Group at the Ecole Polytechnique Federale de Lausanne in Switzerland.

Bibliography

[Glassner84] A. Glassner, "Space Subdivision for Fast Ray Tracing," *IEEE Computer Graphics and Applications*, vol. 4, no. 10, October 1984, pp. 15-22.

[Haines86] E. Haines, D. Greenberg, "The Light Buffer: A Shadow-Testing Accelerator," *IEEE Computer Graphics and Applications*, September 1986, pp. 6-16.

[Heckbert90] P. Heckbert, "Adaptive Radiosity Textures for Bidirectional Ray Tracing," *Computer Graphics*, vol. 24, no. 4, August 1990, pp. 145-154.

[Immel86] D. Immel, D. Greenburg, M. Cohen, "A Radiosity Method for Non-Diffuse Environments," *Computer Graphics*, vol. 20, no. 4, August 1986, pp. 133-142.

[Kajiya86] J. Kajiya, J., "The Rendering Equation," *Computer Graphics*, vol. 20, no. 4, August 1986, pp. 143-150.

[Shirley90] P. Shirley, "A Ray Tracing Method for Illumination Calculation in Diffuse-Specular Scenes," Proceedings of Graphics Interface '90, May 1990, pp. 205-212.

[Wallace89] J. Wallace, K. Elmquist, E. Haines, "A Ray Tracing Algorithm for Progressive Radiosity," *Computer Graphics*, vol. 21, no. 4, July 1989, pp. 315-324.

[Ward88] G. Ward, F. Rubinstein, R. Clear, "A Ray Tracing Solution for Diffuse Interreflection," *Computer Graphics*, vol. 22, no. 4, August 1988, pp. 85-92.

[Whitted80] T. Whitted, "An Improved Illumination Model for Shaded Display," *Communications of the ACM*, vol. 23, no. 6, June 1980, pp. 343-349.

An Adaptive Sampling Technique for Multidimensional Integration by Ray-Tracing

Christophe Schlick

1 Introduction

Creating pictures with photorealistic effects (diffuse and specular interreflections, shadow and penumbra, depth of field, motion blur...) implies to solve multidimensional integral equations — so-called *rendering equations*. Ray-tracing is a well-known powerful and flexible tool that has been used successfully for solving these rendering equations, either in an image-oriented process (distributed ray-tracing [COOK84], path tracing [KAJI86]) or in a scene-oriented one (deterministic radiosity [WALL89] [LESA90], stochastic radiosity [MALL89] [SHIR90]).

Unfortunately, ray-tracing is a point sampling technique and thus is prone to aliasing artifacts. Moreover, because the signal to sample is generally not bounded in the frequency domain, aliasing cannot be totally eliminated but only reduced. Antialising methods that have been proposed for ray-tracing usually combine an increase of the sampling rate and a strategy for samples selection. But the most powerful of them remain expensive because of the great number of rays needed; thus some improvements have to be found to reduce that number.

A classical technique — adaptive sampling — consists in sending more or less rays, according to the way the signal varies. The drawback of the existing adaptive sampling methods for multidimensional integration by ray-tracing (**MIRT**) is that they *don't have all the wanted characteristics* (irregular and importance sampling, uncorrelation between the dimensions to sample, complete stratification at each refinement level, efficient reconstruction). In this paper, an adaptive sampling technique is presented, that fulfills all these conditions and can be totally implemented with look-up tables for efficiency considerations.

Section 2 presents some notations and terminology, section 3 describes the characteristics any successful sampling strategy should have, section 4 gives an overview of the method while section 5 details it, section 6 presents the results obtained and section 7 concludes.

2 Terminology

The process of generating realistic images can be viewed as the resolution of two equations : the *Image Rendering Equation* (IRE) and the *Scene Rendering Equation* (SRE)

$$I(i,j) = \frac{1}{T\pi R^2}\int_{i-1/2}^{i+1/2}\int_{j-1/2}^{j+1/2}\int_0^R\int_0^{2\pi}\int_0^T L(P_{xyr\theta t}, V_{xyr\theta t})\, dx dy r dr d\theta dt \qquad \textbf{(IRE)}$$

$$L(P,V) = L_E(P,V) + \int_{V'\in\mathcal{V}_{or}\, P'\in\mathcal{P}} R(P,V,V')A(P,P')G(P,P')L(P',V')\, dV'_{or} dP' \qquad \textbf{(SRE)}$$

- $I(i,j)$: Illumination of pixel (i,j)
- R : Lens radius — F : Focal length — T : Shutter time
- $L(P,V)$: Radiance leaving point P in direction V

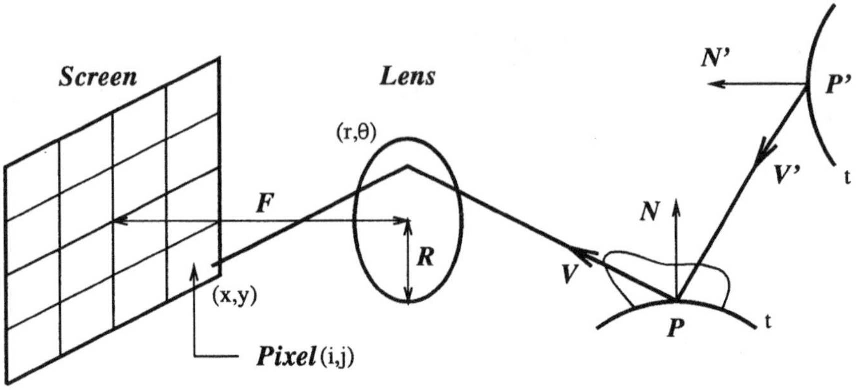

Fig 1. Image rendering and scene rendering equations

- $L_E(P, V)$: Self-emitting radiance leaving point P in direction V
- $R(P, V, V')$: Bidirectional reflectance at point P between direction V and V'
- $A(P, P')$: Absorption factor between P and P' (ie influence of the medium surrounding the two points). This absorption term is given by the Bougher law and by convention, $A(P, P') = 0$ when P is not visible from P'
- $G(P, P')$: Geometrical factor between point P and P' (ie influence of the distance and orientation of the two points). This geometrical term is given by : $G(P, P') = \frac{N \cdot V' N' \cdot V'}{D^2}$ where D is the distance from P to P'
- $\mathcal{V}$: Set of directions for incoming light — $\mathcal{P}$: Set of origins for incoming light

The IRE expresses the illumination of a given pixel in the image. It is essentially a mean over several integration dimensions (pixel width, pixel height, lens radius, lens angle, shutter time) of the light reaching the screen. The SRE expresses the radiance for each couple (P, V) in the scene. It is a Fredholm equation of the second kind that represents the light transport in the environment (see Figure 1). It should be noticed that both IRE and SRE are monochromatic equations and so they should theoretically be solved for each wavelength of the visible spectrum. However some efficient wavelength selection schemes have been developped that achieve the same results with a limited (either constant [MEYE88] or adaptive [BOUV88]) number of spectrum samples.

The two main families of current image synthesis methods differ by the way the rendering equations are solved :

Image-oriented Methods The IRE is solved first and the SRE is evaluated, on the fly, for each couple (P, V) needed for the computation of the IRE. (Examples : early direct illumination methods, ray-tracing, distributed ray-tracing, path tracing...)

Scene-oriented Methods The SRE is solved first, for a finite set of couples (P, V). Then the IRE is computed for each pixel, taking and interpolating the radiance values computed with the SRE. (Examples : basic radiosity, progressive radiosity, extended radiosity...)

The adaptive sampling technique presented in this paper is particularly well-suited to image-oriented methods. But it could be easily adapted for scene-oriented methods using ray-tracing (especially the stochastic ones).

3 Desired Characteristics for Sampling

The rendering equations are multidimensional integral equations and could be solved numerically by every quadrature method. But in [COOK84], the authors showed that **MIRT** is an efficient tool for solving them, despite the aliasing artifacts that plague every point sampling method.

Indeed, according to Shannon's theorem, for a given sampling rate, every signal which has frequencies beyond the Nyquist limit will alias. In our case, the signal $L(P, V)$ has generally no upper bound in the frequency domain. So it can be extremely expensive to sample at a rate were the aliasing of $L(P, V)$ will be sufficiently small (several hundred rays per pixel). Thus, powerful sampling strategies have to be found to reduce aliasing while maintaining a low cost. For that purpose, some characteristics of an *optimal antialiasing method* have been brought out by several authors :

Adaptivity One way to reduce the number of rays is to send them according to the way the signal varies. The algorithm for a given pixel is to send rays until a certain refinement criterion is fulfilled. Several criteria have been proposed in the past; the basic idea is to create a metric to estimate the current error and to stop when it falls below a predefined threshold. The metric can come either from statistics (variance [LEE85], confidence [PURG88]), from signal theory (signal to noise rate [DIPP85]) or better, take into account some characteristics of the human eye (constrast [MITC87]).

Irregularity As shown in [COOK86], irregular sampling achieves better results than regular one, because it replaces coherent aliasing patterns by uncoherent broad-band noise that is much less objectionable for human eye. Several irregular sampling techniques exist in the litterature (Poisson sampling [COOK86], Jittering [COOK86], N-Rooks separated sampling [SHIR90]).

Complete Stratification Stratification is a well-known sampling technique. When N samples are to be taken in an interval of length L, complete stratification consists in taking exactly one sample in each stratum (interval of length L/N). Mixing adaptivity and stratification rises some difficulties because one has to insure that there is a complete stratification at each refinement level.

Importance Sampling When there is a weighting function for the signal to sample (for instance, the lens radius dimension has a weighting function r in IRE), it is more efficient to sample with a non uniform density. As noted by Shirley [SHIR90], such an importance sampling can be obtained from a uniform sampling by transforming the samples by the inverse of the distribution function associated with the weighting function.

Uncorrelation Uncorrelated sampling [KAJI86] is a powerful technique that allows multiple integrations at sensibly the same cost as simple ones. The idea is to create a bijection between the strata of a dimension and those of another. To avoid visual artifacts, the bijection must be different for neighbouring dimensions and for neighbouring pixels — here is the reason of the term *uncorrelation*. In Figure 2, such an uncorrelated sampling is shown for four dimensions (A,B,C,D). Note that the data between two neighbouring dimensions are in bijection and that the bijection is different for each couple of dimensions.

Fast Reconstruction When sampling has been done with uniform density, reconstruction is only a convolution with a limited-support filter. When the sampling is not uniform (adaptive sampling, for instance) reconstruction becomes more complex and needs more expensive filters (multi-stage reconstruction [MITC87])

None of the previous antialiasing methods proposed for **MIRT** does include all those six points. In fact, the last five conditions can be obtained relatively easily, but adaptivity is somewhat conflicting. For instance, the refinement criterion uses the notion of neighbourhood that is broken by uncorrelation — neighbours in a dimension are not necessarily neighbours in another —, complete stratification is hard to be insured at low cost for each refinement level and finally, adaptivity leads to non-uniform sampling that needs expensive reconstruction.

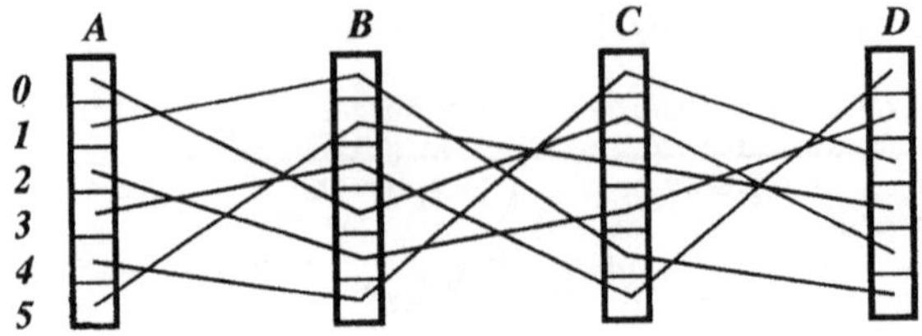

Fig 2. Uncorrelated sampling of four dimensions (A,B,C,D)

In fact, we will show that these contradictions can be removed by reconsidering the antialiasing process and mixing some results from the image synthesis field with some from the signal processing field.

4 Overview of the Technique

The classical antialiasing technique, used in the signal processing field, is to convolve the signal with a low-pass filter in order to bring its highest frequency below the Nyquist limit for the wanted sampling [FOLE89]. The filtering method supposes to know the continuous signal, on which the filter will be applied. In image synthesis, the signal is only virtual and is not available under its continuous form, thus the low-pass filtering method cannot be used as it is; our technique can be considered as an adaptation of that method.

The smallest feature that can be displayed on the screen is the pixel, so the highest frequency of the signal should be the pixel frequency. If we are able to filter the signal in that way, Shannon's theorem states that four samples per pixel will be sufficient to reconstruct it. The principle of our technique is to subdivide pixels into four sub-pixels and to compute a mean value of the light reaching each sub-pixel. This value will become more and more precise, as long as the refinement progresses.

Because the six conditions of an optimal antialiasing method cannot be achieved by a unique sampling, the idea is to use a *double sampling*. The first is regular (one value per sub-pixel) and serves both for the refinement criterion and the reconstruction. The second is adaptive, irregular, stratified, uncorrelated (several samples per sub-pixel) and computes the value for the first sampling by doing a kind of low-pass filtering of the signal at the Nyquist frequency.

5 Description of the Technique

5.1 Algorithm

As shown in the pseudo-code of the algorithm, there are four different modules to detail : Initial Evaluation, Refinement Condition, Refinement of the Evaluation, Reconstruction...

begin
    foreach sub-pixel **do**
        begin
            Initial Evaluation
            while *Refinement condition* is true **do**
                Refinement of the Evaluation
        end
    Reconstruction
end

5.2 Initial Evaluation

The sampling rate taken by default is one ray per sub-pixel. That is the cheapest sampling that guarantees not to miss any pixel-size feature. The construction of the ray follows the process described in [COOK84] and improved in [SHIR90] : take one sample for each integration dimension of the IRE (pixel width, pixel height, lens radius, lens angle, shutter time) and compute the origin, the direction and the time of the ray according to the chosen samples.

When the ray hits an object, the potential contributions of the light sources on the intersection point are computed and used to create a distribution function called *illuminance*. Then a sample is taken according to the illuminance weighting function and the corresponding ray is built. The same scheme is used for reflected and/or transmitted rays, taking reflectance and/or transmittance as distribution function. The scheme can even be improved by tracing only one ray (either reflected or transmitted) according to the Fresnel coefficient.

5.3 Refinement Condition

The refinement condition that we have selected uses the *contrast* notion defined in [MITC87] which has the advantage of following approximatively the sensibility curve of the human eye. The contrast between two pixels (or sub-pixels) with illumination I_a and I_b respectively, is :

$$\frac{Max(I_a, I_b) - Min(I_a, I_b)}{Max(I_a, I_b) + Min(I_a, I_b)}$$

When the illumination is calculated for several wavelengths — RGB, for instance — contrast has to be computed for each wavelength and the *global contrast* between the two pixels will be the maximum of the different contrasts. That leads to the following condition :

Refinement Condition *Evaluation has to be refined while the global contrast between a sub-pixel and its neighbourhood is above a predefined threshold.*

The neighbourhood can be either the four or the eight sub-pixels surrounding the center sub-pixel, whether you consider 4-connectivity or 8-connectivity. In the same spirit, the illumination value used to compute the contrast can either be the brute value stored in the sub-pixel, or the result of a convolution with the reconstruction filter. The latter solution gives better results for mainly two reasons. First, the illumination value used in the computation is the one that will finally be displayed, so the contrast really seen by the eye is computed. Second, the reconstruction filter tends to smooth the variations of the signal, so the refinement condition will be fulfilled earlier and less rays will be sent.

After having computed the contrast and found that the evaluation needs some refinement, the sub-pixel to refine has to be selected, using the following process :

Refinement Selection *The central sub-pixel and the one with the maximal contrast are taken. Their refinement levels are compared and the sub-pixel with the lowest one is refined.*

In Figure 3, some sub-pixels are shown, with their illumination value and their refinement level. The sub-pixel which has the greatest contast with the center is the lower-left one. After comparison of their levels, the lower-left sub-pixel is chosen for the refinement.

5.4 Refinement of the Evaluation

The module that computes the evaluation of the IRE/SRE has to be considered as a *black box* which takes as entry, a refinement level for a given sub-pixel, and returns an illumination value for that sub-pixel, more or less accurate, according to the current level. The implementation of the module is an improvement of the N-Rooks sampling described in [SHIR90], in order to include adaptativity, irregularity, uncorrelation and complete stratification. Moreover, the technique is totally implemented using a unique look-up table for efficiency considerations.

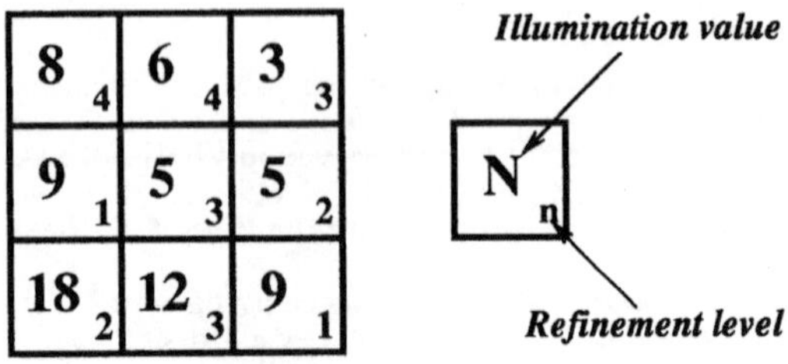

Fig 3. Sub-pixel neighbourhood and refinement selection

At the beginning of the program, a table T of length 2^N is created[1], that will store a permutation σ_N on the set $\{0...2^N-1\}$ (see Figure 4). All the samples needed during the processing will be extracted from the table, and converted from the range $[0...2^N-1]$ to the range of the corresponding integration interval, by an affine transformation. After that, conversion to importance sampling can eventually take place as done in [SHIR90]. For instance, if a sample for the time integral is needed (range $[0, T]$), a number n is extracted from the table and is simply scaled by $T/2^N$. If a sample for the lens radius integral of IRE is needed (range $[0, R]$ with weighting function r), the extracted number n is transformed by $R\sqrt{n/2^N}$ in order to get an importance sampling for the given weighting function.

By working modulo 2^N, the table T contains in fact 2^N different permutations, depending on the element of T taken as entry point. So, if neighbouring dimensions and neighbouring sub-pixels are associated with different entry points, a multidimensional uncorrelated sampling will be created *de facto* (see Section 3). The entry point of a given sub-pixel and a given dimension can be obtained by every classical hash-coding function[2]. When several samples are needed (ie when the sub-pixel is refined), they are read sequentially from the table, starting from the entry point. For instance, in Figure 4, the four first samples to be extracted are 4, 10, 1, 13.

Entry* ↓ *point

2	**15**	**4**	**10**	**1**	**13**	**7**	**8**	**3**	**14**	**5**	**11**	**0**	**12**	**6**	**9**
0	*1*	*2*	*3*	*4*	*5*	*6*	*7*	*8*	*9*	*10*	*11*	*12*	*13*	*14*	*15*

Fig 4. Permutation and entry point

At this point of the description, we have got an adaptive, unregular, uncorrelated, N-Rooks multidimensional sampling. The last characteristic to include is the stratification. A sampling with a complete stratification whatever the refinement level means that every set of n consecutive elements of T realizes such a stratification. In our case, it means that every n adjacent elements of T have to belong to a different stratum of length $2^N/n$.

We propose a permutation σ_N that does not exactly fulfill that condition, but an approximation of it. σ_N is created using the following recursion law :

$$\sigma_0 \ : \ \{0\} \longmapsto \{0\} \quad \text{and} \quad \forall i \in \{0, ..., 2^n-1\} \begin{cases} \sigma_{n+1}(i) = 2\sigma_n(i) + \rho(i) \\ \sigma_{n+1}(i+2^n) = 2\sigma_n(i) + 1 - \rho(i) \end{cases}$$

where $\rho(i)$ is a number randomly chosen among the set $\{0, 1\}$, for each i.

---

[1] A table of 256 elements is sufficient and allows each element of T to be a byte instead of an integer.

[2] We use the function : $n = 17i + 73j + 131d$ for dimension d of sub-pixel (i, j) where $d \in \{0...6\}$ (0 : time, 1 : pixel width, 2 : pixel height, 3 : lens radius, 4 : lens angle, 5 : illuminance, 6 : reflectance/transmittance)

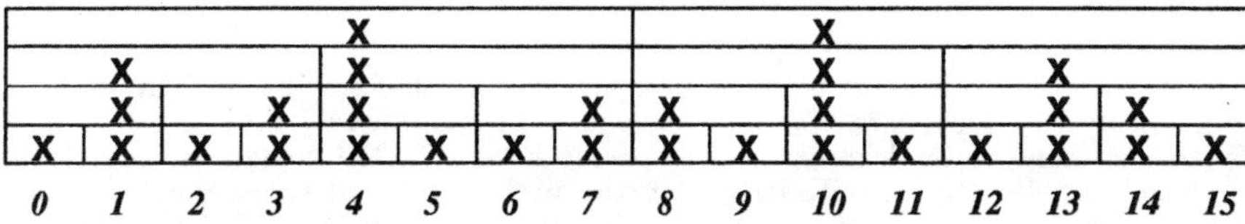

Fig 5. Four steps of the refinement process

Figure 4 shows an example of a permutation σ_N (limited at $N = 4$) obtained using that recursion law. The entry point is the third element of the table, and the samples are taken sequentially from that point. In Figure 5, the location of the samples for different steps of the refinement process are shown. Several remarks can be made concerning the figures. First, note that the stratification is complete when the number of samples is a power of 2, and varies gracefully between two adjacent powers of 2, always preserving a uniform distribution of the samples and not favouring a particular region of the table. Second, when the refinement is carried on up to the table resolution (16 samples in our figure), the sampling is no longer irregular. This problem can be avoided by putting an upper bound for the number of samples to be extracted from a given table[3] — that will create a kind of minimal distance Poisson sampling. And finally, the number of refinement levels can be bounded too, by grouping several samples in one step, in order to compute the refinement criterion less often[4].

5.5 Reconstruction

The result of the evaluation loop is an image signal, filtered at the pixel frequency and sampled at the Nyquist rate. The last stage consists in reconstructing the signal from the samples and resampling it at the display resolution. Shannon's theorem states that the signal can be perfectly reconstructed by convolving the samples with a *sinc filter of infinite support.* Because the pure sinc filter is impracticable, several filters with finite support have been developped in the signal processing field, to reduce post-aliasing due to bad reconstruction.

The advantage of our method over previous adaptive sampling methods is that the final samples are evenly spaced, so the reconstruction is only a convolution with a weighting window. Moreover, if the weightings are carefully chosen, the convolution can be done using only additions, substractions and shifts on integers. Such an improvement is particularly valuable when the reconstruction filter is also used during the refinement step (see Section 5.3). Examples of such weighting windows of various support size are given in Figure 6.

$$\frac{1}{4}\begin{bmatrix}1 & 1\\ 1 & 1\end{bmatrix} \qquad \frac{1}{64}\begin{bmatrix}2 & 3 & 3 & 2\\ 3 & 8 & 8 & 3\\ 3 & 8 & 8 & 3\\ 2 & 3 & 3 & 2\end{bmatrix} \qquad \frac{1}{256}\begin{bmatrix}3 & 4 & 5 & 5 & 4 & 3\\ 4 & 7 & 8 & 8 & 7 & 4\\ 5 & 8 & 20 & 20 & 8 & 5\\ 5 & 8 & 20 & 20 & 8 & 5\\ 4 & 7 & 8 & 8 & 7 & 4\\ 3 & 4 & 5 & 5 & 4 & 3\end{bmatrix}$$

Fig 6. Examples of weighting windows for reconstruction

[3] For our table of 256 elements, an upper bound of 64 works well. Indeed, 64 samples per sub-pixel gives 256 samples per pixel, which suffices for the most complicated pictures.

[4] We use a 12 steps refinement scheme, taking each time 1 to 8 samples to get the following progression : 1, 2, 4, 8, 12, 16, 24, 32, 40, 48, 56, 64 samples per sub-pixel.

6 Results

Four pictures have been used to test the efficiency of the technique. CHECK represents simply an infinite checkboard. BALLS represents the same checkboard with two mirror balls (one in front of the viewer and one above him). GLOSS represents the classical sphereflake modelled by E.A. Haines, with the middle sphere having glossy reflection. FOCUS represents the sphereflake using a finite lens camera model. All the pictures have been computed at a 512x512 resolution (see Figure 7, on color plates).

The method has not been compared with previous adaptive antialiasing techniques ([LEE85], [DIPP85], [MITC87], [PAIN89]), because none of them fulfilled all the desired characteristics, and the effects of such a lack have already been studied in [SHIR90]. Instead, we wanted to test the benefit of our method compared to non-adaptive methods that include these characteristics ([SHIR90], for instance).

Two metrics can be used for comparison of sampling techniques :

- Pictures with the same quality : *Compare rendering times*
- Pictures with the same rendering time : *Compare quality*

We have used two different tests to evaluate our technique for both metrics.

For the first test, the four pictures have been rendered using our sampling stategy with no adaptivity at a rate of 256 rays per pixel (ie 67108864 rays per image). Then the adaptivity has been added with an initial rate of 4 rays per pixel and an upper bound of 256. The pictures were undistinguishable from the ones without adaptivity and table 1 shows the speed-up obtained[5]

Image Name	Rays per Image	Rays per Pixel	Speed-up
CHECK	1468324	5.6	45.7
BALLS	1924336	7.3	34.8
GLOSS	3351728	12.8	20.0
FOCUS	7137980	27.2	9.4

TABLE 1. Improvement in rendering time for the same quality

For the second test, the non-adaptive method has been used, taking as sampling rate the average number of rays per pixel found during the first (respectively 6, 7, 13 and 27). The difference of quality is shown in Figure 8 (see color plates), where close-ups of a part of the images can be seen — the left picture represents the adaptive method and the right one the non-adaptive with the same rendering time.

7 Conclusion

A sampling technique for multidimensional integration by ray-tracing has been described. It is as powerful as the most recent non-adaptive sampling strategies [SHIR90] because it includes all their characteristics (irregular and importance sampling, uncorrelation between the dimensions, complete stratification). But it is also more efficient because it includes adaptivity and fast reconstruction. These features were easily added by reconsidering the antialiasing process and using a double sampling, one for the evaluation and one for the refinement criterion and the reconstruction.

The method can be totally implemented using look-up tables — both for the sampling and for the reconstruction — built once, at the beginning of the program. The look-up table used for the sampling realizes a mixing of N-Rooks and minimal distance Poisson

[5]The overhead due to evalution of the refinement condition is negligible, so the speed-up in time is in fact the savings of rays.

samplings, while the table used for the reconstruction approximates of a sinc filter using only integer arithmetic.

Another nice thing with the method is that it uses no random choice, a fact that can be advantageous for different reasons. For instance, one can imagine doing a first pass on the image with a low upper bound on the number of samples per sub-pixel. If we store the current refinement level for each sub-pixel (in the alpha channel, for example) the image can be refined locally, using a pixel stream strategy a la Perlin. Because there is no randomness, it is insured that the refinement will be the same — with the same desired characteristics and no redundant work — as if it had be done during the first pass. So a fast rough image will converge to the final antialiased one, using simply a multi-pass rendering, at no extra cost.

Acknowledgements

Special thanks to Peter S. Shirley and Don P. Mitchell for their valuable comments on an earlier version of this paper.

References

[BOUV88] C. Bouville, J.M. Dubois, F. Marcal, G. Viaud *Monte Carlo Integration applied to an Illumination Model*, Proc. of Eurographics 88, p483-497

[COOK84] R.L. Cook, T. Porter, L. Carpenter, *Distributed Ray Tracing*, Proc. of SIGGRAPH 84, p137-145

[COOK86] R.L. Cook, *Stochastic Sampling in Computer Graphics*, ACM Transactions on Graphics, v5, n1, p51-72, 1986

[DIPP85] M.A. Dippé, E.H. Wold, *Antialiasing through Stochastic Sampling*, Proc. of SIGGRAPH 85, p69-78

[FOLE90] J.D. Foley, A. VanDam, S.K. Feiner, J.F. Hughes, *Computer Graphics, Principles and Practice*, Addison-Wesley Publishing Company, 1990

[KAJI86] J.T. Kajiya, *The Rendering Equation*, Proc. of SIGGRAPH 86, p143-150

[LEE85] M.E. Lee, R.A. Redner, S.P. Uselton, *Statistically Optimized Sampling for Distributed Ray Tracing*, Proc. of SIGGRAPH 85, p61-67

[LESA90] B. LeSaec, C. Schlick, *A Ray-Tracing Based Progressive Radiosity with General Reflectance Functions*, Proc. of Eurographics Workshop on Rendering (Rennes), p103-116, 1990

[MALL89] T.J. Malley, *A Shading Method for Computer Generated Images*, Ms Thesis, University of Utah, 1989

[MITC87] D.P. Mitchell, *Generating Antialiased Images at Low Sampling Densities*, Proc. of SIGGRAPH 87, p65-72

[PAIN89] J. Painter, K. Sloan, *Antialiased Ray-Tracing by Adaptive Progressive Refinement*, Proc. of SIGGRAPH 89, p281-287

[PURG86] W. Purgathofer, *A Statistical Method for Adaptive Stochastic Sampling*, Proc. of Eurographics 86, p145-152

[SHIR90] P.S. Shirley, *Physically Based Lighting Calculations for Computer Graphics*, PhD Thesis, University of Illinois at Urbana-Champaign, 1990

[WALL89] J.R. Wallace, K.A. Elmquist, E.A. Haines, *A Ray-Tracing Algorithm for Progressive Radiosity*, Proc. of SIGGRAPH 89, p335-344

The Simulation of Radiant Light Transfer with Stochastic Ray-Tracing

Brigitta Lange

Abstract

In the following an implementation of the Rendering Equation according to Kajiya (Kajiya 86) is presented, that models the radiant light transfer in order to generate photorealistic images. By invoking a probabilistic model of the radiation exchange process and applying Monte Carlo sampling techniques, it is possible to utilize an already existing ray-tracer by modifying it. A view-dependent solution is presented that can handle diffuse and specular reflections and that can also model the illumination from area sources of circular shape. Moreover, a weighting factor for the sampled intensities has been developed that provides not only a correct treatment of the intensity variations of illumination in dependence of the distance, but can also be used to accelerate the image generation process by controlling the number of paths to be traced for each pixel individually as required by the particular lighting situation for the corresponding point in the scene, without producing a loss in accuracy.

1. Introduction

The production of realistic images requires the ability to simulate the propagation of light in an environment, i.e. the ability to completely account for the global illumination arising from complex interreflections within the environment.

In principal it should be possible to trace the progress of radiant energy through any system by applying Maxwell's equations and the associated boundary conditions. In practice, however, this is often an impractical if not impossible task. Each of the incoming directional intensities is a reflected directional intensity leaving another surface in the surrounding. This leads to a complex expression since the outgoing directional intensities must be solved for all surfaces simultaneously. The problems are made tractable by making certain simplifying assumptions which will be described in the following.

An alternative approximation to the Maxwell equations for the simulation of radiant light transfer has been presented by (Kajiya 86) and is referred to as Rendering Equation. This equation is helpful in providing a physical picture of the light transport and is useful in Monte Carlo methods. This approach was chosen because it is the only method modeling diffuse interreflections that allows to fall back on an already existing ray-tracer by modifying it. Furthermore such a ray-tracing based access does not limit or burden the geometric model with illuminance information as required by the surface discretization of the radiosity method.

2. The Implementation Concept

In the Monte Carlo process radiation is presented by a number of rays whose origins and directions are selected from a series of random numbers. For a point on a reflecting surface the total hemisphere of incoming directions has to be accounted for in order to calculate the emitted energy. But it is impossible to follow all of the rays needed to simulate the radiative exchange process. Thus we sample the hemisphere of incoming directions for a point on a surface by a finite set of randomly selected rays. The goal is to obtain an optimum sample set increasing the accuracy with the fewest number of samples possible.

Instead of branching a tree of rays at each surface we reduce the number of higher generation rays by selecting only one random ray at each surface point (*path-tracing* (Kajiya 86)). For a scene with a lot of reflection this will cut down the number of intersection tests. A path then corresponds to a discrete Markov Chain used in Monte Carlo techniques to compute the solution of simultaneous linear equations.

To calculate the intensity of a pixel we have to trace as many paths as there are samples needed for the hemisphere of the first ray-surface intersection point in the path. Since we do not know in advance how many samples are needed for a particular pixel we trace the same fixed number N of paths for all pixels. For all other ray-object intersection points encountered during the path-tracing process the total hemisphere can only be sampled in one direction due to the path structure. The path length has to be finite, therefore the path depth has a fixed maximum level for recursion.

In order to reduce the variance we have to select the random numbers in such a way that a lot of rays are send in directions where the most light is arriving and relatively few rays in directions where the incoming light is sparse. Therefore, we should not renounce the precious information the lighting model provides us. If the light sources are sampled directly for each surface point this will increase accuracy and reduce the number N of rays that have to be selected randomly.

If the lighting model of a scene consists of more than one light source then, for computational efficiency, we select only one of the light sources randomly. That means that at each level of the path two rays are traced where the one belongs to a Markov Chain and samples the indirect contributions and the other is found deterministically and samples the direct contribution of a light source. To conserve energy the contributions of the light sources have to be weighted thus preventing that a point reflects more energy than there is incident.

The *indirect illumination* of a point visible to a pixel can be calculated by approximating the integral over the hemisphere of incoming directions (θ_j,φ_j), $j = 1,\ldots,N$ with a discrete set of N samples that do not intersect the light sources. The intensity reflected from the point on a surface in the direction (θ_r,φ_r) of the incident ray is thus:

$$I(\theta_r,\varphi_r) = 1/N \sum_{j=1}^{N} [(1-w)\cdot I(\theta_j,\varphi_j) + w\cdot I(\theta_l,\varphi_l)\, f(\theta_l,\varphi_l)] \cdot sr , \qquad (2.1)$$

where $I(\theta_j,\varphi_j)$ is the intensity of a random sample obtained by recursion, $I(\theta_l,\varphi_l)$ is the intensity from the light source direction that varies with a function $f(\theta_l,\varphi_l)$ of the incident angles, w is a weighting factor to conserve energy, and $sr \in [0,1]$ is the surface reflection coefficient.

To implement this path-tracing procedure the Monte Carlo distribution of the samples has to be determined and an appropriate weighting factor w has to be found.

3. The Random Ray Distribution

To simulate the radiant light transfer by path tracing a method of assigning values to individual samples is desired so that the distribution of values for all the samples will agree with a known probability distribution that occurs in a physical process. At every ray-surface intersection point a new ray must be "emitted" from the same point to sample the incoming energy, thus functions are required to determine the direction of emission.

The Monte Carlo method allows a physical process to be represented by the selection of a series of random numbers. The characteristic of a uniformly distributed random variable can be applied to the selection of the direction of an emitted bundle of energy, a ray. The direction of emission can be defined in terms of the circumferential angle and the cone angle. The angles of emission are then obtained by the following general equation:

$$R = P(\xi) = \frac{\int_{l1}^{\xi} f(x)\,dx}{\int_{l1}^{l2} f(x)\,dx} \quad . \tag{3.1}$$

That is the cumulative distribution function, describing the transformation of a set of uniformly distributed random numbers R of the unit interval [0,1) into a set of usually non uniformly distributed random numbers ξ of an interval [l1, l2). f(x) is the probability density function, that is the average distribution that must be satisfied by the ξ values determined by the simulation scheme or simply the proportion of values that lie in the region $\Delta\xi$ around ξ. Each value ξ is then obtained by choosing an R value at random and using the functional relation $P(\xi)$ to determine the corresponding value of ξ. By choosing a suitable function f(x) we achieve a description of the desired physical process.

We want to determine the intensity reflected by a point on a surface, therefore we have to sample the total hemisphere of incoming directions. By using the reflection model we can perform importance sampling and concentrate the distribution of the sample rays in the parts of the hemisphere that are of major "importance" in terms of contribution to the reflected intensity. Thus variance reduction can be achieved and the probability density function f(x) can be specified according to the surfaces reflection properties. The surfaces that have to be considered are either diffuse or specular reflecting.

3.1 Sample Distribution for a Diffuse Surface

In case of an ideal diffuse reflecting surface the incident energy from a particular direction is reflected according to Lambert's cosine law. Thus we should distribute the samples according to the intensity distribution for diffuse reflection. The energy $e'(\lambda,\theta,\varphi)$ emitted from a diffuse surface in a given direction (θ,φ) per unit area dA, per unit time, within a unit wavelength interval centered around the wavelength λ, and into a unit elemental solid angle $d\omega$ centered around the direction (θ,φ) can be expressed as

$$e'(\lambda,\theta,\varphi)\, dA\, d\omega\, d\lambda = i'(\lambda)\, dA \cos\theta\, d\omega\, d\lambda\ , \tag{3.2}$$

where $i'(\lambda)$ is the spectral intensity of radiation in an interval $d\lambda$ around a single wavelength (Figure 1). Consequently, there exists the relation

$$e'(\lambda,\theta,\varphi) = i'(\lambda) \cos\theta\ , \tag{3.3}$$

which is Lambert's cosine law. For a diffuse surface the directional spectral emissive power $e'(\lambda,\theta,\varphi)$ does not depend on φ.

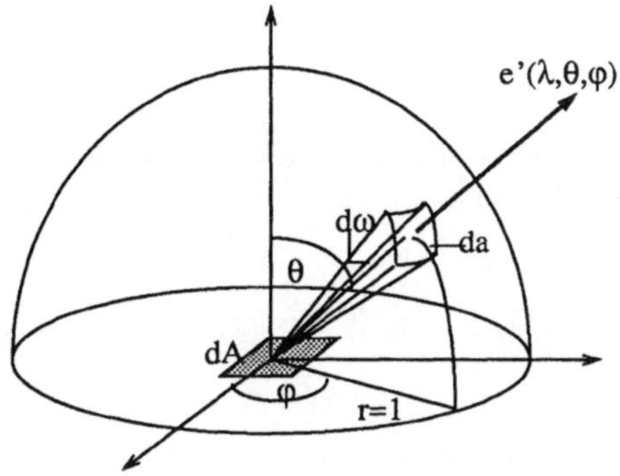

Figure 1: Geometry for radiant energy calculations

In the calculation of total radiant energy rejection by a surface one needs the spectral emissive power integrated over all solid angles of a hemisphere located over the surface. This quantity is called the hemispherical spectral emissive power $e(\lambda)$ of a diffuse surface. It is the energy leaving a diffuse surface per unit time per unit area dA at the center of the base of a unit hemisphere. By definition, a solid angle anywhere above dA is equal to the intercepted area on the unit hemisphere. An element of this hemispherical area is given by

$$d\omega = da / r^2 = r^2 \sin\theta \, d\theta \, d\varphi / r^2 = \sin\theta \, d\theta \, d\varphi \,. \tag{3.4}$$

Hence, the spectral emission from dA per unit time and unit surface area passing through the element on the hemisphere is given by

$$e'(\lambda,\theta,\varphi) \, d\omega = i'(\lambda) \cos\theta \sin\theta \, d\theta \, d\varphi \,, \tag{3.5}$$

which is the probability density function. To obtain the emission passing through the entire hemisphere, equation (3.5) has to be integrated over all solid angles to give

$$e(\lambda) = \int_0^{\pi/2} \int_0^{2\pi} i'(\lambda) \cos\theta \sin\theta \, d\theta \, d\varphi \,. \tag{3.6}$$

Principally we now have all we need to know in order to choose the angles θ and φ of emission. The energy emitted in $d\theta$ (or in $d\varphi$) is found by integrating the other variable φ (θ). Expressed in terms of the RGB primaries, equation (3.5) becomes

$$e'(\theta,\varphi) \, d\omega = I \cos\theta \sin\theta \, d\theta \, d\varphi \,, \tag{3.7}$$

where I is the intensity of either red, green, or blue light. The probability of emission in a an angular interval around one of the angles of emission is then the energy emitted in the interval around the angle divided by the total energy (equation (3.6)). Using the cumulative distribution function (3.1) gives the emission angles in terms of a random number of the unit interval.

The distribution function for the circumferential angle φ can be obtained by relating φ which lies in the interval $[0, 2\pi)$ to a random number R_1 of the unit interval $[0,1)$ by

$$\varphi = 2\pi R_1 . \tag{3.8}$$

From function (3.8) we see that the circumferential angle φ will be uniformly distributed over the range $[0, 2\pi)$.

Similarly the cone angle θ which lies in the interval $[0, \pi/2)$ can be related to a random number R_2 of the unit interval. Thus

$$\cos\theta = \sqrt{1 - R_2} \quad .$$

Since R_2 is uniformly distributed over $[0,1)$, $(1 - R_2) = R_3$ is also uniformly distributed and we can write

$$\cos\theta = \sqrt{R_3} \quad \Rightarrow \quad \theta = \arccos(\sqrt{R_3}\,). \tag{3.9}$$

Thus we distribute our samples over a total solid angle of 2π steradian, this is the entire hemisphere, but sample the incoming light more dense at small angles since these directions are most important for a point on a visible surface in terms of variance in the reflected intensity (Figure 2).

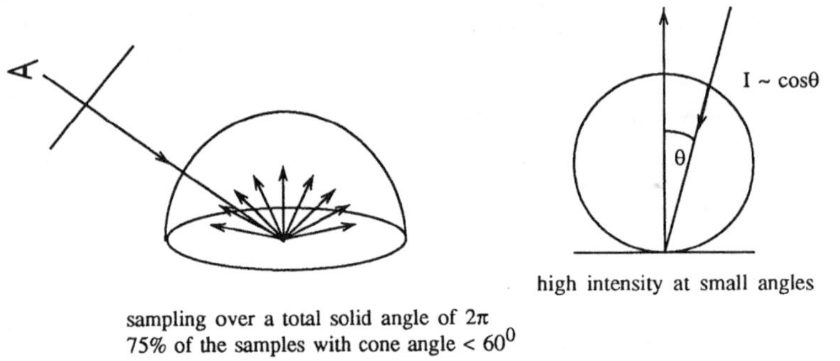

Figure 2: Sample distribution for a diffuse surface according to Lambert's Law for reflection

We have found a way to sample the incoming intensities for a point on a diffuse surface. Similar the perfect specular reflecting surfaces obey well-known laws for reflection that can be used for the distribution of samples.

3.2 Sample Distribution for a Specular Surface

For a perfect specular reflector the entire hemisphere of incoming directions has only to be sampled in the direction of reflection of the incident ray in order to reduce variance since only light from that direction is reflected back in the direction of the incident ray.

But most real surfaces are no perfect specular reflectors. And we can distribute the samples using the simple Phong reflection model. Thus we do not have to consider the entire hemisphere of incoming directions but only a smaller solid angle around the mirror direction. Only the energy arriving through this small solid angle makes a significant contribution to the intensity. By varying the size of the sampling cone, centered around the mirror direction, we can determine the surfaces glossiness. Due to the Phong model the directional emissive power for a specular surface can be expressed as

$$e'(\upsilon,\varphi)\, d\omega = I \cos^n \upsilon \sin\upsilon\, d\upsilon\, d\varphi, \tag{3.10}$$

where I is the intensity of light of either red, green, or blue color, the exponent n determines the glossiness of the surface, and υ is the angle between the mirror direction **m** and the direction of reflection **xe** (Figure 3). We can easily determine the distribution functions for the angles determining the sample direction if we change the geometry. That means we have to rotate the tangent plane T in such a way that the mirror direction **m** becomes perpendicular to it and thus transform the local spherical coordinate system with unity vector **k** as normal **n** into a new local coordinate system with origin at the intersection point P and the mirror direction **m** as unity vector **k**, normal to the tangent plane T'.

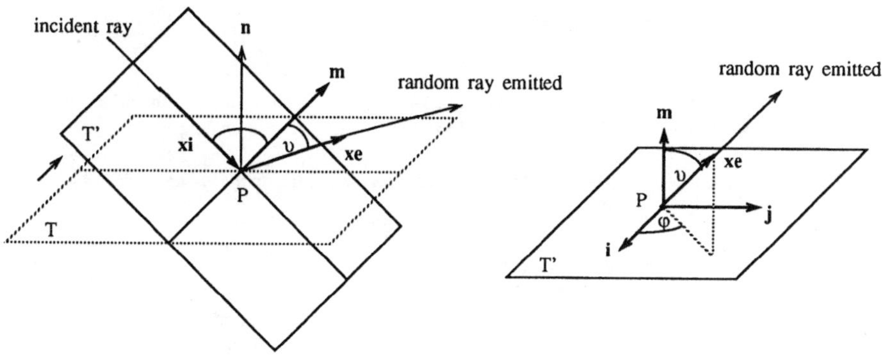

Figure 3: Geometry for the calculation of a random ray emitted by a specular surface

This transformation has the advantage that the emission from a specular surface in this description is not dependent on the circumferential angle φ, and thus we can use the same random ray selection process as described in section 3.4.1 for a diffuse surface. By substituting equation (3.10) into equation (3.1) we obtain the distribution functions for the circumferential angle φ and the cone angle υ describing the ray direction, relative to the new local coordinate system depicted in Figure 3.

To obtain the distribution function for the circumferential angle φ, we relate $\varphi \in [0, 2\pi)$ to a random number $R_4 \in [0,1)$ by

$$\varphi = 2\pi R_4 . \tag{3.11}$$

The cone angle υ which lies in the interval $[0, \pi/2)$ can be obtained by selecting a random number $R_5 \in [0,1)$.

$$\cos\upsilon = \sqrt[n+1]{1 - R_5} \Leftrightarrow \upsilon = \arccos(\sqrt[n+1]{1 - R_5}\,) . \tag{3.12}$$

The distribution function (3.11) indicates that the circumferential angle φ is uniformly distributed over the range of $[0, 2\pi)$.

The distribution function (3.12) for the cone angle υ determines the size of the sampling cone centered around the mirror direction **m**. If the exponent n of the cosine is large the sampling cone is small and we have a nearly mirror like surface.

According to the Phong reflection model the intensity of light reflected decreases as the angle υ between the mirror direction **m** and the direction of the incoming light (- **xe**) increases. Thus we have to cast our samples more frequent in the peak region of the sampling cone (Figure 4). Distributing the samples for a specular surface around the mirror direction enables us to model reflections that are no more sharp but soft. Thus also aliasing effects can be avoided.

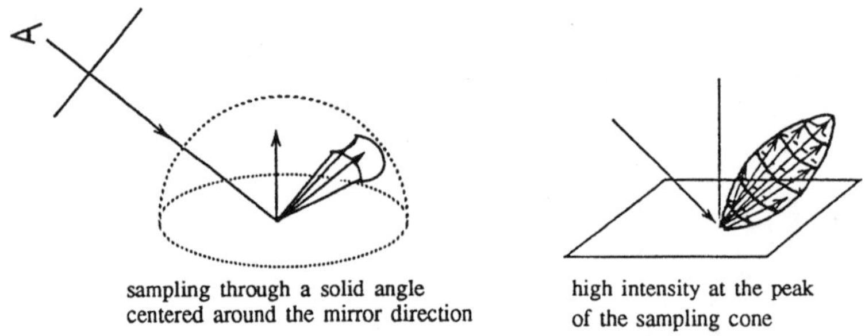

Figure 4: Sample Distribution for a specular surface according to the Phong model

3.3 Sample Distribution for Non "Ideal" Surfaces

Most real surfaces are not ideal diffuse nor ideal specular instead they incorporate both reflection properties. We can render these surfaces without having to find new sample distribution functions by simply combining the sampling process for diffuse and specular surfaces. The fraction of the reflectance that is specular is determined by the specular reflection coefficient ks and the fraction of light that is diffusely reflected is determined by the diffuse reflection coefficient kd, where kd + ks = 1. This corresponds to approximating the surfaces reflection properties as the sum of a diffuse portion which is independent of the view point and a specular portion which depends on the view direction.

These reflection coefficients are usedto determine the percentage of samples that are distributed according to the distribution functions for ideal diffuse surfaces and the distribution functions for ideal specular surfaces. That means, from the total number of N samples for a visible point on a surface, we distribute kd·N samples according to the diffuse distribution functions (3.8), (3.9), and ks·N samples according to the specular distribution functions (3.11) and (3.12) (Figure 5).

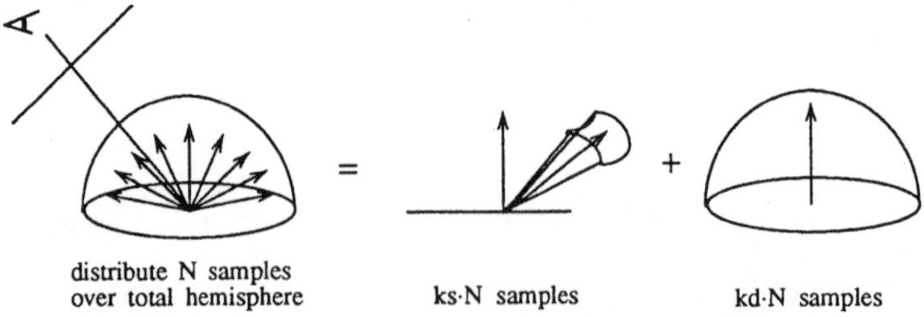

Figure 5: Sample distribution for a non ideal surface according to the diffuse and specular reflection properties

But for all higher-level ray-surface intersection points we only have one random ray to sample the incoming light due to the path structure. In such cases we choose one of the two distribution functions per random.

The sample distributions described in the previous sections provide two great advantages: First, we do not have to weight the intensity of a sample according to its direction in order to account for the reflection by Lambert's law and by the Phong model, because we do not distribute the samples uniformly over the hemisphere of incoming directions. Instead, we perform importance-sampling. Thus, we already account for the contributions of the incoming intensity to the reflected intensity by sampling with the distribution functions (3.9), (3.12).

This also improves the performance of path-tracing. The number of samples necessary can be essentially reduced since the contribution of each path to the final intensity is increased and thus also the variance is reduced.

Second, we need not to reduce the incoming intensity of a point P1 from a point P2 by multiplying it with the inverse square of the distance r between those points according to the law of inverse squares [BAR 25]. Since we distribute our samples radially from a point the factor $1/r^2$ is automatically included in the ratio of sample density to intercepted surface area. Thus at each ray-surface intersection point in the path the intensity reflected in direction (θ_r, φ_r) is calculated by

$$I(\theta_r, \varphi_r) = I(\theta_i, \varphi_i) \cdot sr , \tag{3.13}$$

where $I(\theta_i, \varphi_i)$ is the incoming intensity from the random sample direction (θ_i, φ_i), and sr is a surface reflection coefficient. Thus the intensity is reduced after each reflection. The path is then followed until a sufficient number of reflections have occurred to reduce the intensity below some predetermined level. This level is chosen so that the effect of the ray in a succeeding reflection would be negligible. The path then is terminated.

It has been found by trial that the rendering of an image with the sample distribution functions described in the previous sections in general requires a constant 40 paths per pixel if the direct contributions of the light sources are considered also to guarantee a bound on the variance. Thus, the hemisphere of incoming directions for each point visible to the eye is sampled by 40 random rays plus one shadow ray for the light source. A comparison of images of the same scene rendered by the same process only using different random numbers of the unit interval to determine the sample directions reveals that the deviation in intensity of individual pixels is maximally 10%. But for the eye those differences are not perceptible. The path length has been fixed to a maximum of 16 rays.

4. Light Source Contributions

We made the assumption that light sources are the most significant sources of illumination and thus provide us a basis for importance sampling. In order to reduce the variance we let the light sources participate in a separate direct component calculation (equation (2.1)) similar to the shading model of an usual ray-tracer. It seems rather unlikely that a point light source will be hit by a random ray, and since we know already from the geometry of the scene where the light sources are we should not neglect that valuable information. How important this is even when we have extended the lighting model to area sources is demonstrated in Figure 6. A series of images is shown that were generated without calculating the illumination in an extra component, merely stochastic sampling was performed. An object of the scene acted as a spherical light source with uniform intensity. The images from left to right were generated with 40, 100, 200, 400 paths per pixel. A lot of speckles occur due to stochastic sampling. It is obvious that the number of samples is not sufficient to guarantee the same results as an image generated by calculating the light source contributions in an separate component. Compare with Figure 7 where the light source contribution has been calculated in a separate direct component.

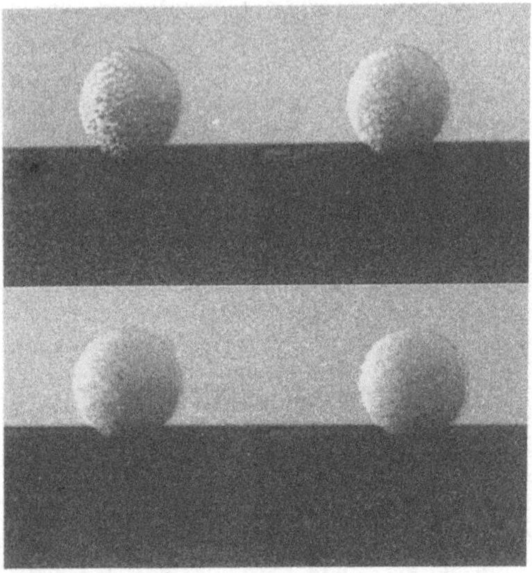

Figure 6: Random sampling without direct component calculation of light source contributions

Figure 7: Random sampling with direct component calculation of light source contributions

At each ray-surface intersection point in the path we go explicitly to one of the light sources by means of a shadow ray. If the point lies in shadow of the light source we have no contributions to the illumination from the light source and thus only the contribution of the random sample (3.13). If the point is illuminated by the light source then we have to account for the intensity contribution of the light source also, but we have to weight the incident energy according to the angle of incidence (Lambert's law or Phong model), and the distance to the source. To prevent us from counting the intensity contributions to a point by the light source twice the random samples are such that they do not intersect the light sources.

Since we have two samples for the incoming energy to a point their contributions must be weighted suitable to obtain the proper sample mean value (2.1). The weighting factor we have found was developed on the extension of the simple lighting model consisting of point light sources with no shape to radiant area sources of circular shape. It includes an automatic attenuation of the intensity by the distance the light travels.

4.1 Weighting Factor for a Diffuse Surface

The contribution a light source makes to the total hemisphere of incoming directions for a point on a surface can be measured by means of the solid angle the light source occupies. The solid angle ω is the projected area of the light source normal to the view direction from the surface point divided by the square of the radius R of the hemisphere.

$$\omega = A / R^2 \tag{4.1}$$

Thus we should weight the intensity contribution of the light source by $\omega/2\pi$ which is the ratio of the solid angle ω to the total hemisphere. The intensity of the random sample has then to be weighted by $1 - \omega/2\pi$. Since the solid angle varies inversely to the square of the distance R the law of inverse squares is already accounted for. If we want to weight the intensity contributions of a light source by the solid angle we have to change our lighting model. For a point light source we cannot determine the solid angle since it has no associated area. Therefore we extended the lighting model to area sources by trying to stick to the simple concept of a point light source so that area sources can be implemented to a ray-tracer with little additional cost.

The illumination produced by a sphere of any size is inversely proportional to the square of the distance measured from its center, and is equal to the intensity of a point source having the same total amount of light divided by the square of the distance (always, of course, assuming the cosine-law to hold for the spherical surfaces, and the brightness to be uniform over the sphere). Thus we can extend the point light sources to spheres of uniform intensity.

Principally, we regard the light source as a circular disk of radius a and uniform intensity I which always is assumed to be perpendicular to the direction **l** from the point to the center of the light source. Thus we have an idealized light source model that requires three parameters as specification: the center of the light source, a radius, and the intensity in RGB values. The solid angle of the circular disk source at a point P on a surface can now be calculated by

$$\omega = 2\pi\,(1 - \cos\alpha) = 2\pi\,(1 - r / \sqrt{r^2 + a^2}\,), \tag{4.2}$$

where α is half the opening angle of the light source, which can be determined by the radius a of the source and the distance r of point P to the center of the source. The weighting factor wd for the direct contribution of a light source to a point on a diffuse surface then is

$$wd = 1 - r / \sqrt{r^2 + a^2}. \tag{4.3}$$

The intensity reflected from a point on a diffuse surface then gives

$$I(\theta_r, \varphi_r) = (wd\, I(\theta, \varphi) \cos\theta + (1 - wd)\, I(\theta', \varphi')) \cdot sr, \tag{4.4}$$

where (θ_r, φ_r) is the direction of reflection, (θ, φ) is the direction **l** from the point to the center of the light source, (θ', φ') is the direction of the random sample.

Since we always calculate the full solid angle ω we get problems in those cases where the angle θ is almost 90 deg. and ω is large. There will be no intensity from the light source and the intensity from the random sample is also very small. This results in dark spots at the transition from lighted to non-lighted area of an object. We may not multiply the intensity by $\cos\theta$ because the light source is still visible from point P at angles $\theta \in [0, \pi/2 + \alpha]$ and thus contributes to the illumination of point P. Substituting angles $\theta \in (\pi/2, \pi/2 + \alpha]$ in the cosine function would result in negative values. Since there cannot be a negative intensity we have to find a function $f(\theta)$ that transforms the interval $[0, \pi/2 + \alpha]$ into the interval $[0, \pi/2)$, so that $\cos(f(\theta)) \in [0,1]$:

$$f(\theta) = \pi/2\ \theta / (\pi/2 + \alpha). \tag{4.5}$$

The weighting factor wd (4.3) is correct for $\theta \in [0,\pi/2 - \alpha)$ but for $\pi/2 - \alpha < \theta \le \pi/2 + \alpha$, the solid angle is smaller than ω. To calculate the surface of the spherical segments at those angles involves quite complicated integrations. In order to provide a fast calculation approximating functions have been determined. Since we know that the solid angle of a light source $g(\omega)$ with opening angle α is

$$g(\omega) = \begin{cases} \omega/2\pi, & \text{if } \beta = \alpha \\ 1/2\ \omega, & \text{if } \beta = 0, \\ 0, & \text{if } \beta = -\alpha \end{cases} \qquad \text{where } \beta = (\pi/2) - \theta. \tag{4.6}$$

We can model the continuous decrease of the solid angle by increasing angle θ with

$$g(\omega) = \sin^2(h(\beta)) \cdot \omega, \tag{4.7}$$

where

$$h(\beta) = \pi/4 + \pi/4 \cdot \beta/\alpha. \tag{4.8}$$

Thus the final weighting factor wd becomes

$$wd = \begin{cases} \omega/2\pi, & \beta \ge \alpha \\ \sin^2(\pi/4 + \pi/4 \cdot \beta/\alpha) \cdot \omega/2\pi, & \text{if } \beta \in [-\alpha, \alpha). \\ 0, & \text{else} \end{cases} \tag{4.9}$$

The weighting factor wd provides a proper calculation of the solid angle a light source has seen from a point on a diffuse surface. A light source only contributes to the illumination of a point if the angle $\theta \leq \pi/2 + \alpha$, and if the point is not in shadow of the source.

4.2 Weighting Factor for Specular Surfaces

The reflection of light from a specular surface is described by the Phong model, thus only light incident from a small solid angle around the mirror direction is considered to be relevant for the illumination. This also holds for the direct contributions of light sources. Therefore we have to weight the light source contributions by

$$ws = \cos^n \upsilon, \tag{4.10}$$

where υ is the angle between the light source direction and the mirror direction and n determines the sampling cone size. The contribution of the random sample is then weighted by (1 - ws). If the light source lies in the small solid angle around the mirror direction ws will be close to one and it can be assumed that the contributions from the indirect illumination is negligible. Therefore weighting with $ws = \cos^n \upsilon$ seems to be reasonable in the case of a point light source. Weighting with (4.10) causes reflections of point light sources. Since we extended the lighting model to area sources we have to modify the weighting factor ws in order to achieve correct reflections of area light sources, i.e. the correct highlight size.

An intensity contribution to a specular surface will only occur if an illumination ray lies within the narrow sampling cone. But for area light sources it may happen that they contribute to the intensity of a point although the center of the light source already lies outside the sampling cone. Thus we have to test if the angle υ between the mirror direction and the illumination ray is smaller than α, where α is the angle defined by

$$\alpha = \arctan (r/R)\,, \tag{4.11}$$

where r is the radius of the light source and R is the distance from the point on the specular surface to the center of the light source. If the angle υ is smaller than α, then a part of the light source can be seen in the sampling cone. Thus we have to weight the intensity contribution of the light source by $ws = 1$. If α is smaller than υ we generate a Phong like decrease of intensity by

$$ws = \cos^n \upsilon'\,, \quad \text{where} \quad \upsilon' = \upsilon - \alpha\,. \tag{4.12}$$

Weighting the Intensity contribution of an area source by

$$ws = \begin{cases} 1, & \text{if } |\upsilon| < \alpha \\ \cos^n (\upsilon - \alpha), & \text{if } |\upsilon| \geq \alpha \end{cases} \tag{4.13}$$

allows a correct rendering of the reflections of area sources in specular surfaces as far as circular shaped light sources of uniform intensity are concerned.

The indirect contributions to the illumination of a point on a specular surface can then be weighted by (1 - ws). Therefore, the intensity reflected by a point on a specular surface is

$$I(\upsilon_r, \varphi_r) = (\, ws\, I(\upsilon, \varphi) + (1 - ws)\; I(\theta, \varphi)\,) \cdot sr\,, \tag{4.14}$$

where $I(\upsilon, \varphi)$ is the incoming intensity of the light source, $I(\theta, \varphi)$ is the incoming intensity of a random sample casted in the sampling cone and sr is a surface reflection coefficient.

Differing from the usual shading model of a ray-tracer we do not combine the specular and diffuse reflected components of the light source contributions for each point on a non-ideal surface. At those points we either sample according to the diffuse distribution function or according to the specular distribution function. In the first case the surface is treated as ideal diffuse thus the direct light is reflected diffusely and the intensity can be calculated according to (4.4). In the second case we treat the surface as ideal specular and calculate the reflected intensity according to (4.14). Since we distribute $ks \cdot N$ samples according to the specular distribution function and $kd \cdot N$ samples according to the diffuse distribution function the ratio of diffuse and specular reflected light remains correct.

5. Shadow Test Extension

Before calculating the contribution to the illumination of a point by a light source a shadow test is performed that determines whether the light source is visible from that point. But if an area source is partially obscured by an object penumbras will occur. We cannot test whether a point lies in shadow of an area source by a *single* shadow ray. We have to trace shadow rays to all different parts of the light source. The first ray-surface intersection point is encountered N times. Thus we have a number of N shadow rays that can be distributed in the solid angle of the light source. Two unit vectors i, j of the tangent plane perpendicular to a vector l from the point on the surface to the center of the light source have to be determined. These vectors can be used to generate a sample direction d for the extended shadow test. In the simplest case the shadow rays are distributed uniformly over the boundary of the light source. With this method soft shadows can be generated (Figure 8), but small objects obscuring parts of the light source will not be found. Therefore we have to distribute the shadow rays uniformly over the entire light source area to make sure that all parts of the light source are tested.

We tried two different methods of distributing samples uniformly over a circular area of radius r: A two-dimensional random distribution (Lange 89), and a segmented distribution, where only the distance from the center of the light source was selected randomly, i.e. we used a random number R_i of the unit interval [0,1) and determined the ray directions by

$$\mathbf{d}_i = \mathbf{l} + \mathbf{i}\ r \sqrt{R_i}\ \cos\frac{i\,360}{N} + \mathbf{j}\ r \sqrt{R_i}\ \sin\frac{i\,360}{N}\ , \qquad i = 1,\ldots,N\ . \tag{5.1}$$

Although the results of this method are not of vast difference to the previous method, the segmented distribution provides the most satisfying results (Figure 8). This is due to the fact that we can make sure that every sector of the light source is sampled exactly once.

At all higher level intersection points only one sample for the light source is available, therefore we decided to calculate the shadow at those points by tracing the shadow ray still to the center of the light source, otherwise we have to add samples to the path structure. This is a simple solution which of course is not the best in terms of realism (Figure 8).

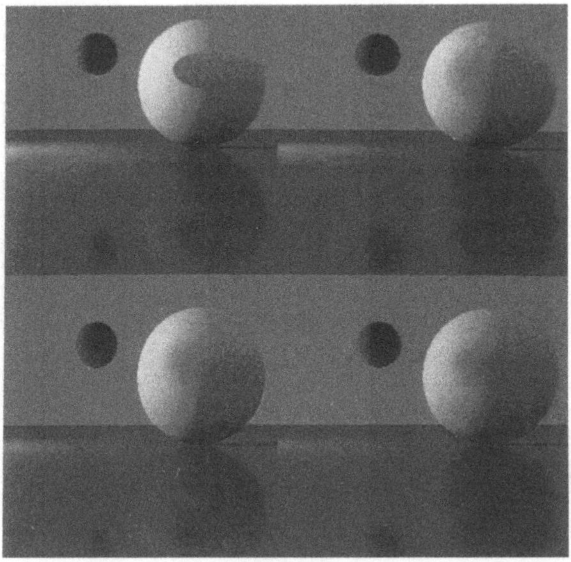

Figure 8: Four different shadow test performances

Upper left: Center of light source was sampled as in usual ray-tracing procedures. The shadow is sharp and of uniform intensity.
Upper right: Samples were distributed uniformly over the boundary of the light source. Soft shadow but no umbra.
Lower left: Samples were distributed uniformly over the area of the light source by the two-dimensional distribution. Umbra and penumbra areas are visible.
Lower right: Segmented distribution, the light source was sampled with (5.1). Umbra area more distinct.

All images have a resolution of 256 × 256. The light source has a radius 4, the yellow sphere has radius 7, and the purple sphere has radius 2. In the reflected image of the sphere it is recognizable that the shadow the small sphere produces is not reflected correctly. Since we sample for all higher level intersection points only the center of the light source, the reflection will always show a shadow as in the upper left image.

6. Reduction of Sample Set Size

The weighting factors wd and ws described above provide a basis for the reduction of paths per pixel. Up to now we have sampled the indirect contributions to a point visible to the eye by the same fixed number N of samples for every pixel. But if a light source occupies a large solid angle of the hemisphere, then the contribution of a random sample is negligible. Thus we can reduce the number of paths for each pixel in dependence of the weighting factor. That means, we first trace only one path through a pixel and determine the weighting factor wd. We then multiply the total fixed number N of paths per pixel by the factor (1-wd).

A comparison of the intensity diagrams of images generated with a fixed number of 40 paths per pixel and an image with variable number of paths per pixel determined by wd has shown that the maximum deviation in intensity values of single pixels is less than 5%. Similarly, we can reduce the number of samples per pixel by the weighting factor ws. If ws equals one we do not have to distribute samples in the reflection cone since their contribution is (1-ws) = 0. How much computation time can be saved by that method cannot be determined with universal validity since this strongly depends on the scene. The number of paths per pixel for the generation of Figure 7 could have been reduced from 40 to an average of 26 paths per pixel.

7. Conclusion

A stochastic ray-tracing method has been presented that models the radiant light transfer in order to generate realistic images. This method can be implemented to basic ray-tracing algorithms so that the deterministic shading model of a ray-tracer can be extended to a stochastic model. The implementation can be achieved by extending the light transport model, i.e. by adding random samples. Importance sampling has been performed by distributing the samples for the indirect illumination according to the reflection properties of the surfaces and by calculating the light source contributions in a separate direct component. Area sources of circular shape have been introduced. A weighting factor for the direct and indirect contributions to the illumination has been found that provides a basis of reducing the sample set size. The method renders objects of ideal diffuse, ideal specular, and of both reflection properties. Transmission has not been included yet, but is straightforward.

Acknowledgments

I am grateful for support and encouragement from all members of the Graphics & Vision group at the Philips Research Laboratories Eindhoven, the Netherlands where I have been occupied with the implementation of a stochastic ray-tracing algorithm for the simulation of radiant light transfer. All images presented have been generated there in an UNIX environment for a SUN workstation.

References

Barrows, W.E. (1925) Light, Photometry and Illuminating Engineering. McGraw-Hill, New York

Cook, R. (1985) Antialiasing by Stochastic Sampling. ACM SIGGRAPH'85, Course Notes of the 12. Annual Conference&Exhibition on Computer Graphics and Interactive Techniques, July 1985

Dippé, M.A.Z.; Wold E.H. (1985) Antialiasing Through Stochastic Sampling. Proceedings of SIGGRAPH'85, In Computer Graphics, Vol. 19, No. 3, July 1985, pp. 69-78

Kajiya, J.T. (1986) The Rendering Equation. Proceedings of SIGGRAPH'86, In Computer Graphics, Vol. 20, No. 4, August 1986, pp. 143-150

Lange, B. (1989) Die Simulation Optischer Strahlungsprozesse durch Stochastisches Ray-Tracing. Diplomarbeit, RWTH Aachen, pp. 79-86

Lee, M.E.; Redner, R.A.; Uselton S.P. (1985) Statistically Optimized Sampling for Distributed Ray Tracing. Proceedings of SIGGRAPH'85, In Computer Graphics, Vol. 19, No.3, July 1985, pp. 61-65

Rubinstein, R.Y. (1981) Simulation and the Monte Carlo Method. John Wiley & Sons

Unbiased Variance Reduction for Global Illumination

David Kirk, James Arvo

Abstract

We describe the application of two standard Monte Carlo variance reduction techniques to the problem of global illumination. First, we examine the technique known as importance sampling using a simple example of integration, and show how the principle can be applied in the context of solving the rendering equation. Secondly, we show that stratification can be performed by partitioning incident illumination into approximate strata representing direct and indirect illumination. By applying both importance sampling and stratification we can significantly increase the statistical efficiency of the estimators obtained through path tracing without introducing bias.

1 Introduction

The rendering equation, introduced by Kajiya [Kajiya 86], is a mathematical model of radiative transfer which subsumes nearly all existing illumination models used in image synthesis. Algorithms such as classical ray tracing [Whitted 80], radiosity [Goral 84], or the "Utah approximation" [Phong 73], [Gouraud 71] can be viewed as approximations in which specific simplifying assumptions lead to tractable methods of solution. As such, these approaches are not "competing" or "alternative" techniques to the rendering equation; they are algorithms for an approximate or partial solution. In fact, in [Kajiya 86], Kajiya proposes Monte Carlo path tracing as one potential solution technique.

The algorithms mentioned above present different trade-offs between optical effects and efficiency and address restricted forms of the rendering equation. We will concentrate on distribution ray tracing [Cook 84] and Monte Carlo methods to approach an accurate solution of the rendering equation in its full generality.

In this paper we describe the use of two powerful Monte Carlo techniques to increase the statistical efficiency of Monte Carlo solutions. This is accomplished by reducing the variance of the estimators while leaving the mean unaffected. That is, when applied correctly, they will reduce the noise without introducing statistical bias. This increase in efficiency allows us to attain a desired accuracy with dramatically fewer samples than would otherwise be necessary. Zero bias ensures us that the solution to which we converge is the correct solution.

Our approach to variance reduction is to combine importance sampling and stratification. Both techniques are applied to the problem of selecting a new sampling direction at each of the scattering events comprising a Markov chain, or *random walk*. Through stratification, the integral of direct and indirect illuminance are estimated independently. The role of importance sampling is to distort the probability densities of the directions chosen for the scattering in order to emphasize those of high incident illuminance. These are the "important" directions with respect to the point in question, so we wish to sample them more frequently.

Both of these techniques are applied in such a way that the random variable estimating the incident illuminance remains unbiased. This fact ensures that we can approach the correct answer arbitrarily closely by employing a sufficiently large number of samples. Many adaptive sampling plans fail to meet this criterion, so there is no guarantee that the correct answer can be reached. In [Kirk 91], we discussed bias from adaptive sampling in screen space and described an algorithm for avoiding such bias.

2 Importance Sampling

In the approach known as *analog Monte Carlo*, simulated events are assigned probabilities that are faithful to those in the physical process being modeled. *Importance sampling* is a departure from this which alters the sampling distribution in order to invest most of the computational effort in collecting samples that contribute significantly to the solution. By weighting these contributions appropriately, we can maintain the original expected value while reducing the variance.

The principle is simple to illustrate with a finite-dimensional problem, such as estimating the value of a one-dimensional definite integral. For example, suppose we wish to estimate the value of I, where

$$I = \int_a^b g(x)\,dx \tag{1}$$

First, define a function $f : [a,b] \rightarrow R$ such that $f(x) \geq 0$ for all $a \leq x \leq b$ and

$$\int_a^b f(x)\,dx = 1. \tag{2}$$

Then f is a valid *probability density function* (henceforth abbreviated PDF) over the interval $[a,b]$. Let $\mathbf{X}$ be a random variable distributed according to f. That is

$$\Pr(x_0 \leq \mathbf{X} \leq x_1) = \int_{x_0}^{x_1} f(x)\,dx. \tag{3}$$

Now consider another random variable, ξ, defined in terms of $\mathbf{X}$ by $\xi = g(\mathbf{X})/f(\mathbf{X})$. The expected value of ξ, denoted $E(\xi)$, is given by

$$E(\xi) = \int_a^b \frac{g(x)}{f(x)} f(x)\,dx = I. \tag{4}$$

Thus, we can pose the problem of evaluating a definite integral in terms of finding the expected value of a random variable. Because $E(\xi) = I$, the random variable ξ is an *unbiased estimator* of I and, following the nomenclature in [Halton 70], we shall called it the *primary estimator* of I. Any PDF, f, which is non-zero almost everywhere on the support of g will result in a primary estimator which is unbiased, although the variance of this estimator will typically depend strongly on the choice of f.

If $\xi_1, \ldots, \xi_n$ is a random sample of ξ, then the random variable $\mathbf{Y}$ given by

$$\mathbf{Y} = \frac{1}{n}\sum_{i=1}^{n} \xi_i \tag{5}$$

is a statistic called the sample mean of ξ. This is another unbiased estimator of I, but of lower variance than ξ. We shall call this the *secondary estimator* [Halton 70]. As an example, if f is the uniform density function on the interval $[a,b]$, that is, if $f(x) = 1/(b-a)$ for all $a \leq x \leq b$, then $\mathbf{Y}$ reduces to

$$\mathbf{Y} = \frac{b-a}{n}\sum_{i=1}^{n} g(\mathbf{X}_i) \tag{6}$$

where $\mathbf{X}_1, \ldots, \mathbf{X}_n$ are drawn uniformly from $[a,b]$. This is the well-known *sample-mean Monte Carlo* method [Rubinstein 81] for estimating a definite integral. By the law of large numbers the variance of the secondary estimator approaches zero as n approaches infinity. However, because the standard error varies only as $1/\sqrt{n}$, this can be a very costly means of achieving the desired accuracy.

Where the cost of sampling the primary estimator is large, it may be vastly more efficient to first reduce its variance through a judicious choice of the PDF rather than employing an enormous number of samples in the secondary estimator. This strategy is called *importance sampling*. We shall refer to a PDF which is used to reduce the variance of a primary estimator in this way as an *importance sampling probability density function* and abbreviate it ISPDF. Because sampling in the context of solving the rendering equation entails the fairly expensive process of stochastically tracing paths through the environment, it can benefit greatly from importance sampling.

If the integrand, g, in the previous example is non-negative, it is possible to obtain a zero-variance estimator through importance sampling. In particular, if we let $f(x) = g(x)/I$, this results in an estimator ξ which is identically I. Thus, if we "already know the answer," importance sampling returns it infallibly [Rubinstein 81]. In practice, however, the value of importance sampling lies in the fact that partial information about the solution can be used to reduce the variance. Intuitively, the greater the similarity between the ISPDF and the integrand, the greater the statistical efficiency of the resulting primary estimator. Importance sampling makes use of *a priori* information about the solution by increasing the probability of sampling where the magnitude of the integrand is large.

In Sec. 4 we describe how to construct ISPDFs that are useful in the context of solving the rendering equation.

3 Stratification

Stratification is a common Monte Carlo variance reduction technique [Hammersley 64] [Halton 70] [Rubinstein 81] [Kalos 86] which has been widely used in computer graphics to improve image quality [Cook 84] [Cook 86] [Kajiya 86] [Lee 85]. It is an effective variance reduction technique for Monte Carlo integration if the domain of integration can be partitioned into strata within which the variance is smaller than the difference between their means [Hammersley 64]. In such a case, estimating the integral of each stratum independently is superior to integrating over the entire domain as a whole.

Although stratification has been applied to a variety of problems in computer graphics, the published literature has thus far not addressed the roles that this technique can play in solving the rendering equation efficiently. In this context we can perform a very useful form of stratification by partitioning the incident illuminance into approximate strata consisting of direct and indirect illumination.

If the majority of the surfaces in the environment are far from ideal reflectors, the intensity of indirect illumination will tend to be significantly less than direct illumination. Identifying the strata corresponding to first and subsequent generations of light scattering can be accomplished using the geometry of the light sources. This information is too valuable to ignore. A bounding volume around a light source is very useful importance information for our ISPDF. We can be *certain* that first-generation light rays will come only from emitters, so restricting the search for these rays to domains closely approximating the visible portions of emitting surfaces can be of great value.

4 ISPDFs for the Rendering Equation

When evaluating a definite integral via Monte Carlo, importance sampling is typically done by selecting as an ISPDF some analytic function which shares important features of the integrand. For instance, a good choice is one with singularities that coincide with those of the integrand. In solving integral equations, importance sampling alters the probabilities of the Markov process used as the primary estimator [Albert 56] [Kalos 86]. In the context of the rendering equation, this means altering the probabilities associated with selecting the direction in which to cast the next ray at each scattering event of the random walk. Given such a direction, the subsequent point along the random walk is found deterministically via ray tracing. At each such point we wish to construct an ISPDF which allows us to select "important" directions with high probability; that is, directions in which the incident illuminance is large. The features required of the ISPDF are as follows:

1) Because "important" directions are dependent upon position, the ISPDF must itself be a function of position. That is, we must be able to associate an ISPDF with any surface point.

2) The ISPDF is actually a PDF over the hemisphere of directions defined by the surface normal at the given point. (We insist that the ISPDF be non-zero over the entire hemisphere, since it is difficult to find the support of the incident illuminance function. We would consider a sphere to include transmittance.)

3) We must be able to efficiently generate random direction vectors distributed according to the ISPDF. If this operation is too costly, it may be less effective than simply using the primary estimator.

4) We must be able to efficiently compute the ISPDF density at any chosen direction in order to appropriately weight the resulting estimator (by dividing by this density).

5) The ISPDF constructed at a given surface point should capture the most important features of the incident illuminance at that point.

6) We should be able to construct the ISPDFs in any environment, without placing restrictions on geometry or reflectance properties. This allows the algorithm to operate in any environment which can be ray traced.

During rendering, primary estimators are computed via Markov chains originating at the eye, as in [Kajiya 86]. At each point of the random walk, we construct an ISPDF which possesses properties 2 through 5 above. Once an ISPDF is constructed, we use it to obtain a new direction vector for continuing the random walk. Then we weight the resulting contribution by the reciprocal of the ISPDF density of this direction. One such mechanism that allows us to perform these operations efficiently is described in Sec. 5.

5 Constructing an ISPDF

In this section we describe a simple mechanism for associating an ISPDF with any given point in space, typically a point on a surface. The role of each ISPDF is to approximate the most important features of the illuminance distribution at the given point. As discussed above, in order to be effective, the ISPDF must be easy to generate, sample from, and evaluate.

Because each ISPDF must incorporate some knowledge of the light sources, we construct an approximate light source geometry by surrounding each emitter with a bounding sphere. From this we construct bounding cones, each with its apex at the point of scattering. The ratio of the solid angle subtended by the cones to the solid angle of the remainder of the hemisphere determines the probability that a uniformly distributed ray direction would sample the light. For our ISPDF, we increase that probability, since the light is very likely to be important. We then must normalize the result by the distorted probability, that is, by dividing by the density of the ISPDF in the direction of the sample chosen. Sampling the light is then a matter of sampling the cone uniformly (which is described in Appendix A) and normalizing the result. For multiple light souces this becomes more complicated when the cones overlap. This overlap must be taken into account when computing the ISPDF density. Failing to do so will favor the areas of overlap by sampling them more frequently without attenuating the contribution appropriately. This would bias the estimator.

We must be careful in selecting the cone spread. If it is too small, we may often miss the important information which is outside the cone. A small cone may also attenuate the important information too much when we hit it while drawing a sample from within the cone. Cones which are too large are not as objectionable since this simply diminishes the effect of the ISPDF. A too-large cone simply fails to focus "attention" closely on the important features. Erring either way will increase the variance without introducing bias.

Since the difference between hitting the light and not hitting the light is likely to be very great, we want to sample the light often. However, if we create an ISPDF which is inappropriately distorted, we may increase the variance rather than reducing it. For instance, an "unexpected" event, such as finding an emitter when sampling outside the bounding cones, can produce a sharp "spike" in the estimator owing to a large contribution being divided by a low probability density. Sampling at a higher rate in areas of low contribution causes less harm, so it is safer to err in this direction.

One variance reduction approach which has these features is stratification. We will *always* sample the light(s), and in addition, we will always sample the remainder of the hemisphere. We will combine the two by using the ISPDF density to weight them appropriately. It is important to note that Kajiya's technique of "always sampling the light" [Kajiya 86] is only appropriate when the light is treated as an independent stratum and weighted appropriately. Simply adding an extra sample to the estimator without renormalization may well reduce the variance, but at the expense of considerable bias.

We can also use any other knowledge that we possess to construct the ISPDF. In an approach similar to [Arvo 86] or [Heckbert 90], we can use information from an earlier "illumination" pass to determine where important objects are located. This can help to find bright indirect illumination. Any such information can be used to create an ISPDF, as long as we normalize by the density to avoid bias.

Examples of statistically biased approaches can be found throughout image synthesis. As discussed in [Arvo 90], adaptive tree depth control [Hall 83], or weight cutoff, can cause a systematic bias. [Kirk 91] describes how common practices of adaptive sampling in screen space can cause bias. Although biased estimators are sometimes convenient, they frequently appear in image synthesis unnecessarily. Often the bias can be eliminated by carefully designing the stochastic process through which the estimate is obtained. Viewing the operations as forms of stratification or importance sampling can be very helpful when doing so.

Anytime we depart from analog Monte Carlo, if we wish to avoid bias, we must take care to appropriately normalize the result. In the case of adaptively sampling the incoming illuminance over a hemisphere, the error due to bias can be very great. In the screen space regime, the gamut of values obtained is frequently restricted to the range of $[0, 1]$. When sampling incoming illuminance, the brightness can be far greater and more concentrated. Consequently, the opportunities for error are far greater.

6 Conclusions

We have described an approach for improving the efficiency of Monte Carlo solution of the rendering equation, employing both importance sampling and stratification. Although our technique for sampling direct illumination is very similar to [Kajiya 86], we have formally placed it in the framework of importance sampling and ISPDFs. Therefore, the same techniques can be applied to near-ideal reflectors or other "important" objects in the environment, without biasing the estimate. The computational cost of importance sampling is potentially far less than the cost of uniformly sampling bright features using a vast number of rays. With proper weighting, statistical bias can be guaranteed to be zero.

We expect that future work can use these results to formulate adaptive importance sampling plans which can increase both the quality of images and the efficiency with which they are generated. A careful design of these sampling plans can ensure zero bias, as well.

Acknowledgements

Much of this research was performed while the authors were employed at Apollo Computer and Hewlett-Packard. The authors also wish to thank the anonymous reviewers for their comments.

References

[Albert 56] Albert, G. E., "A general theory of stochastic estimates of the Neumann series for solution of certain Fredholm integral equations and related series," in Symposium on Monte Carlo Methods, edited by M. A. Meyer, J. Wiley, New York 1956, pp. 37-46.

[Arvo 86] Arvo, James, "Backward Ray Tracing," Developments in Ray Tracing, Siggraph Course Notes, Vol. 12, August 1986.

[Arvo 90] Arvo, James and David Kirk, "Particle Transport and Image Synthesis," Computer Graphics 24(4), August 1990, pp. 63-66.

[Cook 84] Cook, Robert L., Thomas Porter, and Loren Carpenter, "Distributed Ray Tracing," Computer Graphics 18(3), July 1984, pp. 137-145.

[Cook 86] Cook, Robert L., "Stochastic Sampling in Computer Graphics," ACM Trans. Graphics, 5(1), January 1986.

[Goral 84] Goral, Cindy M., Kenneth E. Torrance, Donald P. Greenberg, and Bennett Battaile, "Modeling the interaction of light between diffuse surfaces," Computer Graphics 18(3), July 1984, pp. 213-222.

[Gouraud 71] Gouraud, H., "Continuous Shading of Curved Surfaces," IEEE Transactions on Computers, 20(6), June 1971, pp. 623-628.

[Hall 83] Hall, R. A. and D. P. Greenberg, "A testbed for realistic image synthesis," IEEE Computer Graphics and Applications, 3(10), November, 1983, pp. 10-20.

[Halton 70] Halton, I. H., "A retrospective and prospective survey of the Monte Carlo method," SIAM Review, 12, 1270, pp. 1-63, 1970.

[Hammersley 64] Hammersley, J. M. and D. C. Handscomb, "Monte Carlo Methods," Chapman and Hall, 1964.

[Heckbert 90] Heckbert, Paul, "Adaptive Radiosity Textures for Bidirectional Ray Tracing," Computer Graphics 24(4), August 1990, pp. 145-154.

[Kajiya 86] Kajiya, J. T., "The Rendering Equation," Computer Graphics 20(4), August 1986, pp. 143-150.

[Kalos 86] Kalos, M. H. and Paula A. Whitlock, "Monte Carlo Methods, Volume I: Basics," J. Wiley, New York, 1986.

[Kirk 91] Kirk, David and James Arvo, "Unbiased Sampling Techniques for Image Synthesis," Computer Graphics 25(4), August 1991, pp. 153-156.

[Lee 85] Lee, Mark E., Richard A. Redner, and Samuel P. Uselton, "Statistically Optimized Sampling for Distributed Ray Tracing," Computer Graphics 19(3), July 1985, pp. 61-68.

[Phong 73] Phong, B. T., "Illumination for Computer-Generated Images," Ph.D. Dissertation, Department of Computer Science, University of Utah, Government ordering #AD-A0008-786, Salt Lake City, Utah, July 1973.

[Rubinstein 81] Rubinstein, R. Y., "Simulation and the Monte Carlo Method," J. Wiley, New York, 1981.

[Whitted 80] Whitted, Turner, "An Improved Illumination Model for Shaded Display," Communications of the ACM 32(6), June 1980, pp. 343-349.

APPENDIX A: Uniform Sampling from Cones

To generate directions uniformly distributed over a cone with axis (A_x, A_y, A_z) and spread angle α, we first sample from the cone centered on the z-axis with spread angle α, then transform the vector to obtain the desired orientation. This requires two uniform random variates as follows:

$$\text{sample } z \text{ uniformly from } [\cos\alpha, 1]$$
$$\text{sample } \theta \text{ uniformly from } [0, 2\pi]$$

$$D = \begin{bmatrix} \cos\theta\sqrt{1-z^2} \\ \sin\theta\sqrt{1-z^2} \\ z \end{bmatrix}$$

To map D into the cone with the desired orientation we apply the Householder reflection matrix, H, which maps the z-axis into the cone axis. That is,

$$H = I - 2NN^T$$

where N is orthogonal to the plane bisecting the z-axis and the cone axis. For very efficient storage and computation we need only store $V = \sqrt{2} * N$ which, for any cone axis, A, different from $(0, 0, -1)$, is given by

$$V = \sqrt{2}\frac{W}{||W||}$$

where

$$W = \begin{bmatrix} -A_x(1 + A_z) \\ -A_y(1 + A_z) \\ A_x^2 + A_y^2 \end{bmatrix}$$

For a cone axis of $(0, 0, -1)$ we set $V = (0, 0, 1)$. We then transform D into D' by

$$D' = D - V(V^T D)$$

which requires only 5 adds and 6 multiplies.

Direct Lighting Calculation by Monte Carlo Integration

Peter Shirley, Changyaw Wang

1 Introduction

Many rendering algorithms separately calculate direct and indirect lighting for visible surfaces. In early rendering algorithms, the direct lighting calculation was carried out on point luminaires, and the indirect lighting was approximated by an "ambient" term [3, 10]. This idea was later used by Whitted, but shadowing was added using visibility rays [15]. Cook extended this idea to area luminaires by using Monte Carlo integration [1]. Since then many researchers have extended Cook's method to include adaptive sampling [8, 2, 11, 5, 9].

One problem with Cook's method is that a direct lighting calculation generates a shadow ray for every luminaire, which is excessive for some environments. Kajiya pointed out that it would be better to send some number of rays to luminaires chosen probabilistically, though he did not propose a specific selection method [5]. Shirley implemented a method where one shadow ray was generated regardless of the number of luminaires, but did not supply many analytical details [12, 13]. In this paper, we derive an unbiased estimator for direct lighting using Monte Carlo integration. The method builds on work from [12], but will work for arbitrary collections of area luminaires.

An important characteristic of the direct lighting estimate discussed in this paper is that, for the multiple luminaire case, it is lower in cost and higher in variance than traditional methods. For this reason, the new method will work best when many samples are taken in each pixel. This would be true in complex scenes where the dimensionality of the integration is high[1]. If few samples are to be taken, culling methods [14] would undoubtably work better.

2 Direct Lighting for One Luminaire

Suppose we want to calculate the direct lighting component at a point $\mathbf{x}$ viewed from direction ψ. This quantity will be a *spectral radiance* [6], and can be written as a function of wavelength, λ: $L(\mathbf{x}, \psi, \lambda)$. Given a luminaire, S, the direct lighting resulting from S can be written:

$$L(\mathbf{x}, \psi, \lambda) = \int_{\mathbf{x}' \in S} g(\mathbf{x}, \mathbf{x}')\rho(\mathbf{x}, \psi, \psi', \lambda) L_e(\mathbf{x}', \psi', \lambda) \cos\theta \frac{dA' \cos\theta'}{||\mathbf{x}' - \mathbf{x}||^2} \tag{1}$$

[1]In ray tracing the dimensions could include pixel area (2), lens area (2), reflection direction (2), shadow ray direction (2), time, wavelength, and polarization; a total of 11 dimensions.

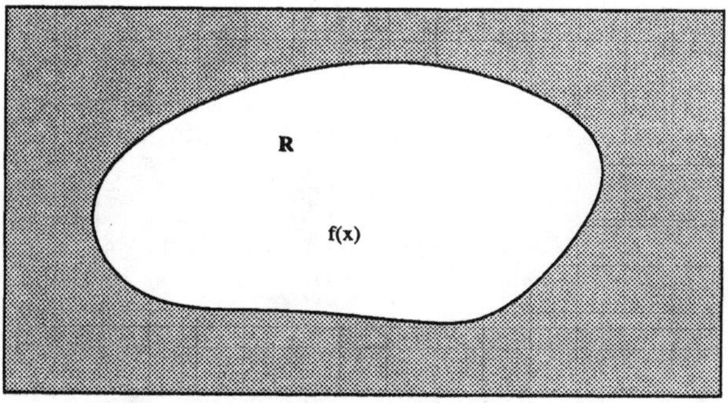

Fig. 1: A function **f** defined over a region **R**.

where $g(\mathbf{x}, \mathbf{x}')$ is the *geometry term*, which is zero if there is an obstruction between $\mathbf{x}$ and $\mathbf{x}'$, and one otherwise [5]; $\rho(\mathbf{x}, \psi, \psi', \lambda)$ is the BRDF [6]; ψ' is the direction from $\mathbf{x}'$ to $\mathbf{x}$; θ is the angle between ψ' and the surface normal at $\mathbf{x}$; θ' is the angle between ψ' and the surface normal at $\mathbf{x}'$; dA' is the differential area of $\mathbf{x}'$.

Any integral over a region $\mathbf{R}$ (see Figure 1) can be approximated using Monte Carlo methods [4]:

$$\int_{x' \in \mathbf{R}} f(x') d\mu(x') \approx \frac{f(x)}{p(x)} \tag{2}$$

where the point x is a random variable with probability density p. For this formula to be valid, p must be positive where f is nonzero. Equation 2 gives a *primary estimator* which might have a high variance. A lower variance *secondary estimator* can be generated by averaging several primary estimators (each with a different x).

Equation 2 can be applied to the direct lighting integral (Equation 1) if we have a density p with which to sample, and a method to choose x_i with density p on the surface of the luminaire:

$$L(\mathbf{x}, \psi, \lambda) = g(\mathbf{x}, \mathbf{x}')\rho(\mathbf{x}, \psi, \psi', \lambda) L_e(\mathbf{x}', \psi', \lambda) \cos\theta \frac{\cos\theta'}{p(\mathbf{x}')\|\mathbf{x}' - \mathbf{x}\|^2} \tag{3}$$

Once a random point $\mathbf{x}'$ has been chosen on S, evaluating this expression is straightforward except for the geometry term $g(\mathbf{x}, \mathbf{x}')$, where a visibility ray must be sent.

An important issue is the selection of p on the luminaire. At a minimum, p must be a valid probability density on S, and p must be non-zero for all points on S that are both visible to $\mathbf{x}$ and have non-zero L_e.

If S is a finite polygon, choosing p(x) is straightforward. We can use a simple uniform density ($p = 1/A$) function defined on the polygon surface. If the polygon is large, then we might be better making p drop with the square of the distance from $\mathbf{x}$ (essentially integrating with respect to the solid angle).

In our implementation, we break a polygon into triangles. Since it's very difficult to compute the exact solid angle of a triangle, we evenly sample points on an imaginary triangle, which covers the same solid angle as the triangle light source and whose vertexes have distance 1 to $\mathbf{x}$. This imaginary triangle is very close to the solid angle and the uniform samples on the imaginary triangle yields a better approximation to uniform samples through a solid angle than uniform sampling of the luminaire.

Figure 2 shows three possible strategies for sampling a triangle. On the left is uniform sampling. In the middle is the sampling method used by Lee et al. [8]. On the right is the method just described.

If S is a sphere, making p a uniform density would yield an unbiased estimator, but that estimator would have an unnecessarily large variance. This high variance arises

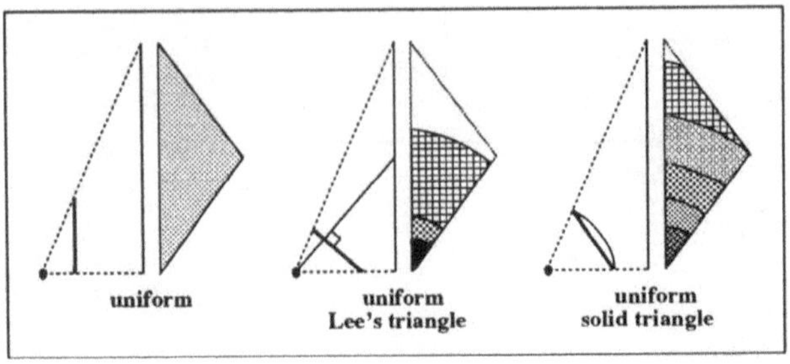

Fig. 2: Sampling strategies for a triangle.

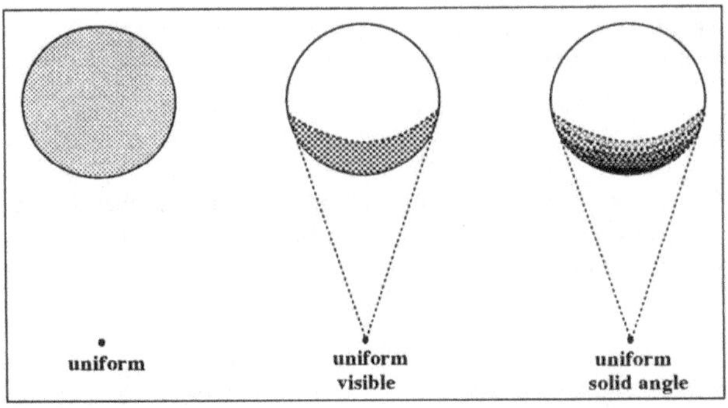

Fig. 3: Sampling strategies for a sphere.

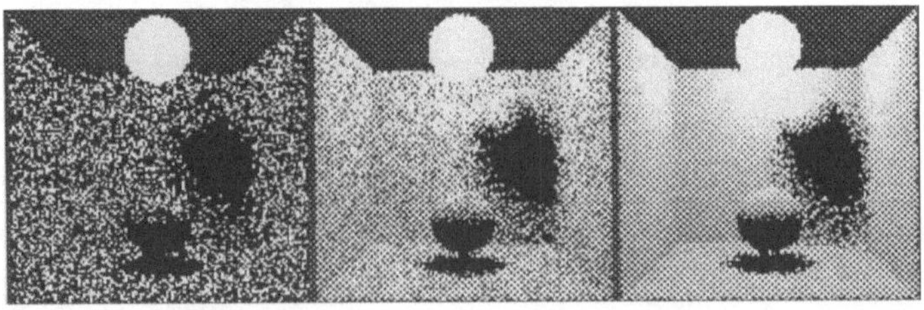

Fig. 4: Luminaire sampling, left to right: uniform, uniform visible, uniform solid angle.

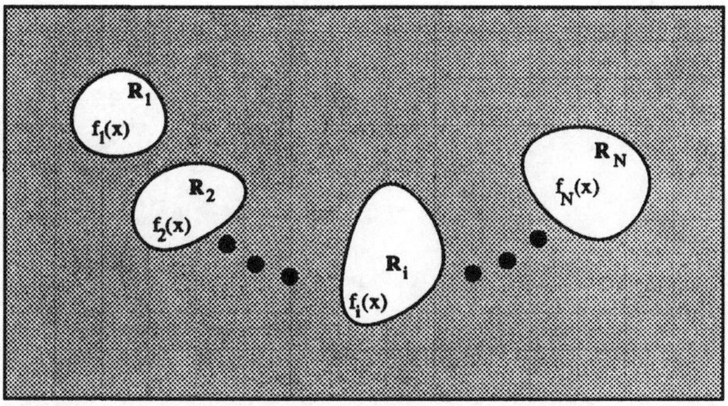

Fig. 5: Several functions f defined over several disjoint regions R.

because there is at least a one half chance that we will pick a sample point not facing x. The closer the object is to a spherical luminaire, the larger the chance that a point invisible to **x** will be chosen. A better way is to uniformly select a sample point from the part of the sphere that is visible to **x**. Better still, we could select points uniformly with respect to the solid angle as seen from **x**. Figure 3 shows these three ways of sampling a sphere. Figure 4 shows thre results of using these strategies in a simple picture at one ray per pixel.

3 Direct Lighting for Multiple Luminaires

If there are N luminaires S_1 to S_N, then the direct lighting will be given by the same integral as Equation 1, but the domain of integration will have to be extended to the union of the areas of every luminaire. Assuming we can construct a probability density function that covers all luminaires, then we can use a Monte Carlo estimator for direct light, and thus use only one or a few shadow rays.

We usually view the direct lighting as I, a sum of N integrals, where each integral represents the radiance contribution from a single luminaire. In an abstraction of this problem we have N integrals over N domains $\mathbf{R}_1$ through $\mathbf{R}_N$ (see Figure 5):

$$I = I_1 + I_2 + \cdots + I_N \tag{4}$$

where each integral I_i is defined by:

$$I_i = \int_{x' \in \mathbf{R}_i} f(x') d\mu(x') \approx \frac{f_i(x)}{p_i(x)} \tag{5}$$

where $p_i(x)$ is a probability density on $\mathbf{R}_i$

This can be extended using N separate Monte Carlo integrations:

$$I \approx \frac{f_1(x)}{p_1(x)} + \frac{f_2(x)}{p_2(x)} + \cdots \frac{f_N(x)}{p_N(x)} \tag{6}$$

This corresponds to sending N shadow rays to N luminaires. Instead, we can define a region $\mathbf{R}$ to be the union of all $\mathbf{R}_i$, and define a function $f(x)$ to be whatever f_i is appropriate for the point being evaluated. This comes from the simple observation that Equation 2 can be applied even if $\mathbf{R}$ has holes or is not fully connected. We have a estimator as soon as we can develop a valid density function p on $\mathbf{R}$. An easy way to do

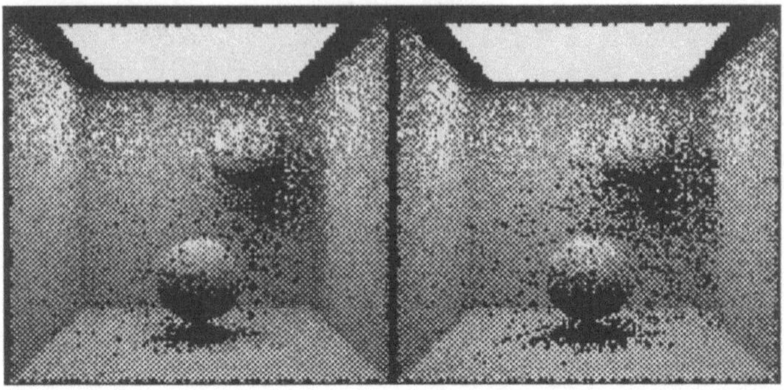

Fig. 6: Two triangle luminaires. Left: 2 shadow rays. Right: 1 shadow ray.

this is to combine the known p_i:

$$p(x) = \begin{cases} \alpha_1 p_1(x) & \text{if } x \in \mathbf{R}_1 \\ \alpha_2 p_2(x) & \text{if } x \in \mathbf{R}_2 \\ \vdots & \vdots \\ \alpha_N p_N(x) & \text{if } x \in \mathbf{R}_N \end{cases} \tag{7}$$

where the α_i sum to one, and where each α_i is non-zero if I_i is non-zero. We can define f in a similar manner:

$$f(x) = \begin{cases} f_1(x) & \text{if } x \in \mathbf{R}_1 \\ f_2(x) & \text{if } x \in \mathbf{R}_2 \\ \vdots & \vdots \\ f_N(x) & \text{if } x \in \mathbf{R}_N \end{cases} \tag{8}$$

We can use the same types of p_i for luminaires as used in the last section. The question remaining is what to use for α_i. We could just make $\alpha_i = 1/N$ for all i, but this would produce a high variance. Instead, we should make α_i large for bright or nearby luminaires. To obtain better α_i we can get an estimated contribution L_i at $\mathbf{x}$ by evaluating Equation 3 for S_i (with a random $\mathbf{x}'$) with the geometry term set to one. These L_is can be directly converted to α_i by scaling them so their sum is one. Sometimes, an estimated L_i can be a false zero, because we happen to select an invisible point, which is blocked by the object or the luminaire. To get an unbiased estimation, the α_i *must* be assigned a positive value in this case. We check to make sure no point on the ith luminaire is above the tangent plane of the object at $\mathbf{x}$ before we allow α_i to be set to zero.

4 Examples

A simple system with both conventional and one ray shadow testing was implemented in C++. All figures we run at 128 by 128 pixel resolution to allow noise to be clearly visible.

Figures 6 and 7 show a two luminaire scene with one viewing ray per pixel and 100 viewing rays per pixel. On the left of each figure, conventional shadow testing is used, and on the right, only one shadow ray is fired for each viewing ray. In this case, the number of shadow rays is only halved, so the optimization makes little difference.

In Figures 8 and 9 a scene with 32 luminaires is shown. In the one ray per pixel images (Figure 8), the conventional shadow ray test generates 32 shadow rays at each pixel, so the variance of the estimate of the shadow is much lower than the variance of pixels containing edges. In the 100 viewing ray images (Figure 9), both scenes end up

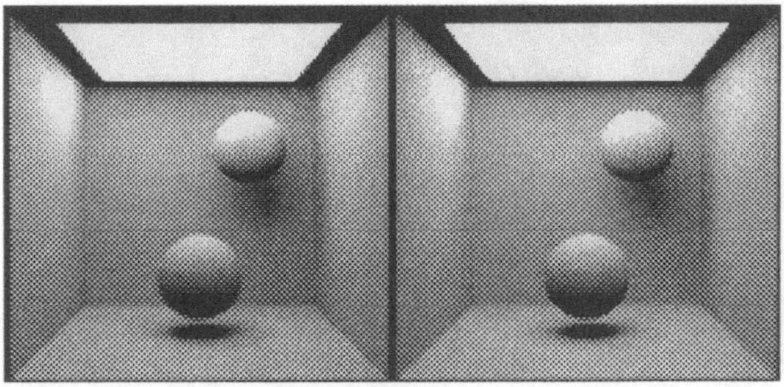

Fig. 7: Same case as Figure 6 with 100 rays per pixel.

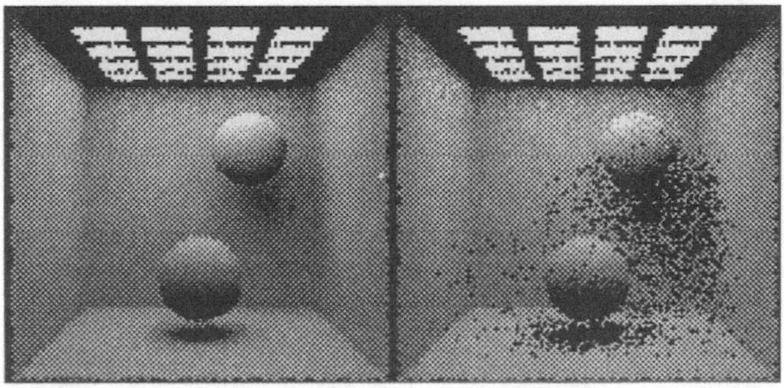

Fig. 8: 32 triangle luminaires. Left: 32 shadow rays. Right: 1 shadow ray.

with low variance shadows and edges. Because conventional shadow testing generates 3200 shadow rays, that shadow is probably overly sampled.

In the left image of Figure 10, the α_i of Equation 7 are all set to be the same $(1/32)$. This is considerably noiser than the image on the right, where α_i is made proportional to the potential influence of the ith luminaire.

On the left image in Figure 11, the α_i are chosen based on the potential contribution to luminaire centers. The incorrect banding on the sphere results from cases where the center of a luminaire is not visible, but off-center regions of the luminaire are visible. This emphasizes the rule that α_i must not be set to zero unless the ith luminaire contributes no light to the point being tested.

5 Conclusion

We have presented a method of direct lighting calculation that uses only one shadow ray for any number of area luminaires with arbitrary directional properties. Any number of shadow rays can be sent if more accuracy is required by using secondary estimators. The method will work for planar and non-planar luminaire surfaces, as long as a method for selecting uniform random points from the luminaire surface is known.

The method could behave poorly in the presence of very bright luminaires that do not contribute radiance to the visible points in the scene. An example of when this could happen is a room at noon with the window shades fully closed. In this case it might

Fig. 9: Same case as Figure 8 with 100 rays per pixel.

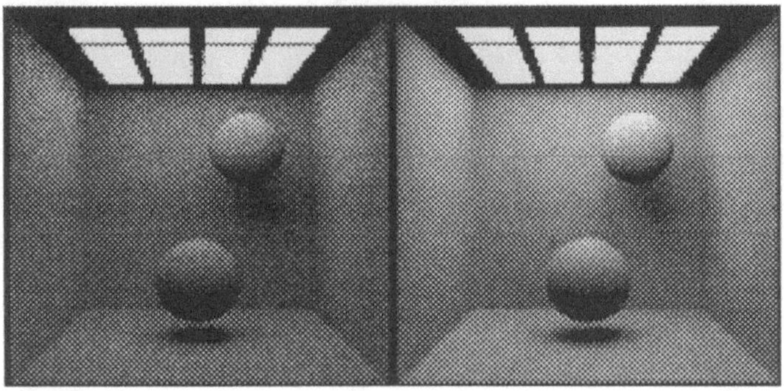

Fig. 10: Left: uniform α_i. Right: variable α_i.

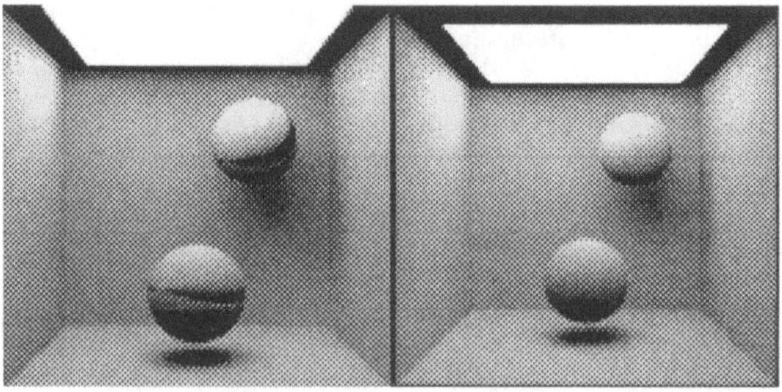

Fig. 11: Left: inappropriate zeros for α_i. Right: correct α_i.

be wise to send one shadow ray to the sun, and one to the union of interior luminaires. The overall reason this case is a problem is because the α_i are bad estimates. Bad α_i estimation can happen because the geometry term, $g(\mathbf{x}, \mathbf{x}')$, is not accounted for (as in the sun example), or because the radiance estimate for the luminaire is bad, as might happen when using a complicated directional luminaire.

Future work should include more sophisticated ways to construct probability densities on luminaires, and fast estimates of luminaire contributions for the assignment of α_i.

The sampling methods presented in this paper could also be used to gather estimates of indirect lighting from patches whose radiance characteristics are known from a zonal precomputation [7].

The basic rationale for this method is that direct lighting should not be calculated to a higher accuracy than necessary. This is very similar in concept to Kajiya's argument that we should not expend much work for deep parts of the ray tree [5]. It is certainly not true that one shadow ray per viewing ray is optimal, however. For the 100 luminaire case, one shadow ray is better than 100 shadow rays, but two or three might be better still. This issue requires further investigation.

Acknowledgements

Thanks to William Brown, Jean Buckley, Greg Rogers, Kelvin Sung, and Greg Ward for their help and input, and to William Kubitz, the thesis advisor for the project in which this work began.

References

[1] Robert L. Cook, Thomas Porter, and Loren Carpenter. Distributed ray tracing. *Computer Graphics*, 18(4):165–174, July 1984. ACM Siggraph '84 Conference Proceedings.

[2] Mark A. Z. Dippe and Erling Henry Wold. Antialiasing through stochastic sampling. *Computer Graphics*, 19(3):69–78, July 1985. ACM Siggraph '85 Conference Proceedings.

[3] H. Gouraud. Continuous shading of curved surfaces. *Communications of the ACM*, 18(6):623–629, June 1971.

[4] J. M. Hammersley and D. C. Handscomb. *Monte Carlo Methods*. Wiley, New York, N.Y., 1964.

[5] James T. Kajiya. The rendering equation. *Computer Graphics*, 20(4):143–150, August 1986. ACM Siggraph '86 Conference Proceedings.

[6] John E. Kaufman, editor. *The Illumination Engineering Society Lighting Handbook, Reference Volume*. Waverly Press, Baltimore, MD, 1984.

[7] A. Kok and F. Jansen. Source selection for the direct lighting calculation in global illumination. In *Proceedings of the Second Eurographics Workshop on Rendering*, 1991. This Proceedings.

[8] Mark E. Lee, Richard A. Redner, and Samuel P. Uselton. Statistically optimized sampling for distributed ray tracing. *Computer Graphics*, 19(3):61–68, July 1985. ACM Siggraph '85 Conference Proceedings.

[9] James Painter and Kenneth Sloan. Antialiased ray tracing by adaptive progressive refinement. *Computer Graphics*, 23(3):281–288, July 1989. ACM Siggraph '89 Conference Proceedings.

[10] Bui-Tuong Phong. Illumination for computer generated images. *Communications of the ACM*, 18(6):311–317, June 1975.

[11] Werner Purgathofer. A statistical method for adaptive stochastic sampling. *Computers & Graphics*, 11(2):157–162, 1987.

[12] Peter Shirley. *Physically Based Lighting Calculations for Computer Graphics*. PhD thesis, University of Illinois at Urbana-Champaign, November 1990.

[13] Peter Shirley. A ray tracing algorithm for global illumination. *Graphics Interface '90*, May 1990.

[14] Greg Ward. Adaptive shadow testing for ray tracing. In *Proceedings of the Second Eurographics Workshop on Rendering*, 1991. This Proceedings.

[15] Turner Whitted. An improved illumination model for shaded display. *Communications of the ACM*, 23(6):343–349, June 1980.

Structure-Directed Sampling, Reconstruction and Data Representation for Global Illumination

George Drettakis, Eugene Fiume

Abstract

The structure of the sampling and reconstruction process, as well as that of the scene and its data representation, can be utilised to effect significant benefits in global illumination algorithms. As an introduction, previous methods that have utilised this structure are reviewed and some possible improvements are suggested. A new framework is then introduced that utilises the concept of an *oracle* as an abstraction to provide information about the complex computational processes of global illumination. This framework is based on the recognition of the relevant aspects of scene structure, the description of the problems to be solved in terms of oracles, the use of structure to create approximations to the oracles, and finally techniques to monitor the cost, error and quality of the approximations. Each aspect of the framework is discussed in detail, presenting certain oracles that are useful for global illumination algorithms. Two examples of its application are presented. The first example is a simple local illumination problem, that allows us to consider several aspects of the oracle methodology. The second example uses the framework to enhance the progressive refinement radiosity based solution to global illumination. Both examples illustrate the benefits of the proposed framework. Finally, some suggestions are made for future research directions.

1. Introduction

The sampling theorem for the digital representation of continuous signals places exact constraints on the class of signals that are expressible when using a fixed set of regularly-spaced samples. For our purposes, the "signal" will be an illumination function defined throughout a scene. Many signals can interpolate any finite set of sample values, and the "least complex" of these signals, which is usually an alias for a more desirable signal, is unlikely to be visually satisfactory, unless we are careful to use prefiltering to reduce the class of signals that we intend to represent. The additional facts that, in computer graphics, reconstruction mechanisms in displays are often beyond our control, and that illumination signals inherently have significant high-frequency content, implies that the sampling rate should be even further increased; if this is not possible, then an even smaller class of signals can be faithfully represented. This gloomy news is of course simply a paraphrasing of well-known results from signal theory. However, for several reasons, the news is not nearly as bad as the theory suggests. First, results from texture mapping have shown that careful filtering can dramatically improve image quality using a fixed sampling rate. This suggests that current sampling rates are generally visually adequate. Second, the human visual system appears to be fairly adept at coping with some forms of noise. Third, our *knowledge* of scene structure and of the rendering process can be used to reduce the set of samples required to represent a given illumination signal. If we are careful, therefore, a much smaller set of illumination samples (associated to objects in space) than the set of image samples can be taken. The problem of creating a rectangular set of pixel samples then reduces to performing an appropriate resampling of the set of illumination samples.

Global illumination models and their corresponding algorithms typically require a large amount of storage and computation to simulate light-object interactions, and to finally generate a two-dimensional image. In this paper we introduce the idea of using *structure-directed* rendering to effect significant computational and storage savings. There are two relevant aspects of structure in the rendering process. The first is the structure of the actual scene, and that of its data representation. The second is the structure that exists in, or is imposed on, the sampling and reconstruction process during image generation.

We will propose a framework that utilises these forms of structure, with two main goals in mind. First, to achieve better understanding of global illumination and the rendering process, and second, to develop a set of suitable abstractions and conceptual mechanisms that will lead to better algorithms for rendering. The framework that we will introduce hinges on the following four components: (i) the identification of the aspects of the scene and sampling/reconstruction structure that are relevant for the specific problem being solved, (ii) the development of the *oracle* concept, which is a convenient abstraction allowing us to reason about illumination and rendering, (iii) the development of approximations to these oracles that will lead to algorithms and practical solutions, and (iv) error and quality metrics allowing us to evaluate the quality of the approximations developed.

Before entering a detailed discussion of this framework, we will try to give an intuitive feeling for how knowledge of structure can be used in rendering. The first example is the selection of samples to be used in illumination calculations. If we have a totally diffuse environment, we only require a small number of samples, placed at locations on the surfaces where there is a significant change in illumination. With appropriate reconstruction kernels to interpolate these sample values, accurate pictures can be produced from very little information. If we compare this to the amount of information required in a traditional radiosity-based solution, where the surfaces have to be blindly subdivided into patches, and possibly elements, we see that there are significant savings to be had (see Fig. 1). If a process existed that could choose the best sample locations and the best filters for reconstruction, it would in effect be an *oracle* for image synthesis. In its pure and perfect form, such a process is infeasible, since it would require previous knowledge of the distribution of illumination, a way to minimise the number of samples selected and also an inverse-reconstruction solution to choose the appropriate filters. Nevertheless, a number of meaningful approximations to an ideal oracle can be made, using known properties of a given scene, and structuring the sampling and reconstruction process.

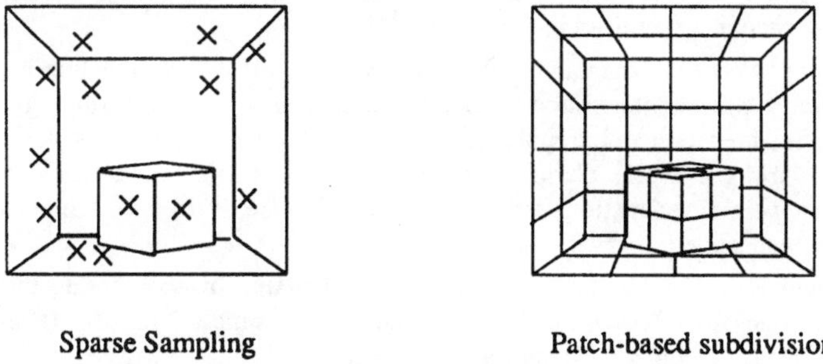

Sparse Sampling — Patch-based subdivision

Fig. 1: Gains from Sparse Sampling

A second example illustrating a similar concept is that of *importance sampling* in Monte Carlo methods. In effect, an ideal importance sampling technique attempts to concentrate the samples taken to approximate an illumination integral, based on visibility and light power information, on those regions that contain the largest amount of information (Hammersly and Handscomb 1964). Again, given an appropriate *oracle*, the computation of an integral over a hemisphere around a point collecting or shooting illumination, could be sped up considerably. The oracle would identify which parts of the hemisphere contribute the most to the integral, and the sampling operation of space would then be limited to these regions. Utilisation of knowledge about scene structure, sample placement information and corresponding reconstruction techniques, can allow an approximation of such an oracle.

The concept of structured rendering has appeared in various forms in previous work. In the following section we review such techniques, and discuss how further use of structure could enhance these methods. In the next section we introduce our proposed framework. This involves a discussion of scene properties used in the rendering process, identifying which of these are more important for global illumination. We then discuss the oracle concept in more detail and how such abstractions can be approximated. Some error and quality measures are then proposed. We continue with two examples. In the first, a fully detailed case-study is presented, in which the framework is applied to solve a simple shading problem. To give an indication of how the framework applies to global illumination problems, the second example shows how a set of approximations can result in certain benefits to the progressive refinement, radiosity-based algorithm. We close with a discussion and suggestions for future research. Throughout the text we refer to the sampling operation performed by a grid superimposed on the image-plane as *image-sampling*; the term *space-sampling* refers to the operation of sampling the hemisphere above a point or area to determine illumination.

2. The Use of Structure in Previous Approaches

Structure, either in the form of the data-representation scheme, or in the sampling and reconstruction process, has been previously utilised to varying degrees. We will concentrate on a limited number of approaches, in which some effort has been made to *combine* more than one aspect of structure to attain a certain goal, thus allowing us to illustrate the importance of the concept of using an integrated approach. We examine cases from both the ray-tracing and the radiosity-based global illumination model.

2.1. Ray-Tracing Approaches

Ward et al. in (Ward et al. 1988) and (Ward and Rubinstein 1988) discuss a sparse sampling technique for the calculation of diffuse illumination in a ray-tracing environment. Four instances of structure are utilised in this method.

Sparse Sample Selection. When a ray intersects a surface, diffuse illumination is calculated in one of two ways: the neighbourhood is searched to determine if there are enough diffuse samples nearby, in which case an interpolation is performed, otherwise a hemispherical integral is approximated using a Monte Carlo technique. An error metric based on distance between samples and a weighting function, derived from this distance, is used for sample selection and interpolation respectively.

Stratified Sampling. The Monte Carlo integral is calculated by segmenting the hemisphere into equal regions to effect faster convergence than a simple "crude" (Hammersly and Handscomb 1964) or "blind" Monte Carlo solution.

Data Structure Utilisation for Sample Location. The sparse samples described above are located using a data-structure that maintains a list of such samples allowing faster access, for interpolation and neighbourhood sample determination.

Simple Reconstruction. For reconstruction, the weights used in the interpolation of the diffuse sample values are based on the distance of the point to be shaded to the various sample locations.

In this method, the scene structure, encoded in the octree, is used in the sample selection and interpolation calculations, resulting in significant computational savings. The sample selection process utilises structure in the following sense: (a) new samples are chosen based on an error approximation metric, and (b) when the new values require an integration to be performed, the nature of the hemisphere sampling operation is taken into account to allow the usage of stratified sampling. In addition, the sample selection metric is used in the specification of the weights for reconstruction, albeit in a somewhat arbitrary fashion.

Thomas et al. (Thomas et al. 1989) introduce the *cover* data representation scheme to enchance the ray-tracing algorithm, that allows efficient determination of shadow boundaries and contours. This is a form of structured data representation for sparse image-sampling.

Cover Data Structure. Each object is covered by an inside and an outside cover that are then intersected with the ray to determine proximity to silhouette contours and shadow boundaries.

Sparse Pixel Sampling. Instead of sampling every pixel in a scan-line, pixels are sampled in a uniformly sparse fashion. When a region of interest is identified by the covers, denser sampling is employed.

The algorithm just described introduces a new data-representation scheme (the covers), specifically built to aid in the determination of the required image sampling rate. This is an example of a new structure being imposed on an existing rendering process (ray-tracing), as opposed to the recognition of underlying or previously existing structure.

2.2. Radiosity-Based Approaches

In the radiosity-based approaches, data-representation structure exists in all scenes, due to the subdivision of the surfaces into planar patches, that are essentially area samples in the rendering process (Goral et al. 1984). Moreover, the hemi-cube (Cohen and Greenberg 1986) defines the way in which the three-dimensional space is sampled. Both these aspects are simple forms of structure, but are not specific attempts to achieve benefit from either recognising, or imposing new, structure in the rendering process of the specified algorithm. Nevertheless, there have been cases that do introduce or utilise such information, and we discuss them in what follows.

Substructuring. The substructuring technique (Cohen et al. 1986) introduces an additional level of subdivision that is used to sample the patches in the hemi-cube computation. The additional subdivision, in essence a hierarchical sampling approach, occurs when deemed necessary by examining the variation of the radiosity values calculated on a surface. An iterative approach is used until desirable convergence is achieved. Space supersampling is thus achieved by the introduction of subsamples, called *elements*. One way of viewing the substructuring process, is that the knowledge of scene structure conveyed by the radiosity gradient is used to determine the sampling rate and to improve image quality.

Ray-Tracing Sampling. In (Wallace et al. 1989) an alternative to the hemi-cube sampling approach is introduced, as applied to the progressive refinement solution (Cohen et al. 1988, Chen 1989). All elements in the scene receive power, but an adaptive sampling approach is used when such sampling of a source (an emitting patch), is required. This adaptive technique is based on the magnitude of the contribution of each region of the source to the form-factor, that depends on the geometrical configuration. The use of adaptive sampling imposes a different structure in the sampling process, utilising knowledge of scene geometry information.

In these methods, there has been some interaction of the structure of the underlying data representation (patches or elements) and the process of space sampling to determine form-factors. Here, as in the ray-tracing approaches, reconstruction has received little attention. This is due to the fact that simple interpolation schemes usually suffice for the purposes of the diffuse environments modelled, at least for the image-sampling case. Finally the view-independence constraint has limited the introduction of new data representation schemes (such as space subdivision) that could be utilised in the space-sampling process.

2.3. Additional Structure Utilisation in Existing Techniques

We now discuss some possible enhancements to the methods described above, that will give a broader view of how the knowledge of structure can be beneficial. For illumination models based on ray-tracing or on radiosity, significant benefits can be achieved, either with the introduction of different data representation, or by taking advantage of existing aspects of the rendering process. We discuss a few such possible improvements to existing methods.

2.3.1. Ray Tracing

Sparse Sampling. In Ward et al.'s approach, space-sampling for diffuse illumination could be enhanced by taking geometry information into account. This information could come from the octree structure, or an enhanced hierarchical representation. The stratified sampling approach for the Monte-Carlo space sampling could be improved by utilising what would amount to importance sampling methods. Techniques similar to the shadow acceleration for octrees (Woo and Amanatides 1990) could be used to determine which part of the hemisphere contributes the most significant part of illumination. This would be especially useful for the case of occluded complex environments, where the existing method is noted to have problems (Ward and Rubinstein 1988). Additional scene information, pertaining to geometry and shadows could be added to the criteria for placing samples, or calculating weights for interpolation. This would enhance the image quality. These operations would constitute a first order approximation to a sample selection oracle (see also Section 3.2).

Covers. Another approach to selecting the sparse grid from which to start shooting rays could be based on a scan-conversion first pass (in the spirit of (Weghorst et al. 1984)), augmented with a shadow volume calculation. The shadow boundaries and first-level contours could then be immediately located. This would allow the avoidance of all shadow casting operations for diffuse objects. In addition, the information recovered from the cover intersection operations could determine the choice of the different filters to achieve better effects, linking the cover structure to reconstruction (see also Section 3.2).

2.3.2. Radiosity-Based Approaches

Hemicube. The hemicube calculations can benefit significantly from a different approach to patch subdivision. A pyramid of varying resolution could be used instead of maintaining one or two levels (patches and elements), similar to the idea of MIP maps (Williams 1983). A related method has been proposed in (Houle 1991) to sample linear and area light sources in an adaptive fashion. If the view independence constraint is relaxed, then an octree-type structure can be utilised to sample adaptively, concentrating again on the regions of the hemisphere that contain the most powerful and large light sources.

Ray-Tracing Solution. Similarly to the hemicube approach, hierarchical structures could be used to reduce computation and storage. The pyramid approach can drastically reduce the number of elements visited by each patch emission operation, since the elements will be in effect adaptively collapsed depending on projected size (onto the light-source).

In both cases, data representation would be enhanced and linked more closely to the sampling and reconstruction process. Additional information can be used in the image-sampling phase, by utilising the tagged sample concept introduced in Section 3.3.

3. Oracles and Structure: a New Framework for Global Illumination

The discussion on previous work gave some intuition on what is meant by identification of scene structure, and by the imposition of structure in the sample/reconstruction processes. In addition, indications were given on how the knowledge of such structure can be beneficial to global illumination and rendering algorithms.

In this section we attempt to establish a methodology that will allow the development of systematic structure-directed approaches. Initially, we examine some of the scene properties with respect to illumination and the rendering process. We attempt to isolate those aspects of scene structure that are important, and identify appropriate methods for their representation. We view this as an initial proposal that requires future research, that will enhance our understanding of global illumination and rendering by isolating the salient structural properties of scene representation as well as sampling and reconstruction. We proceed with a discussion of oracles, that are a convenient conceptual device allowing us to reason about global illumination and rendering. By discussing ways to approximate these oracles, we show how the framework can lead to practical implementations. Finally we present some initial thoughts on determining the quality and measuring the error of these approximations.

3.1. Aspects of Scene Structure

Many properties of scene structure influence the complexity of image synthesis. They can be grouped into four major categories: geometry, viewing parameters, light sources and local illumination.

Geometry. A great deal of scene information resides in the geometric models of the individual objects in the scene, as well as in their subsequent interaction for the purposes of determining occlusion and visibility. Geometric primitives usually include objects such as polyhedra, polygons, free-form surfaces, algebraic surfaces, etc.; the "normal form" for many of these objects at rendering time is a polygonal decomposition. Object descriptions sometimes include an explicit, or have a natural, parameterisation, as is the case for bicubic surfaces or patches. Others can be assigned an implicit parameterisation, to aid in such applications as texture mapping. Surface information is usually encoded in the object structure, that may include textures or complex shading model parameters. One important aspect of geometrical representation is the storage of bounding volume information, and an optional hierarchical "level of detail" structure (for example that in (Rubin and Whitted 1980)). Finally, the geometry of the scene can be enhanced by structuring information such as relative visibility or object proximity relationships, that are crucial, especially in light transfer calculations.

Viewing Parameters. The set of viewing parameters that define the contents of the final image are of extreme importance in any rendering model. A fundamental structure representing this information is the viewing pyramid, based the definition of the viewpoint, the screen size and the near and far planes. This volume allows clipping to be performed when appropriate, provides projected screen size information for objects in the scene, and allows optimisations to be performed, such as backface removal. The use of these parameters has found less use in global illumination models, such as ray-tracing or radiosity-based methods, than it has in traditional scan-line visible surface determination algorithms (Sutherland et al. 1974).

Light Sources. The description of light sources for a computer synthesized scene is extremely important. Primary illumination accounts for a very large number of the important visual effects that we perceive, and it has the benefit that it is somewhat simpler to calculate than secondary lighting. There are three main light source characteristics that need to be described: physical geometry, spatial and spectral light power distributions. There are a number of ways that have been proposed to represent these, concentrating mainly on the latter two. Storage of simple descriptions has been used in (Verbeck and Greenberg 1984) while recent work attempts more compact representations that capitalise on scene structure (including relative light-source size, surface properties, etc.)(Houle 1991).

Local Illumination. Despite the increasing focus on realistic global illumination models, it is the local illumination models that characterise much of the interaction of light with objects. A local illumination model is responsible for providing low-level, often high-complexity detail. Such models typically simulate several classes of surface microstructure, anisotropy, surface self-shadowing, as well as a wide variety of isotropic spectral and spatial reflectance functions, and many forms of texture detail (such as intensity-, bump-, displacement-, environment-, and reflectance-mapping). At first blush, the problem of exploiting the structure of local illumination models may seem impossibly difficult, but this is not so. Many aspects of local surface variation are represented in compact analytic, tabular, or probabilistic forms. This suggests that approaches based on multiresolution or levels-of-detail techniques, low-pass (or scale-space) filtering, or vector quantisation would be effective.

Recognising that the above scene properties are influential is only a small part of the battle. It is now critical to develop representations that facilitate efficient evaluation where possible, and that identify problem areas of the scene. In most of the above cases, we have pointed to possibilities for efficient representations, and one goal of our research is to demonstrate how these representations are useful to the implementation of oracles.

3.2. Oracles for Global Illumination

In theoretical computer science, oracles provide an elegant way of relating problems to one another. For our purposes, an *oracle* is a fictional black box that given a request, produces an exact answer for certain aspects of the sampling and reconstruction process. They are used as decision making processes that give answers to (usually complex and difficult) questions. Oracles allow us to encapsulate complex decision making into isolated objects or entities. These objects have the ability to access information on scene structure and all aspects of the rendering process, and can therefore reach approximate or sometimes exact decisions. This abstraction, as we shall see in Section 4.1, need not be limited solely to global illumination, but can also be used in other traditional problems in computer graphics.

We present three oracles directly related to the global illumination problem, thus giving an indication of how such a mechanism can encapsulate some of the difficult selection procedures.

Optimal Sample Selection. The most obvious oracle that would be useful in a global illumination algorithm is one that selects the minimal set of samples located on the visible surfaces of scene S, from which all the pixel colours could then be satisfactorily reconstructed. Since we are determining a set of samples, a reconstruction method r is implied. As we shall see, the selection of r can be the result of a different oracle, but for the purposes of our definition, we assume that r is given. We informally define oracle O_{SaSel}:

$$O_{SaSel}(S, r, q) =_{df} \textit{ a sample set } \{S_i\} \textit{ s.t.}$$
$$//\{S_i\}// = \textit{minimum, w.r.t. image quality metric } q.$$

Optimal Space Sampling. A second important oracle would allow us to determine the minimal set of illumination samples in the 3D environment required to accurately compute the illumination at a given point, from the surrounding hemisphere of space. Given an area A for which illumination is being calculated, a hemi-sphere of integration above A, h, we define oracle O_{SpSa}:

$$O_{SpSa}(S, h, A, \tau) =_{def} \textit{a sample set } \{S_i\} \textit{ s.t. } \|\{S_i\}\| = \textit{minimum and calcluated illumination value is within tolerance } \tau$$

Optimal Image Reconstruction. A third oracle permits efficient reconstruction, and chooses the correct reconstruction kernel for each specific pixel. In this fashion the final image, reconstructed from the minimal sample set (created for example by oracle O_{SaSel}), is equivalent to the image that would be generated by a given, much more dense, image sampling operation. We define oracle O_{IR}, given sample set $\{S_i\}$ and pixel intensity set $\{P_i\}$ (to be calculated):

$$O_{IR}(S, \{S_i\}, \{P_i\}, \tau) =_{def} \textit{a filter set } \{f_i\} \textit{ s.t. the time to generate } \{P_i\} \textit{ to tolerance } \tau \textit{ is a minimum.}$$

Simple brute-force implementations of these oracles either require the solution of the problem beforehand, or are simply impossible. However, even brute-force approaches may lead to the synthesis of effective approximate oracles based on observations of an ensemble of similar problems. One reason that the formulation in terms of these oracles is actually useful is that it allows us to reason about the complex processes of illumination and rendering in a meaningful way. As such it allows us to isolate the parts that require complicated calculation and knowledge, thus enhancing our understanding of the underlying process. In addition this formulation permits the development of approximations to these oracles in an elegant fashion, achieving modularity and encapsulation of certain complex aspects of the procedures modelled, since in this way only the oracle need have access and knowledge of structure.

3.3. Approximating Oracles Using Scene Structure

When defining a problem in terms of our framework, one would typically first define the set of oracles that suit the solution. The next step would be the identification of the scene and rendering process structure that is relevant, and the salient aspects of this structure. One would attempt to identify these aspects and quantify their effect on the selection or decision making process that the oracle is designed to perform. Based on these observations and this formulation an algorithm can be created that approximates the ideal oracle, which we call an *oracle approximation*. For the purposes of discussion, from now on we will use the terms *oracle* and *oracle approximation* interchangeably, since it will be clear from the context which is meant. To clarify the concept of oracle approximations we present a high-level formulation of approximations for the oracles discussed in the previous section.

To approximate optimal sample selection oracle O_{SaSel}, information from all the scene properties should be used. Light source and geometry information can be used to efficiently predict where higher sampling rates are required due to primary illumination phenomena, such as shadow boundaries or variations due to geometry or local illumination characteristics. Sample selection for secondary illumination is a much more difficult problem and must necessarily be related to the techniques for optimal space-sampling, since the space-sampling approach will determine where in space illumination will be distributed to, or collected from. This can be achieved in part by the adaptive methods described in Section 3.4.

To approximate the space-sampling oracle O_{SpSa}, geometry and viewing information must be utilised. View dependent rendering is critical in achieving low space-sampling rates, since numerous optimisations can be made due to specific geometrical configurations, relative visibility and projected size relationships between objects and the screen. "Level of detail" hierarchical structures as well as proximity information can result in large savings by avoiding significant computation (and storage) for parts of the scene that have little or no visible effect. It must be emphasized here that sacrificing pure view independence does not mean that the ability for motion or animation is forfeited. To the contrary, a necessary addition to the structured approach to global illumination would be that of temporal structure. Such a discussion will not be included in this paper however.

For a efficient reconstruction oracle O_{IR} we introduce the notion of *tagged samples*. When the samples are chosen (by O_{SaSel}) during the sample selection stage, this choice is made due to a scene property, such as geometry, or a preprocessing step, such as shadow boundary determination. The reason causing this choice is stored with the samples, so that it can be accessed in the image reconstruction process. Efficient access mechanisms can be built to determine which samples need to be consulted, and by utilising the tags for each sample, appropriate filters can be chosen, that approximate the optimal. For example, for a set of samples resulting from slowly varying diffuse illumination, wide-support blurring filters can be used, while for boundaries, smaller-support, anisotropic, or even nonlinear filters can be used to maintain, or possibly enhance, the boundaries.

There is an important property of oracles that must be noted here: the computational cost of the oracle (i.e., the decision making process), itself. To make our proposed framework worthwhile, the computation time of the oracle itself must be very small compared to the computation time of the task we are attempting to improve. It may be possible to precompute the cost of the oracle, in which case it can be determined beforehand whether the method has any benefit. Alternatively, if the cost can not be precomputed, then it must be monitored during execution. In either case the implementation must always include a cost calculation mechanism that will allow the system to determine whether use of the oracle is beneficial.

3.4. Error and Quality Assessment for Oracles and their Approximations

A fourth important aspect of the proposed framework is the assessment of oracle quality. The quality of the approximation and of the benefits achieved by using the oracle must be continuously monitored. This checking procedure can then be utilised to tune the selection/decision criteria appropriately, achieving higher quality results. This is in effect a true adaptive technique, since the decision logic adapts, depending on performance. Error and quality measures can be derived from the selection criteria and the approximation techniques used. Typically a problem will have an exact solution, and the approximation can be defined in terms of the error resulting from the use of the approximation. The error can then be calculated during execution, and it can be determined whether or not certain selection criteria appear to be performing badly, contributing to higher error rates. If such a case is found, the criteria can be tuned to deliver better results. This error/quality assessment code is an integral part of any oracle's functionality.

Apart from the adaptive capabilities that the error/quality assessment provides for, it allows us to define oracles in terms of a *tolerance*. Such tolerance measures are usually defined in terms of the mean error μ and variance σ^2 (other statistics may be used too). This is very useful for many aspects of global illumination and rendering, since a requirement of exact equality (e.g., of pixel intensities) would be in many cases impossible to meet. In addition, a tolerance requirement can be relaxed, allowing us to define "levels of quality" of various oracles.

4. Two Examples: a Case Study and Global Illumination

This section attempts to solidify the proposed framework described in the previous section through the presentation of two examples. The first is a case-study that, though not a global illumination problem, is sufficiently simple to allow development of a number of details of the approach, and indicate how the framework leads to useful, practical solutions. In the second example, we discuss a more relevant global illumination problem, relating it to some of the oracles described in Section 3.2, albeit on a more general and abstract level than for the case study.

4.1. A Case Study: an Oracle for Scan Conversion

In a traditional scan-conversion system, one may use either Gouraud shading, in which a simple interpolation of colour computed at vertices is performed, or Phong shading, in which the normals are interpolated and the local illumination model is applied at every pixel. The difference in the renderings can be significant, mainly with respect to the highlights and the banding artifacts. Phong shading is obviously much more expensive computationally, so it is desirable to use it as rarely as possible. For the purpose of this discussion we will assume that the desired quality of rendering is the one generated purely by the application of Phong shading throughout (though it is questionable whether or not Phong shading always gives the best results). For a given scene, and within a certain tolerance (defined in terms of the mean μ and the variance σ^2), we can create the same image by selectively using Gouraud shading for certain cases. We represent this requirement in terms of the previously described framework as follows. Let $O(\Pi, S, \tau)$ be an oracle with respect to a primitive Π (e.g., polygon, scan-line) a scene S and a tolerance τ, that answers the question: "To generate an image of the desired quality, which shading rule should I use ?" as follows:

$$O(\Pi, S, \tau) = \begin{cases} \text{Phong} & \textit{when quality of alternatives exceeds tolerance} \\ \text{Gouraud} & \textit{when satisfactory} \\ \text{Neither} & \textit{when only simple shading required (e.g., } \Pi \textit{ is in shadow)} \end{cases}$$

The general benefits of the oracle framework has been discussed above. One of these is that of software engineering encapsulation that is provided for free. The oracle can be seen as an "object", that encapsulates the decision making logic for the shading rule selection. The specific oracle O would be used in the following way in a real rendering system:

```
foreach Polygon
     foreach scan-line span S_p of polygon
          answer = O(S_p, S, τ)
          if answer == ''Gouraud''
               ScanConvGouraud(S_p)
          else if answer == ''Phong''
               ScanConvPhong(S_p)
          else
               ScanConvConst(S_p)
```

This oracle is now well defined, and we wish to see how to take advantage of scene structure to develop an approximation, as well as the oracle cost and error/quality evaluation metrics.

4.1.1. Relevant Aspects of Scene Structure

Due to the limited nature of the specific problem we are able to create an extensive list of structural aspects that affect the choice of shading rule. We will construct the list assuming a polygon primitive, since the scan-line case can be derived.

The purely geometric scene properties that are relevant are the following: polygon area, polygon shape, planarity, vertex normals and their variation, rate of change of the angle between the normal and the viewer, the distance of the polygon from the viewer, and the angle between the polygon normal and the viewing direction. Other aspects that are of importance are texture, surface properties and the local illumination model used, as well as light and shadow information. Each of these properties of the scene contributes to the decision of whether it is possible to use Gouraud shading, without significant degradation in image quality.

The properties above convey information that is relevant to the result of the applied shading rule. As an example, if the polygon is planar, and there is no variation in vertex normal, and in addition the angle between the viewer and the normal changes very little, Gouraud can be used without a problem. Obviously one can construct arbitrary combinations of cases that result in situations with varying degrees of certainty as to whether the shading rule can be substituted.

To allow a decision to be made, one can assign a weight to each of the properties, and construct a weighted sum that is then used as the decision variable (e.g., if the sum is larger than .5 one can use Gouraud). In addition one can rank each aspect so that the more important ones are those examined first (for example whether or not the polygon is in shadow), and apply increasingly complex heuristics to detect easy reject/accept cases.

The determination of the weights is initially a static, precomputation phase, followed by dynamic alteration discussed in Section 3.4. The static determination is based on a careful analysis of the effect each parameter or its variation has on the calculated (mean) intensity. Even for a simple case like this, one has to see how each of the properties listed above affects the final result, and express the effect each has by assigning appropriate weights.

4.1.2. Oracle Cost

The issue of oracle cost is of course of major importance. The case we have here is straightforward, since we have a yes/no decision to make. Suppose we have a sequence $Opt_1, Opt_2, \cdots$ of rendering options ordered by (increasing) costs. In our specific example, the rendering options are none, Gouraud, and Phong. We can also define a hierarchy of possible oracles as follows:

$O^1(S,S,\tau)$	*Choose shading rule for entire scene S*
$O^2(O,S,\tau)$	*Choose shading rule for object O*
$O^3(P,S,\tau)$	*Choose shading rule for polygon P*
$O^4(S_p,S,\tau)$	*Choose shading rule for span S_p*

We will also write Opt_j^i to denote that option j is being used at level i of this hierarchy. A number of cost constraints are evident:

$$Cost(Opt_{j+1}^i) > Cost(O^i) + Cost(Opt_j^i),$$

where $Cost(O^i)$ indicates the computational cost of the oracle O^i itself, while $Cost(Opt_j^i)$ is the cost of using rendering option j at level i. For example, if $i=4$ and $j=2$, then the above constraint states that the use of Phong shading across a scan-line span must be more expensive than the cost of evaluating the oracle for this primitive together with the cost of employing Gouraud shading for the span. If it is more expensive to use the oracle to determine that one can shade scan-lines with Gouraud than it is to compute directly with Phong the advantage of the whole technique is lost.

In our case the cost of the computation for the oracle is fixed, and the cost of each option is also easily calculable since the cost of the application of illumination model and of the incremental operations is known, given a set number of pixels †.

† Depending on the algorithm used for visibility, the cost comparison between O^1 and O^2 may prove problematic, but we will assume for now that this is not the case.

4.1.3. Error/Quality Assessment and Adaptivity

As discussed in the previous section an important aspect of the oracle approximation process is determining the success of the selection strategy, which is the choice of shading rule in our case. A first order approach would be for the oracle to perform "random tests" in which when it selects Gouraud it also computes some pixels of Phong and computes the real mean error μ and variance σ^2. A second order approach is to identify those parts of the shading rule equations that are most influential in the determination of the intensity, which can be determined by the weights used in the selection process, and compute an approximation of the error metrics, thus saving computation. When the oracle decides that it is performing poorly, it re-examines the weight assignments. By keeping track of selection history, it will be able to determine which weights are incorrect, and modify them so that they are closer to the correct decision.

4.2. An Example of Structural Diffuse Illumination

In this section we will present an augmented, progressive refinement radiosity-based solution to global illumination. Since this a much more involved and complicated problem than that of the shading rule, the discussion is necessarily less detailed and at a more abstract level. We formulate the problem in terms of two main oracles. The first is an approximation to the sample selection oracle O_{SaSel}, treating patches as illumination samples. The approximation uses knowledge of the scene structure to determine patch size. The second is an approximation to oracle O_{IR}, that, given a set of pixels associated to a patch or set of patches, selects a filter to be used.

4.2.1. Approximating the Sample Selection Oracle

The oracle implementation bases its choice of patches and patch size on a two-step selection process. The first step takes into account the effect of viewing parameters, and the second performs a preprocessing step to determine the influence of geometry and shadow boundary information.

As a first step we allow patch sizes to vary, depending on the size of the projection of each object onto the screen. In this fashion, objects that have a small screen projection are subdivided into a smaller number of patches. The area of each of these patches is calculated at subdivision time, since this is necessary in form-factor calculation. It must be noted that initial patch size, even for large objects, is taken to be significantly larger than in a normal radiosity-based approach.

In a second step, shadow boundary computation is performed in the spirit of (Atherton et al. 1978), from light cast from the initial emitters, onto the remaining objects in the scene. Object proximity is considered in these calculations, since diffuse emitters do not produce sharp shadows beyond a certain distance (or solid angle). These boundaries are then utilised to subdivide the patches that they intersect, so that a finer meshing exists along shadow boundaries. Other considerations, such as geometry information, and proximity, especially to light sources, can be used to further refine the meshing approach. All additional subdivision is subject to the projected screen size constraint: i.e., if the projected object is very small, subdivision will not be performed.

4.2.2. Approximating the Reconstruction Oracle

The reconstruction phase implements a simple form of the "tagged sample" concept. For patches that have not undergone additional subdivision, and thus do not have shadow boundaries crossing them, wide area filters are used to effect smooth blurring. An additional criterion in the kernel selection can depend on surface properties such as the existence of texture. Fast approximations based on table lookups or pyramidal techniques can be used

(Feibush et al. 1980, Williams 1983). For more highly subdivided patches, blurring will not be attempted, allowing shadow boundaries, and intensity variations due to geometry to be more pronounced than in the standard solution.

4.2.3. Using the Oracles in the Enhanced Algorithm

The enhanced algorithm can now utilise these two oracles as black boxes. Instead of the initial static subdivision of the scene that typically occurs in traditional radiosity-based implementations, the first oracle is used, resulting in the generation of an "intelligent" selection of patches and patch sizes. This set of patches is subsequently used in the traditional progressive refinement solution. The number of patches will have been reduced compared to a straightforward patch subdivision (see Fig. 2). As a result the actual computation time for a converged solution will be significantly reduced, since the hemi-cube operation complexity is dependent on the number of patches in the environment.

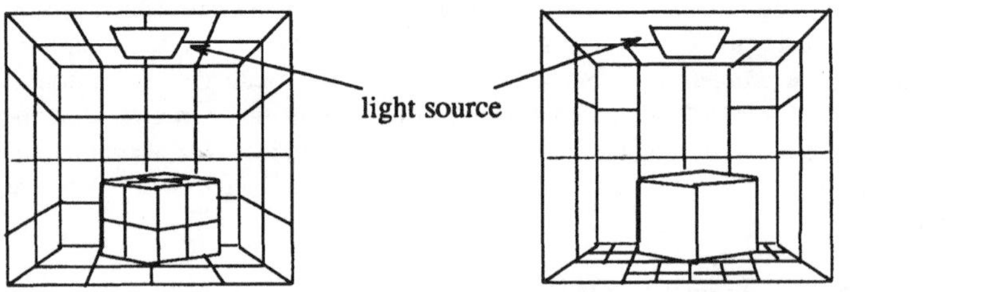

Simple Patch-based subdivision

Adaptive subdivision

Fig. 2: Comparison of Adaptive and Simple Subdivision

At the final rendering stage, the filter selection oracle will be used instead of the usual bilinear interpolation (Goral et al. 1984). This will lead to better defined boundaries for shadows and corners, at a fraction of the cost of higher subdivision that would be required to achieve the same result using the traditional approach.

5. Discussion and Future Directions

As we have seen, structure for the rendering process, both in the scene representation and in sampling/reconstruction, is an important factor that can significantly contribute to global illumination algorithms. The discussion of previous work indicated that, perhaps unknowingly, previous authors have capitalised on certain aspects of structure. In this paper we discussed how conscious and increased structure utilisation can benefit previous approaches.

We then proposed a new structure-directed framework based on the concept of oracles. We discussed initial approaches to approximating these ideal procedures. A discussion of their cost followed, together with simple suggestions on quality and error assessment, their adaptive nature and their definition in terms of tolerance factors.

We presented two examples: a detailed case study on how the framework can be of practical use, and an example for a radiosity-based progressive refinement approach that gives an indication of how this method relates to global illumination.

Further research is being pursued departing from existing global illumination techniques, that will take full advantage of the principles outlined in this paper. As such, algorithms developed will not be retrofitted to accommodate a structural approach but will be designed from the outset with it in mind.

To conclude, we are convinced that a structural approach to global illumination is a worthwhile area for future research. By carefully considering information given in scene descriptions, appropriately tuning data representation, and correctly structuring sampling and reconstruction methods in terms of abstractions such as oracles, it is quite possible that global illumination algorithms will see dramatic improvements both in terms of computational speed and storage requirement reduction.

Acknowledgements

The authors gratefully acknowledge the financial support of National Science and Engineering Research Council of Canada, the Information Technology Research Centre of Ontario, and the University of Toronto, for this research and for travel to the workshop. Thanks also to Pierre Poulin for reviewing an early version of this paper, Gavin Miller for fruitful discussions and also Greg Ward and the reviewers for their useful comments that significantly improved the presentation of this work.

References

Hammersly, John M. and Handscomb, David C. (1964) *"Monte Carlo Methods"*, Methuen, London .

Ward, Gegory J., Rubinstein, Francis M., and Clear, Robert D. (1988) "A Ray Tracing Solution for Diffuse Interreflection," *ACM Computer Graphics (SIGGRAPH '88 Proceedings)*, vol. 22, no. 4.

Ward, Gregory J. and Rubinstein, Francis M. (1988) "A New Technique for Computer Simulation of Illuminated Spaces," *Journal of the Illuminating Engineering Society*, vol. 17, no. 1, pp. 80-91.

Thomas, David, Netravali, Arun N., and Fox, D. S. (1989) "Anti-aliased Ray-Tracing with Covers," *Computer Graphics Forum*, vol. 8, pp. 325-336.

Goral, Cindy M., Torrance, Kenneth E., Greenberg, Donald P., and Battaile, Bennet (1984) "Modelling The Interaction Of Light Between Diffuse Surfaces," *ACM Computer Graphics (SIGGRAPH '84 Proceedings)*, vol. 18, no. 3.

Cohen, Michael F. and Greenberg, Donald P. (1985) "The Hemi-Cube, A Radiosity Solution For Complex Environments," *ACM Computer Graphics (SIGGRAPH '85 Proceedings)*, vol. 19, no. 3.

Cohen, Michael F., Greenberg, Donald P., Immel, David S., and Brock, P. J. (1986) "An Efficient Radiosity Approach for Realistic Image Synthesis," *IEEE Computer Graphics & Applications*, vol. 6, no. 3.

Wallace, John R., Elmquist, Kells A., and Haines, Eric A. (1989) "A Ray-Tracing Algorithm for Progressive Radiosity," *ACM Computer Graphics (SIGGRAPH '89 Proceedings)*, vol. 23, no. 3.

Cohen, Michael F., Chen, Eric S., Wallace, John R., and Greenberg, Donald P. (1988) ''A Progressive Refinement Approach to Fast Radiosity Image Generation,'' *ACM Computer Graphics (SIGGRAPH '88 Proceedings)*, vol. 22, no. 4.

Chen, Shenchang Eric (1989) ''A Progressive Radiosity Method and its Implementation in a Distributed Processing Environment,'' *Masters of Science Thesis, Cornell University.*

Woo, Andrew and Amanatides, John (1990) ''Voxel Occlusion Testing: A Shadow Determination Accelerator for Ray Tracing,'' *Proceedings of Graphics Interface '90.*

Weghorst, Hank, Hooper, Gary, and Greenberg, Donald P. (1984) ''Improved Computational Methods for Ray-Tracing,'' *ACM Transactions on Graphics*, vol. 3, no. 1.

Williams, Lance (1983) ''Pyramidal Parametrics,'' *ACM Computer Graphics (SIGGRAPH '83 Proceedings)*, vol. 17, no. 3, pp. 1-11.

Houle, Caroline (1991) ''Light Source Modelling,'' *M.Sc. Thesis, Department of Computer Science, University of Toronto.*

Rubin, Steve M. and Whitted, Turner (1980) ''A 3-Dimensional Representation for Fast Rendering of Complex Scenes,'' *ACM Computer Graphics (SIGGRAPH '80 Proceedings)*, vol. 14, no. 3, pp. 110-116.

Sutherland, Ivan E., Sproull, Robert F., and Schumacker, Robert A. (1974) ''A Characterization of Ten Hidden-Surface Algorithms,'' *Communications of the ACM*, vol. 6, no. 1.

Verbeck, Channing P. and Greenberg, Donald P. (1984) ''A Comprehensive Light-Source Description for Computer Graphics,'' *IEEE Computer Graphics & Applications*, vol. 3, no. 7, pp. 66-75.

Atherton, P., Weiler, K., and Greenberg, Donald P. (1978) ''Polygon Shadow Generation,'' *ACM Computer Graphics (SIGGRAPH '78 Proceedings)*, vol. 12, no. 3.

Feibush, Eliot, Levoy, Marc, and Cook, Robert (1980) ''Synthetic Texturing Using Digital Filters,'' *ACM Computer Graphics (SIGGRAPH '80 Proceedings)*, vol. 14, no. 3, pp. 294-301.

Source Selection for the Direct Lighting Computation in Global Illumination

Arjan J.F. Kok, Frederik W. Jansen

Abstract

Computation of the global illumination in a scene can be improved by separating the direct component of the lighting, which is received by a patch directly from light sources, from the indirect component, which is received by intermediate interreflection from other patches. The indirect component is calculated during the preprocessing and is stored as the radiosity shading at the patch. The direct component is calculated during the rendering phase by tracing shadow rays like in conventional ray tracing. The number of shadow rays can be reduced by exploiting shadow coherence, and by making a selection for the number of light sources that are taken into account for the direct lighting computation. Different criteria to select these sources are given.

Key words: radiosity, global illumination, source sampling

1 Introduction

Highly realistic images can be generated by two-pass radiosity algorithms (Wallace et al 1987; Sillion and Puech 1989) that calculate, in a preprocessing step, the diffuse component of the light reflection on a patch. During the rendering phase, the specular component of the reflection is added by ray tracing. A problem with these algorithms is to represent accurately the preprocessed radiosity shading, because it often shows much variation over the patch. This is mainly due to the light that is received directly from strong (point) light sources, and that is responsible for sharp shadows. To represent these intensity variations, the patches can be subdivided into elements. However, it is nearly impossible to find a good subdivision scheme that represents the shadows accurately. Methods such as adaptive patch refinement (Cohen et al 1986) and shadow meshes (Campbell and Fussell 1990) require large data structures and fail to provide the dynamic adaptation that is needed, for instance, when one zooms in on a single patch in a very large scene. The use of adaptive radiosity textures (rexes) in combination with a viewpoint dependent preprocessing that restricts the preprocessing to those patches that are prominent in a given view (Heckbert 1990), reduces the problem, but does not solve it.

A solution to the problem is to make a distinction between light that is received directly and light that is received indirectly from light sources (Ward et al 1988; Shirley 1990). Direct light is light that a patch receives from a light source (a light emitting patch or a strong radiant patch) without being reflected by other patches. Indirect light is the light that is received after one or more (diffuse or specular) reflections. In Shirley's approach, during the preprocessing step (illumination ray tracing and radiosity) only the diffuse reflection of the indirectly received light is calculated for each patch and stored as illumination maps and the radiosity shading on the patch. To calculate the radiosity component, first the complete global illumination problem is solved, and then the direct lighting component is subtracted from the total radiosity shading. The resulting radiosity shading will show only minor intensity variations over a patch. Therefore, patch refinement can be relatively coarse.

During the rendering phase, the indirect lighting reflection component for a point on a patch is calculated by interpolating the radiosity shading values that are pre-calculated for the vertices of the patch. The direct component is added by calculating the contribution of the light sources by casting shadow rays to the light sources. For small sources, one shadow ray suffices, whereas for large light sources, more shadow rays are needed to obtain soft shadows. The result of this two pass global illumination method is a realistic image, that combines the best properties of ray tracing and radiosity.

A drawback of the algorithm is that the number of shadow rays can be very large, especially when there are many area light sources. It is clear that the calculation of the direct component in these cases will be expensive and will require extra ray tracing compared to the conventional two pass methods. Therefore methods for reducing the number of shadow rays are needed. Shirley (1990) uses a form of importance sampling (Kajiya 1986), in which only one shadow ray is cast to a light source for each ray-surface intersection. To which light source the shadow ray is cast is determined by the probabilities assigned to the sources. This stochastic method has the drawback, that a large number of samples for each pixel is necessary to get reliable results. Another way to reduce the number of shadow rays is to use shadow coherence. We used an adaptive image refinement method in combination with a shadow image buffer (Woodward 1990; Kok et al 1991). A third approach would be to limit the number of light sources to be considered for direct lighting, and to add the contribution of the less prominent light sources to the indirect component. Proper selection of the most important light sources during the rendering could reduce the number of shadow rays without real loss of shading accuracy. This approach will be explored in this paper.

In section 2 we will discuss criteria to select the light sources. Section 3 gives a description of the implementation of the two pass global illumination method. The results of our experiments are presented in section 4. Finally, conclusions and future directions are given in section 5.

2 Source Selection

In radiosity algorithms, all patches in a scene are considered as possible light sources. A radiosity preprocessing is applied to calculate for each patch the diffuse component of the reflection. However, by treating the very important contribution of the strongest light sources in the same way as the light from the other patches, a loss of shadow accuracy occurs, resulting in pictures with very blurred shadows, which in fact can be very unrealistic. The basic idea of Shirley's algorithm (1990) is, therefore, to handle the most important light sources as in conventional ray tracing, and to handle the other light sources (and other radiant patches) as in the conventional two pass radiosity methods. The light received directly from strong light sources is sampled by shooting shadow rays, whereas the light received from the other patches is calculated during the preprocessing step, and is stored as the radiosity shading on the patches.

The question is: which light sources should be considered for the direct lighting computation? If all patches are considered as light sources, then this would result in far too many shadow rays. If only a limited number of light sources is considered, then the shadow accuracy will be low. A related problem is: how many shadow rays will be needed for area light sources? A large number will be expensive, whereas fewer rays will result in shadows that are too sharp compared to the soft shadows that may be expected from area light sources.

An obvious choice for the light source selection is to take the *n* most emitting or most radiant patches, i.e. the first *n* patches that are selected in the progressive refinement (shooting) radiosity method. However, this *global selection* neglects the fact that a light source can have a strong effect on some patches, but may have a negligible influence on other patches. Therefore it is better to make a *local selection*, i.e. select for each patch the *n* most important light sources. We now give some criteria to select a patch *j* to be a light source for patch *i*.

The first criterion is the *radiance criterion*. This criterion is based on the energy, that is received by a patch from another patch, and on the diffuse reflectivity of the patch. It is, for example, not necessary to cast rays from patch *i* to patch *j* if the energy contribution of patch *j* to patch *i* is negligible. Also, it is not necessary to cast shadow rays for a patch for which the diffuse reflectivity is almost zero (for example a perfect specular reflective patch), although it receives a lot of energy from other patches. So if the reflected intensity of patch *i* caused by the energy received from patch *j* is larger then a certain threshold, then patch *j* is considered to be a light source.

The radiance criterion is applied during the first pass. In the second pass a more strict selection is made out of the *n* sources selected in the first pass. The next criteria can be applied to make this selection during the second pass.

Although the intensity of the received light is important, for finding possible shadows on a patch the distribution of the light over that patch is also important. If the light coming from patch *j* gives large intensity variations over patch *i*, then patch *j* should be considered as a light source for patch *i*. On the other hand, if it can be detected a priori that the light from a strong light source shines on the whole patch *i* or does not reach patch *i* at all, then it is not necessary to cast shadow rays to patch *j*, because no large shading gradients will be found. We will call this the *gradient criterion*. To detect gradients we have to compute the contributions for patch *j* to patch *i* at different points on patch *i*. If these contributions show large differences, then some kind of shadow or shading boundary will appear on this patch, and then the light source should be sampled with shadow rays during rendering. Of course, as with all coherence methods, care should be taken not to miss small shadows.

Another aspect of the light source selection is related to the number of shadow rays that will have to be cast. If a possible light source (patch *j*) is small (e.g. if it is a point light source) or is seen with a small solid angle from patch *i*, then only one shadow ray will be needed. Shadow rays are then effective, and patch *j* is likely to be chosen as a light source. If, however, the possible light source is an area light source, and the solid angle is large, then a soft shadow is expected that can be represented much better with the radiosity shading. So, based on the solid angle it is determined whether shadow rays will be cast, and if so, how many shadow rays will be cast. The larger the solid angle, the more shadow rays are needed. We call this criterion the *solid angle criterion*.

Finally, it may be observed that patches farther away from the viewer are not so critical for the perceived image quality. A more approximated shadow computation can be accepted, and the number of shadow rays can be reduced or the shadow ray casting can be limited to only the most important light sources (the contribution of the other light sources are added to the general radiosity). We will call this the *distance criterion*.

If a local selection is used for the selection of the sources and/or the number of shadow rays, care should be taken that patch boundaries do not become visible due to differences between neighbouring patches in the selection of light sources. It is therefore advisable to apply a selection to a cluster of patches (an object) instead of to a single patch.

Given the described criteria, an optimum combination of these criteria must be derived, taking into account the radiance of all patches, the viewpoint, etc.

3 Implementation

To calculate the indirect reflection component in the preprocessing, both the full-matrix (gathering) method (Cohen and Greenberg 1985) and the progressive refinement (shooting) method (Cohen et al 1988) can be used. If the full-matrix method is used, the choice of the most important light sources can be made after calculation of the formfactors. This can be done as follows:

1. Calculate all formfactors.
2. Solve the complete system of equations.
3. Choose the most important sources.
4. Set formfactors representing indirect light transport to zero (the formfactors F_{ij}, where patch *i* and patch *j* are not chosen as light sources).
5. Solve the system of equations again. This will result in the direct reflection components.
6. Subtract the radiosity values calculated in step 5 from the values calculated in step 2. The resulting radiosity values are the indirect reflection components.

A disadvantage of this approach is that difficulties arise when specular reflections (use of extended formfactors) must be included in the calculation. The progressive refinement (shooting) method is therefore more suited. On basis of the progressive refinement radiosity method, we propose the following source selection algorithm.

During the preprocessing the radiosities for all patches in the scene are calculated. The radiosities are stored at the patch vertices (several points on the patch). To be able to make a selection of light sources for the direct lighting computation, for each patch vertex, in addition to a general radiosity value, a list is maintained that contains the contributions of the *n* most important light sources (based on the radiance criterion). Contributions that fail to enter this list (contributions from less influential sources) are added to the general radiosity value. The number *n* must not be chosen too small, to be able to make a good selection before rendering.

After the shooting is done, a further selection of light sources is made based on the the gradient criterion and the solid angle criterion. The gradient criterion is applied by comparing the contributions of the possible light sources (stored in the list) at two neighbouring vertices. If there is a large difference between the contributions, then the possible light source will be chosen. With the solid angle criterion also the number of shadow rays that will have to be cast from the patch to the light source can be determined.

Prior to the rendering, an additional viewpoint-dependent selection can be made based on the distance criterion (some patches that first were selected as sources can be omitted, or the number of needed shadow rays can be changed). The above selection reduces the number of lightsources to m ($m <= n$, m is not constant for all patches if a local selection is used). The other $n - m$ contributions (of less influential sources) are added to the general radiosity.

For the rendering, an adaptive stochastic ray tracing algorithm (Dippé and Wold 1985; Cook 1986; Mitchell 1987; Painter and Sloan 1989) is used. Viewing rays are traced from the eye through the viewing plane into the scene. If a viewing ray hits a patch, the indirect reflection component is calculated by interpolation from the general vertex radiosity values. The direct component, due to the *m* selected light source patches, is calculated by casting shadow rays. First the scene is sampled at a low resolution. After the first sampling, the screen sampling is adaptively refined. Shadow rays are cast adaptively too, using a shadow image buffer (Woodward 1990; Kok et al 1991) that serves to indicate whether neighbouring samples (viewing rays) have the same shadow properties (sample in shadow or in light; the shadow image buffer stores at each sample one bit for each light source). Shadow rays are sent to jittered points on area light sources to avoid aliasing. The following criteria are successively applied to sample the screen adaptively:

1) Object criterion. If neighbouring samples do not all hit the same object, extra samples are taken to discover the exact object boundaries.
2) Shadow criterion. If neighbouring samples do not show the same shadow property (all light or all shadow), then a shadow boundary exists, and extra samples are taken in this region to find the exact shadow boundaries.
3) Difference (gradient) criterion. If the intensity of neighbouring samples differs more than a certain threshold, then extra samples are taken. By using shadow coherence (by the shadow image buffer), no extra shadow rays are needed in this case.

When all refinement criteria are satisfied, or a maximum sampling rate is reached, the pixel values are calculated using filtering techniques.

4 Experiments

We tested different light source selections to see the influence of the selection on the quality of the image and the number of shadow rays needed. All tests were done on the scene of figure 1. The scene contains two spot light sources A and B (on the corners on the right, pointing to the table) and two area light sources C (on the ceiling) and D (on the left wall).

Several images (see figure 2) were created with different selections of sources, using the same preprocessing. The selections were global, so each patch had the same selection of light sources. The results are given in table 1 (image size 256x256 pixels). The radiosity solution (a) misses shadow details. The shadow of the handle of the teapot and the shadows of the table legs are missing. A very fine subdivision of the patches can solve this, but will cost a lot of memory and processing time. For the two spot light sources, one shadow ray is sufficient (b). For the area light sources, more shadow rays per source are needed (d-h). For light source C, 16 shadow rays are needed to get a good shadow on the table, but for other patches in the environment, smaller numbers (or even 0) will suffice. For light source D, 64 rays are needed to get soft shadows on the floor, but for the table only 16 (or even 4) shadow rays are sufficient. So, local selection criteria to determine the number of shadow rays for each light source can strongly reduce the number of shadow rays.

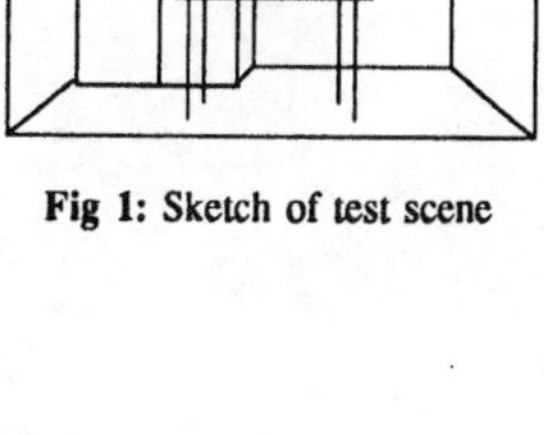

Fig 1: Sketch of test scene

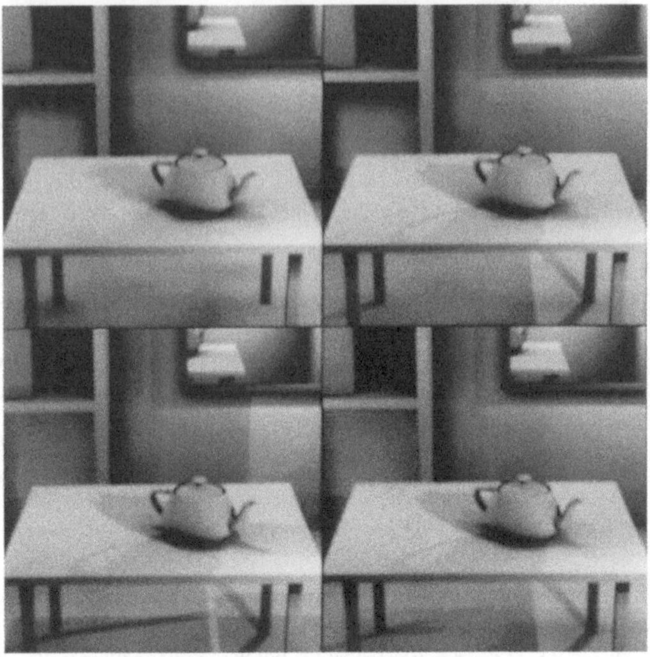

Fig. 2a-d: Influence of selection on image quality.

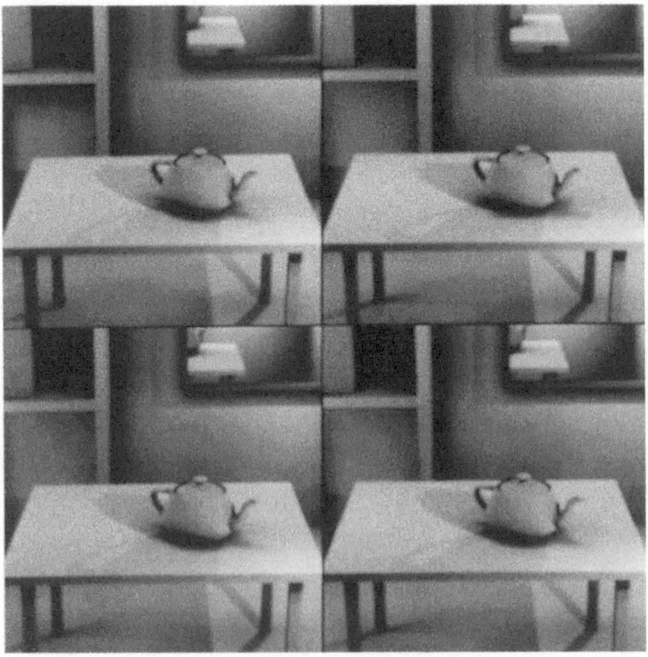

Fig. 2e-h: Influence of selection on image quality.

The number of shadow rays is always much smaller than the number of viewing rays times the number of the shadow rays per sample, because no shadow rays are cast for patches that are backfacing with respect to a light source, and because the program exploits shadow coherence (the reduction approaches a total of 90%).

Table 1: Number of viewing and shadow rays for different source selections

figure 2	number of shadow rays per light source				total rays	
	A	B	C	D	viewing	shadow
a	0	0	0	0	121187	0
b	1	1	0	0	134016	68107
c	1	1	1	1	152325	144786
d	1	1	4	4	382884	641621
e	1	1	16	0	278089	775482
f	1	1	0	16	359244	980414
g	1	1	16	16	454650	1671910
h	1	1	16	64	537437	5073069

We are currently implementing algorithms to automatically apply the described local criteria to obtain high-quality pictures with a minimum of shadow rays.

5 Conclusions and Directions for Future Research

The experiments show that the described method is able to represent sharp shadows, because of the separation of the direct and indirect shading over the two passes. By proper selection of light sources for shadow testing, the images will be generated efficiently.

Different criteria for selecting the most influential light sources were given. All experiments so far were done with a global selection of light sources. Use of local selections will further reduce the number of shadow rays. A disadvantage of the given method is the extra memory required to store the influences of the most important light sources separately at each patch vertex.

Compared to statistical methods to reduce the number of shadow rays (Kajiya 1986; Shirley 1990), for which many viewing rays per pixel are necessary to get reliable results, the presented method will also be useful if only a few viewing rays per pixel are used.

Acknowledgements

This research is part of the radiosity engine project, a cooperation with the Network Theory section of the Faculty of Electrical Engineering at Delft University of Technology. We thank Ed Deprettere and Li-Sheng Shen for the discussions on radiosity methods.

References

Campbell, A.T., Fussell, D.S. (1990) Adaptive Mesh Generation for Global Diffuse Illumination, Computer Graphics 24(4), pp 155-164, Siggraph '90.

Cohen, M. F., Greenberg, D. P. (1985) The Hemi-Cube: A Radiosity Solution for Complex Environments, Computer Graphics 19(3), pp 31-40, Siggraph '85.

Cohen, M.F., Greenberg, D.P., Immel, D.S., Brock, P.J. (1986) An Efficient Radiosity Approach for Realistic Image Synthesis, IEEE Computer Graphics and Applications 6(3), pp 26-35.

Cohen, M.F., Cheng, S.E., Wallace, J.R., Greenberg, D.P. (1988) A Progressive Refinement Approach to Fast Radiosity Image Generation, Computer Graphics 22(4), pp 75-84, Siggraph '88.

Cook, R.L. (1986) Stochastic Sampling in Computer Graphics, ACM Transactions on Graphics 5(1), pp 51-72.

Dippé, M., Wold, E. H. (1985) Antialiasing through Stochastic Sampling, Computer Graphics 19(3), pp 69-78, Siggraph '85.

Heckbert, P.S. (1990) Adaptive Radiosity Textures for Bidirectional Ray Tracing, Computer Graphics 24(4), pp 145-154, Siggraph '90.

Kajiya, J.T. (1986) The Rendering Equation, Computer Graphics 20(4), pp 143-150, Siggraph '86.

Kok, A.J.F., Jansen, F.W., Woodward, C. (1991) Efficient Complete Radiosity Ray Tracing using a Shadow Coherence Method, submitted for publication.

Mitchell, D. P. (1987) Generating Antialiased Images at Low Sampling Densities, Computer Graphics 21(4), pp 65-72, Siggraph '87.

Painter, J., Sloan, K. (1989) Antialiased Ray Tracing by Adaptive Progressive Refinement, Computer Graphics 23(3), pp 281-288, Siggraph '89.

Shirley, P. (1990) A Ray Tracing Method for Illumination Calculation in Diffuse Specular Scenes, Proceedings Computer Graphics Interface '90, pp 205-212.

Sillion, F., Puech, C. (1989) A General Two Pass Method Integrating Specular and Diffuse Reflection, Computer Graphics 23(3), pp 335-344, Siggraph '89.

Wallace, J.R., Cohen, M.F., Greenberg, D.P. (1987) A Two-Pass Solution to the Rendering Equation: A Synthesis of Ray Tracing and Radiosity Methods, Computer Graphics 21(4), pp 311-320, Siggraph '87.

Ward, G.J., Rubinstein, F.M., Clear, R.D. (1988) A Ray Tracing Solution for Diffuse Interreflection, Computer Graphics 22(4), pp 85-92, Siggraph '88.

Woodward, C. (1990) Methods for Computer-Aided Design of Free-form Objects, PhD thesis, Helsinki University of Technology, Acta Polytechnica Scandinavica, Mathematics and Computer Science Series No. 56, Finnish Academy, Helsinki.

The Constant Radiosity Assumption Syndrome

Filippo Tampieri, Dani Lischinski

Abstract

Most current radiosity solutions to the diffuse global illumination problem suffer from inherent inaccuracies because of the built-in assumption that radiosity is constant across the area of each patch. This paper presents an improved progressive refinement radiosity algorithm that makes the above assumption unnecessary. Progressive radiosity repeatedly chooses a light source and determines its contribution to the radiosities on a finite set of locations in the environment. An improved method for numerically approximating these contributions is presented. The method takes into account the non-uniform radiosity distribution across the light source to yield accurate results. In addition to the enhanced accuracy of the resulting solutions, dropping the constant radiosity assumption allows using larger patches, reducing the number of iterations required to obtain plausible results.

1 Introduction

Radiosity is an important image synthesis technique able to provide view-independent solutions to the diffuse global illumination problem. It was first introduced to computer graphics by Goral *et al.* [GORA84] and produced solutions for simple, unoccluded environments. Soon afterwards the method was extended to deal with arbitrary polyhedral environments [COHE85, COHE86]. Further research resulted in *progressive refinement radiosity* [COHE88], capable of producing good approximations to the complete solutions very early in the solution process.

In all the works mentioned above, no attempt was made to objectively evaluate the numerical accuracy of the results. Goral *et al.* qualitatively compared the computed solution with an actual model of a simple environment. Quantitative comparisons were not made because equipment to perform exact measurements was not available. In [COHE88] the quality of intermediate solutions was judged by comparing them to the final converged solution. While such a comparison is useful for illustrating the rate of convergence, it does not measure the accuracy of these solutions, since there is no guarantee that the converged solution itself is accurate. First attempts towards obtaining numerical accuracy were made only recently [BAUM89, WALL89]. In these works, it was shown that the hemi-cube method for computation of form-factors was prone to errors resulting in visible artifacts, particularly in the context of progressive radiosity. Baum *et al.* [BAUM89] suggested enhancing the hemi-cube method with analytical computations, while Wallace *et al.* [WALL89] proposed a new method which utilizes ray tracing to numerically integrate the form-factors. More recently, Hanrahan and Salzman [HANR90] presented a new radiosity algorithm for unoccluded environments that uses a hierarchical subdivision scheme to compute form-factors to a given numerical precision.

The purpose of this paper is to point out another source of inaccuracies in current radiosity implementations. More specifically, we refer to the inaccuracies introduced into radiosity solutions by the built-in "constant radiosity assumption", henceforth referred to as the CRA. The CRA treats patches, which are the basic illuminating units in radiosity algorithms, as having constant radiosity across their areas. Section 2 explains in more detail the role of the CRA in the radiosity formulation and demonstrates how it may result in serious inaccuracies in the computed solution.

In Section 3 an improved ray tracing algorithm for progressive radiosity is described. This method makes no use of the CRA and thus is immune to the associated inaccuracies. More specifically, an improved solution to the problem of determining the contribution of

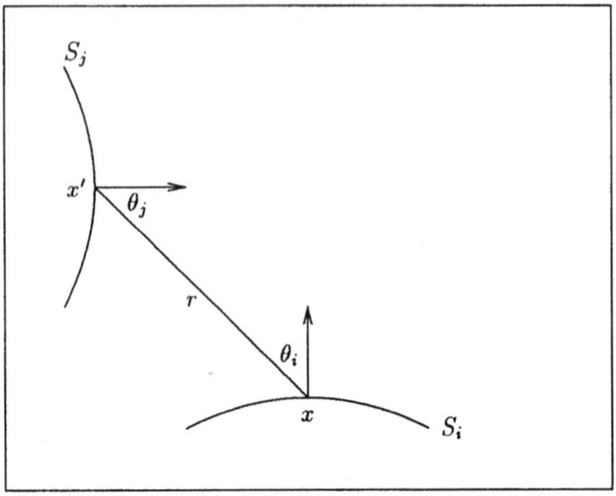

Fig 1. Geometry for Equation (2)

a light source to the illumination of a differential area on a receiving surface is presented. This problem needs to be solved repeatedly in the context of the ray tracing algorithm for progressive radiosity [WALL89]: first, the environment is discretized into a two-level hierarchical mesh of patches and elements [COHE86]; then, a solution is computed incrementally by repeatedly selecting a shooting patch and adding its contribution to the radiosities of all the element vertices in the environment. Rather than assuming a uniform radiosity distribution across the shooting patch, the improved algorithm uses the radiosities of the source element vertices to more accurately represent the actual distribution. The representation is then used to compute a better numerical approximation of the contribution of the light source to the target vertices.

The superior accuracy of the new algorithm is demonstrated in Section 4, where its results are quantitatively compared with an accurate solution obtained through brute-force numerical integration.

2 The CRA and its Implications

The radiosity formulation treats the radiosity B_i at any given point x on a patch i in the environment as being composed of the emitted radiosity $B_i^e(x)$, and the radiosity reflected from that point, where the latter is the sum of radiosity contributions $B_{ij}(x)$ from the other patches in the environment:

$$B_i(x) = B_i^e(x) + \sum_{j=1}^{N} B_{ij}(x) \tag{1}$$

The contribution $B_{ij}(x)$ of patch j to the radiosity at point x on patch i can be easily derived from the rendering equation [KAJI86, BOUV90, SHIR90]:

$$B_{ij}(x) = \rho_i \int_{A_j} B_j(x')v(x,x')\frac{\cos\theta_i \cos\theta_j}{\pi r^2} dA_j(x') \tag{2}$$

where

ρ_i is the reflectivity of patch i;
A_j is the area of patch j;
B_j is the radiosity of patch j;

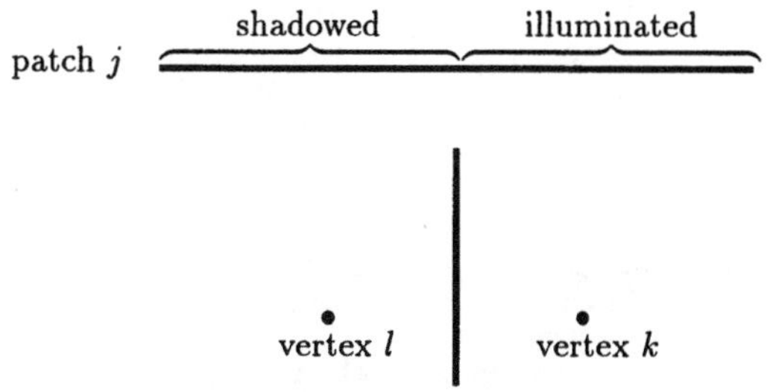

Fig 2. A simple example where the constant radiosity assumption leads to inaccurate results

$v(x, x')$ is a visibility term, 1 if point x "sees" point x', 0 otherwise;
r is the distance between x and x';
θ_i, θ_j are the angles between the surface normals at x and x' and the line connecting the two points (see Fig. 1);
$dA_j(x')$ is the differential area centered around point x' on patch j.

If the radiosity across patch j is assumed constant, B_j can be moved outside the integral:

$$B_{ij}(x) = \rho_i B_j \int_{A_j} v(x, x') \frac{\cos\theta_i \cos\theta_j}{\pi r^2} dA_j(x') \tag{3}$$

The integral on the right-hand side is the differential to finite area form-factor, F_{dA_i,A_j} [GORA84]. Using the reciprocity relationship

$$dA_i F_{dA_i,A_j} = A_j dF_{A_j,dA_i} \tag{4}$$

we can rewrite Equation (3) as follows:

$$B_{ij}(x) dA_i(x) = \rho_i B_j A_j dF_{A_j,dA_i} \tag{5}$$

which is the formula used by Wallace *et al.* [WALL89] to express the contribution of a shooting patch j to a receiving vertex at point x on patch i.

If B_{ij} is assumed constant across a finite area A_i, then integrating both sides of Equation (5) over A_i yields

$$B_{ij} A_i = \rho_i B_j A_j F_{ji} \tag{6}$$

This formula is exactly the one used in the progressive refinement radiosity algorithm [COHE88] to compute the contributions of source patches to all the elements in the environment.

Combining Equations (1) and (6) yields the well-known full matrix radiosity formulation [GORA84, COHE85]:

$$B_i A_i = B_i^e A_i + \rho_i \sum_{j=1}^{N} B_j A_j F_{A_j,A_i} \tag{7}$$

As was shown above, it is the CRA that allows the contribution of patch j to be expressed as a product of two main independent terms: the radiosity, B_j, and the form-factor, F_{A_j,A_i}. The latter term depends exclusively on the geometry of the environment and can be computed without any knowledge of the light distribution in the environment. This decoupling of geometry and radiosity is essential for full matrix radiosity methods since it allows one to precompute all the form-factors and then solve the linear system (7), with the radiosities of the patches as the unknowns. However, as will be shown in the next section, in the context of progressive radiosity, the decoupling is not necessary.

In practice, the CRA rarely holds. Most notably, it is violated by patches crossed by shadow or penumbra boundaries, but even patches unoccluded from the light sources may exhibit considerable variations in radiosity across their areas. In fact, the only patches which truly satisfy the CRA are those which are in total shadow. Primary uniform light sources also come close to satisfying the CRA (the first time they shoot). Thus, no matter how precise the computation of the form-factors, the results may still be far from accurate. Consider, for example, the simple situation illustrated in Fig. 2. Source patch j is an indirect light source whose surface happens to be more brightly illuminated on the right than on the left. Clearly, its contribution to vertex k is larger than that to vertex l. Current radiosity systems, though, would ignore the distribution of the radiosity across patch j, and, because of symmetry, would compute equal form-factors from the patch to the two vertices, resulting in equal contributions to k and l.

The inaccuracies resulting from the violation of the CRA are reduced when patches are small, distances between patches are large, or the radiosity on the shooting patch is relatively constant. Thus, discretizing the environment into a fine mesh of patches will result in an accurate solution. However, increasing the number of patches would have a devastating effect on the time requirements of the method. This is true not only for full matrix methods (where the number of patches determines the size of the matrix), but also in progressive radiosity: the more patches there are, the slower the convergence.

Hanrahan and Salzman [HANR90] proposed a radiosity algorithm that adaptively subdivides pairs of patches to build a hierarchical data structure in which direct light transport interactions are represented by links between nodes (subregions of the original patches). Two subregions are linked only if the form-factors between them fall below a specified threshold. Thus, in effect, only regions that are small compared to the distance between them are allowed to interact, reducing, as was pointed out earlier, the inaccuracies resulting from the violation of the CRA.

In the next section we present another solution to the problem, showing how the ray tracing method of Wallace *et al.* can be modified to take into account nonconstant radiosity across patches, thus allowing one to obtain accurate results even when patches are large.

3 Accurate Computation of Contributions

The exact contribution of a shooting patch j to the radiosity of a point x on a receiving surface i was given in the previous section by Equation (2). The presence under the integral of the arbitrary radiosity distribution B_j and the visibility term $v(x, x')$ suggests that numerical rather than analytical techniques be used to compute the contribution.

If source patch j is broken down into a set of n disjoint areas, its contribution can be expressed as the sum of smaller contributions:

$$B_{ij}(x) = \rho_i \sum_{k=1}^{n} \int_{A_{jk}} v(x, x') B_j(x') \frac{\cos\theta_i \cos\theta_j}{\pi r^2} dA_j(x') \tag{8}$$

If the areas A_{jk}'s are chosen small enough that both visibility and radiosity can be assumed constant across each area, both terms can be taken out of the integral to yield:

$$B_{ij}(x) = \rho_i \sum_{k=1}^{n} v(x, x_k) \bar{B}_{jk} F_{dA_i, A_{jk}} \tag{9}$$

where

x_k is the center of area A_{jk};

$\bar{B}_{jk}$ is the area-averaged radiosity across area A_{jk} (given by $\frac{1}{A_{jk}} \int_{A_{jk}} B_j(x')dx'$).

Note that radiosity is only assumed constant across subregions of patch j whose size is chosen to meet the assumption, and not across the entire patch j as in the CRA.

By multiplying both sides of Equation (9) by $dA_i(x)$ and using the reciprocity property of form-factors given in Equation (4), Equation (9) transforms to:

$$B_{ij}(x) dA_i(x) = \rho_i \sum_{k=1}^{n} v(x, x_k) \bar{B}_{jk} A_{jk} dF_{A_{jk}, dA_i} \tag{10}$$

As pointed out by Wallace *et al.* [WALL89], using straightforward numerical techniques to evaluate the delta form-factor dF_{A_{jk},dA_i} can, in certain cases, result in large errors. As a more numerically stable and cost efficient alternative, they suggest, instead, to approximate each area A_{jk} by a simple finite geometry for which an analytical solution is known. In particular, they use a simple disk of area A_{jk} centered at x_k to derive the following equation:

$$dF_{A_{jk},dA_i} = dA_i(x)\frac{\cos\theta_{ik}\cos\theta_{jk}}{\pi r_k^2 + A_{jk}} \tag{11}$$

Substituting the right-hand side of Equation (11) for dF_{A_{jk},dA_i} in Equation (10), leads to an equation that can be easily evaluated to compute the radiosity contribution $B_{ij}(x)$:

$$B_{ij}(x) = \rho_i \sum_{k=1}^{n} v(x, x_k)\bar{B}_{jk}A_{jk}\frac{\cos\theta_{ik}\cos\theta_{jk}}{\pi r_k^2 + A_{jk}} \tag{12}$$

where $\bar{B}_{jk}A_{jk} = \int_{A_{jk}} B_j(x')dA_j(x')$, is the power leaving delta area A_{jk}.

This equation is very similar to that presented by Wallace *et al.*, but differs from it in a very important aspect. Equation (12) computes the contribution as a sum of delta form factors, *each* weighted by the power leaving the corresponding delta area. The ray-traced form-factor method, instead, computes patch j's overall contribution as the product of its average power and a form-factor. The latter is computed as the area-weighted average of the delta form-factors without any use of the radiosity distribution across the source.

3.1 Adaptive Subdivision

In order to use Equation (12) to compute an accurate estimate of the contribution from a source patch j to a point x on a receiving surface i, patch j must be subdivided into a set of delta areas small enough to satisfy the assumptions made during the derivation of the equation. Furthermore, for computational efficiency, such a set should be as small as possible. As already noted by Wallace *et al.*, a uniform subdivision scheme proves wasteful when different areas of the source contribute different quantities to the overall result. Instead, an adaptive subdivision scheme may provide the flexibility necessary to produce a nearly optimal subdivision for each particular target location x.

Our implementation uses the following approach: for each receiving point x, the source patch j is subdivided in parametric space according to a quadtree partitioning scheme until the delta areas corresponding to the leaves of the quadtree satisfy one of the following criteria:

1. The contribution of the current delta area to x falls below a user specified threshold.
2. The sum of the contributions of the children of the current node equals (within a user specified tolerance) the estimated contribution of their parent.
3. The depth of the current node is equal to a user specified maximum value.

The first criterion is the same as that used in [WALL89], except that it uses the *local* radiosity distribution across the delta area rather than its *average* value across the patch to determine the contribution to the target. The second criterion attempts to limit the subdivision by stopping when further subdivision does not seem to improve the estimated contribution. The third criterion is only intended to give the user an additional means of control in order to avoid excessive subdivision.

3.2 Radiosity Distribution

Throughout this paper, the radiosity distribution of the shooting patch j was denoted by the function B_j, implying that its value could be computed at any point on the surface of patch j. How can this function be represented in current radiosity systems? Let us assume that each patch is divided into a set of smaller elements [COHE86] and that unshot radiosity as well as accumulated radiosity values are computed and stored at each element vertex [WALL89]. Also assume, for simplicity, that quadrilaterals are used

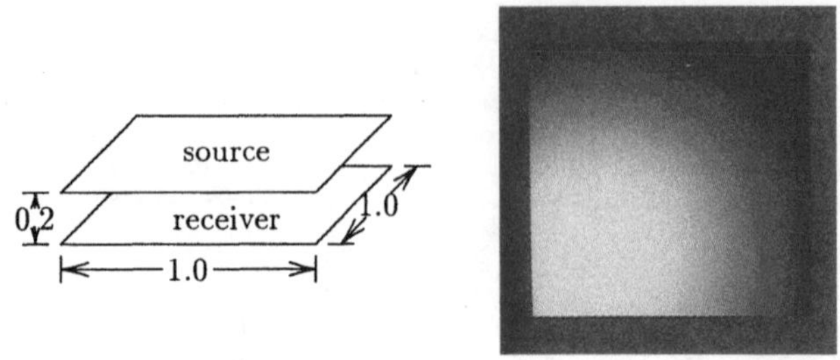

Fig 3. A simple test configuration. The shaded image shows the radiosity distribution of the source

as elements and patches. Then, the radiosity distribution across each element can be approximated by bilinearly interpolating the vertex radiosity values. Thus, the radiosity value at any point x on patch j can be computed by first determining which patch element contains x and then using bilinear interpolation to compute $B_j(x)$ from the element vertex radiosities.

Equation (12) requires that an estimate to $\int_{A_{jk}} B_j(x)dA_j(x)$ be computed. We distinguish two cases, depending on whether delta area A_{jk} is contained in an element of patch j or not. If it is, the radiosity values of each of its corners can be computed by bilinear interpolation as explained above and the integral can then be evaluated analytically. Otherwise, because of the adaptive subdivision scheme, area A_{jk} must cover an integer number of elements and $\int_{A_{jk}} B_j(x)dA_j(x)$ can be computed simply by integrating B_j over each of the contained elements and by summing up the partial results.

Rather than recomputing these values on the fly while computing the contribution of patch j to each target vertex, we use an explicit storage scheme that guarantees that each quantity is computed exactly once.

Once a new shooting patch j is selected, a quadtree structure is build starting from the entire patch j and subdividing until each leaf of the quadtree corresponds to an element of the patch. Each node of the quadtree represents a delta area A_{jk} and stores its center x_k, its area A_{jk}, and its power $\bar{B}_{jk}A_{jk}$ computed as described above.

Once the initial subdivision is created, each target vertex in the environment is processed in turn. If a target vertex requires a finer level of subdivision, new nodes are created and initialized so that other vertices will be able to use the same information, if necessary, without having to recompute it. Notice that, for each target, the quadtree is only traversed down to the necessary depth, not necessarily to the leaves of the data structure.

4 Results

Fig. 3 shows the simple configuration used to test the performance of the new algorithm and to compare it to that of the ray-traced form-factor method. The test environment consists of a single shooting patch placed directly over a receiving surface. The short distance between the two surfaces makes the task of computing an accurate contribution particularly challenging. The shaded square next to the configuration shows the radiosity distribution of the source as seen from above.

Fig. 4 (see also color plates) shows the contributed radiosity on the receiver as computed by the ray-traced form-factor method (left) and the new method (right). In both cases, the receiving surface consists of 32 by 32 elements. Our implementation of the ray-traced form-factor method uses the adaptive source subdivision scheme described by Wallace *et al.* [WALL89] which computes the source-to-vertex form-factor as the area-weighted average of the delta form-factors due to each of the source subregions. The total number of samples (shadow rays) used was comparable for both methods. Given the geometry of the test environment and the particular radiosity distribution of the source, we would expect higher radiosity values in the lower left quadrant of the receiver than in its upper right. While our method (right) correctly computes the expected results, the ray-traced

Fig 4. Radiosity distribution of the receiver due to the shooting source as computed by the ray-traced form-factor method (left) and the new method (right)

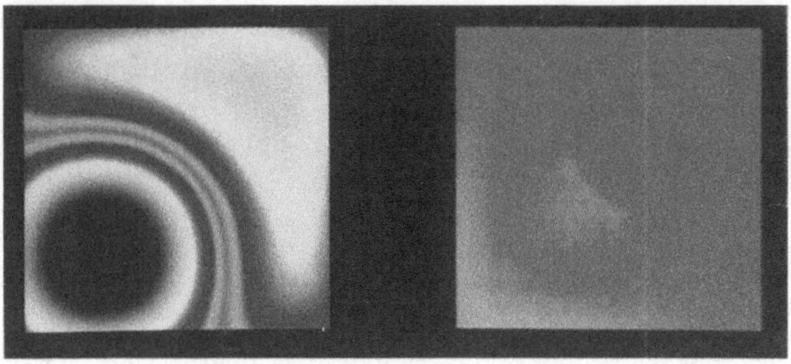

Fig 5. Pseudo-color image of the absolute error in the radiosity distribution across the receiver as computed by the ray-traced form-factor method (left) and the new method (right). The error increases exponentially as the color changes from blue (0), through cyan (0.03), green (0.09) and yellow (0.21) to red (> 0.45)

form-factor method yields a symmetric solution with the brightest spot at the center of the receiver (directly under the center of the source). Notice that this problem stems from the CRA built into the method and not from inaccuracies in the computation of the form-factors. In fact, with the exception of Harahan's and Salzman's scheme [HANR90], any other method currently used for the computation of form-factors [COHE88, BAUM89] could be substituted for the ray-traced form-factor method without any substantial improvement in the solution.

Fig. 5 (see also color plates) shows a pseudo-color image of the absolute error in the radiosity distribution across the receiver when computed by the ray-traced form-factor method (left) and our method (right). A reference solution was computed using a very accurate brute-force numerical integrator. This allows for a quantitative as well as qualitative evaluation of the results. The error image was computed point by point by taking the absolute difference between the radiosity value computed by each of the two methods and the reference solution. The maximum and average radiosity values of the reference solution were respectively 1.0 and 0.35. The absolute error distribution across the receiving surface is shown in the figure in pseudo-color. Since the range of errors produced by the new method was much smaller than that produced by the ray-traced form-factor method, a logarithmic mapping was used to assign colors to error values. More precisely, the color scale interpolates colors from blue to cyan, green, yellow, and red corresponding respectively to absolute error values of 0, 0.03, 0.09, 0.21, and greater than 0.45.

TABLE 1. Error Comparisons

method	max abs err	avg abs err	avg rel err
ray-traced form-factor	0.612	0.2191	63.42%
new method	0.015	0.0034	0.99%

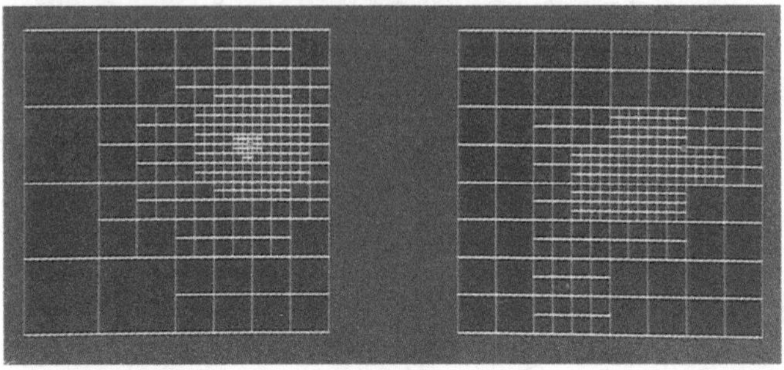

Fig 6. Subdivision of the source for a receiving point placed at (0.74, 0.62) as computed by the ray-traced form-factor method (left) and the new method (right). The location of the target point is marked by a red dot

Table 1 shows the maximum absolute error and average absolute and relative errors over the entire receiving surface corresponding to both the new and ray-traced form-factor methods. Notice the drastic accuracy improvement of the former over the latter.

Fig. 6 (see also color plates) shows the subdivision of the shooting source for one particular receiving differential area produced by the ray-traced form-factor method (left) and our method (right). The receiving point is placed at (0.74, 0.62), where (0, 0) are the coordinates of the lower left corner, and is shown as a red dot in the images. Both methods take into account the magnitude of the contribution as a criteria for subdivision. The ray-traced form-factor method, though, concentrates most of the samples in the area directly above the target. Although this area contributes the most to the form-factor, in this particular case, it adds relatively little to the result. The method proposed by Hanrahan and Salzman [HANR90] would behave similarly since its subdivision scheme is based on geometric considerations only. Our method, instead, takes into consideration the non-uniform radiosity distribution of the source and thus correctly places more samples towards the brightest area of the source. This results in a much more accurate result for the same computational cost. It is worth noting the similarity between the subdivision algorithm and importance sampling techniques; however, while importance sampling is a stochastic process, our method is strictly deterministic.

Finally, Fig. 7 shows a plot of the radiosity distribution along a line segment between the lower left and the upper right corners of the receiver. This segment was sampled at 32 equally spaced locations and the rest of the radiosity distribution was filled in using linear interpolation. The solution generated by the ray-traced form-factor (left, dashed) and by our method (right, dashed) are compared to the reference solution (solid). While the solution provided by the ray-traced form-factor method shows a large error, that provided by our method is almost indistinguishable from the reference solution.

5 Conclusions

An improved solution was presented to the key step in progressive radiosity, i.e. the problem of determining the contribution of a light source to the illumination of a differential

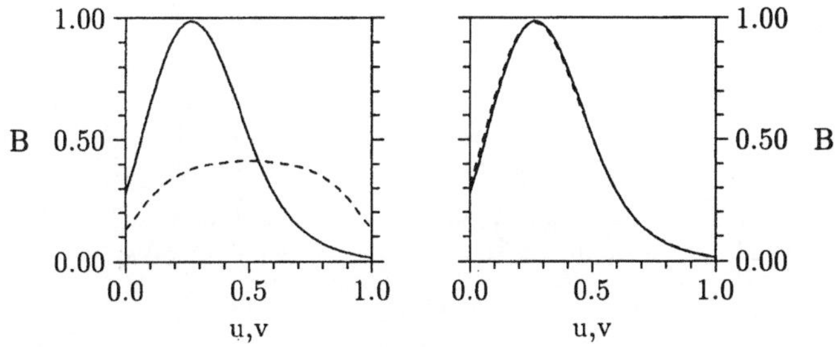

Fig 7. Radiosity distribution for a segment between the lower left and the upper right corners of the receiver. Computed solution (dashed) for the ray-traced form-factor method (left) and the new method (right) vs. the reference solution (solid). Note the tight match of the two curves on the right

area on a receiving surface. The new algorithm computes the contributions more accurately than the form-factor based algorithm of Wallace *et al.*, yet it possesses all the desirable characteristics of the latter.

An important advantage of the algorithm is that it can be used with arbitrarily large patches, whereas current radiosity implementations require the surfaces in the environment to be meshed into small patches in order to come closer to satisfying the CRA. The number of iterations required by a progressive radiosity solution to convergence grows with the number of patches, and thus the new method results in less iterations. Since the CRA is no longer required, the surfaces in the environment can be represented by as little as one patch per surface; no size constraint needs to be satisfied. Note, however, that the problem of determining an optimal subdivision of patches into elements still exists.

The new algorithm uses ray-tracing for the visibility computation, using a single ray to determine visibility across a small area. Since this is a point sampling method, it is inherently prone to aliasing. A more accurate approximation to the integral of Equation (2) may be computed by first identifying the regions on the shooting surface that are visible from the receiving target point and then summing up the contributions of each of these regions as computed by our method, only now omitting the point to point visibility test. Developing an efficient way of determining such regions is essential to the viability of this alternative solution and requires further investigation.

Even if prone to aliasing, we believe that the method presented in this paper represents a significant step forward not only towards more realistic looking images and their efficient computation, but also towards a more physically accurate simulation of the global illumination phenomena.

Acknowledgements

The authors are grateful to the Hewlett Packard Corporation for donating the equipment used in this work. This research was conducted under National Science Foundation grant #CCR-8617880. The presentation of this work at the Second Eurographics Workshop on Rendering was made possible by travel grants generously provided by the Mario Einaudi Center for International Studies and the Department of Computer Science of Cornell University. The authors wish to thank Director of the Program of Computer Graphics, Professor Donald P. Greenberg for making this work possible and James Arvo and François Sillion for reviewing this paper.

References

[BAUM89] Baum, Daniel R., Holly E. Rushmeier, and James M. Winget. "Improving Radiosity Solutions Through the Use of Analytically Determined Form-Factors," Proceedings of SIGGRAPH'89 (Boston, Massachusetts, July 31–August 4, 1989), in *Computer Graphics*, 23(3), July 1989, pages 325–334.

[BOUV90] Bouville, Christian, Kadi Bouatouch, Pierre Tellier, and Xavier Puyeo. "Theoretical Analysis of Global Illumination Models," in Proceedings of Eurographics Workshop on Photosimulation, Realism and Physics in Computer Graphics (Rennes, France, June 11–12, 1990), June 1990, pages 53–66.

[COHE85] Cohen, Michael F. and Donald P. Greenberg. "The Hemi-Cube: A Radiosity Solution for Complex Environments," Proceedings of SIGGRAPH'85 (San Francisco, California, July 22–26, 1985), in *Computer Graphics*, 19(3), July 1985, pages 31–40.

[COHE86] Cohen, Michael F., Donald P. Greenberg, and David S. Immel. "An Efficient Radiosity Approach for Realistic Image Synthesis," *IEEE Computer Graphics and Applications*, 6(2), March 1986, pages 26–35.

[COHE88] Cohen, Michael F., Shenchang Eric Chen, John R. Wallace, and Donald P. Greenberg. "A Progressive Refinement Approach to Fast Radiosity Image Generation," Proceedings of SIGGRAPH'88 (Atlanta, Georgia, August 1–5, 1988), in *Computer Graphics*, 22(4), August 1988, pages 75–84.

[GORA84] Goral, Cindy M., Kenneth E. Torrance, Donald P. Greenberg, and Bennett Battaile. "Modeling the Interaction of Light Between Diffuse Surfaces," Proceedings of SIGGRAPH'84 (Minneapolis, Minnesota, July 23–27, 1984), in *Computer Graphics*, 18(3), July 1984, pages 213–222.

[HANR90] Hanrahan, Pat and David Salzman. "A Rapid Hierarchical Radiosity Algorithm for Unoccluded Environments," in Proceedings of Eurographics Workshop on Photosimulation, Realism and Physics in Computer Graphics (Rennes, France, June 11–12, 1990), June 1990, pages 151–172.

[KAJI86] Kajiya, James T. "The Rendering Equation," Proceedings of SIGGRAPH'86 (Dallas, Texas, August 18–22, 1986), in *Computer Graphics*, 20(4), August 1986, pages 143–150.

[SHIR90] Shirley, Peter. "Physically Based Lighting Calculations for Computer Graphics: a Modern Perspective," in Proceedings of Eurographics Workshop on Photosimulation, Realism and Physics in Computer Graphics (Rennes, France, June 11–12, 1990), June 1990, pages 67–82.

[WALL89] Wallace, John R., Kells A. Elmquist, and Eric A. Haines. "A Ray Tracing Algorithm for Progressive Radiosity," Proceedings of SIGGRAPH'89 (Boston, Massachusetts, July 31–August 4, 1989), in *Computer Graphics*, 23(3), July 1989, pages 315–324.

Some Experiments on Adaptive Subdivision for Progressive Radiosity

Christophe Vedel and Claude Puech

1 Introduction and Related Work

In order to compute the global lighting of a scene, the progressive radiosity method simulates the transport of light between diffusely reflecting surfaces. The scene is divided into patches and the basic step of the simulation is for a patch to shoot its energy to the rest of the environment. During this process, light intensities are stored on the surfaces in object space. This is usually done by choosing a number of sample points and keeping information for each one of them. By using these values and an interpolation scheme, values at all interesting points on the objects are computed before displaying the scene.

We know beforehand that there will be regions in the scene with slowly varying intensity and regions with rapidly varying intensity like shadow borders. To insure uniform accuracy, the latter should be sampled at a higher rate, a rate that for computational cost reasons we cannot afford everywhere. Adaptive subdivision is a twofold process that detects zones which need better sampling and creates new sample points in these zones only. Because all information on the intensity comes from the sampling, subdivision steps are interleaved with form factor computation steps.

The problem of the density and the placement of the sample points did not receive a lot of attention until recently. A possible explanation is that the first methods devised to evaluate form factors were based on projection (hemicube) and not well suited to adaptive subdivision for two reasons. On the one hand, the limited resolution of the hemicube puts a practical lower bound on the size of the patches. When small neighboring patches are projected on the same square of the hemicube, only one of them gets the associated "delta form factor" and "light leaks" occur. On the other hand, form factors for all elements in the environment are computed in one single operation. This makes the interleaving with adaptive subdivision more difficult if one wants to avoid recomputing all the form factors when only a few new sample points are introduced. The significant progress was the introduction of ray tracing to compute form factors by Wallace et al. [Wallace et al., 1989]. The fact that the atomic computation is related to one sample point and that the sampling can be increased locally in order to distinguish arbitrarily close points are the important points.

Results on adaptive subdivision appeared in 1990. Adaptive radiosity textures to store intensity values on surfaces were proposed by Heckbert as an alternative to polygonalization [Heckbert, 1990]. The aim is to uncouple geometry and shading by working in local coordinates on surfaces. Campbell and Fussell have presented a method using BSP trees to solve the hidden surface part of the problem [Campbell and Fussell, 1990]. While it is not as flexible as ray tracing, the method works in object space and avoids the resolution problem of the hemicube. However, it is only good for point light sources and area light sources have to be point sampled. Given the fact that polygons are subdivided along the borders of shadows cast by these point light sources, we see that very precise computations (i.e. the cutting of the polygons) are done based on arbitrary information (i.e. the way we choose to sample the source). These leads us to believe there exists a more economical way to proceed. Finally a paper by Hanrahan and Salzman takes a more global approach of the problem by tightly coupling the sampling of emitters and receptors [Hanrahan and Salzman, 1990]. As a matter of fact, their method is a preprocess and makes no difference between emitters and receptors by treating them symmetrically. This suggests interesting questions on the relationship between the two samplings in a more classical approach.

2 General Adaptive Subdivision Scheme

The progressive radiosity algorithm with adaptive subdivision is outlined below:

```
repeat
       select the most energetic patch
       for every other patch P, process P
until the solution has converged for some criterion

where
      process P consists of
              compute intensities at sample points not already processed
              evaluate the subdivision criterion
              if a subdivision is necessary
                   subdivide P into P1,...,Pn by adding new sample points
                   process patch P1
                   ...
                   process patch Pn
```

In a non adaptive algorithm, processing patch P consists only in computing intensities at the sample points. Here two distinct operations are added: the computation of a subdivision criterion and the subdivision itself with the choice of new sample points. These operations alternate with evaluations of form factors for sample points. Wallace's method to compute form factors using ray tracing is well suited to this because it returns a value for one sample point at a time. In addition to the benefit of not having to store the full matrix of form factors, progressive radiosity allows one to display a partially refined solution at each iteration. By using the above scheme for division, we are able to retain this possibility.

In comparison to Cohen's two pass method [Cohen et al., 1986] that uses substructuring, the drawback of this dynamic approach is that decisions are made based on a partial solution. In some parts of the scene, computation may be done that will be proved useless later on by the rest of the calculation. As the eye is sensitive to contrast rather than to intensity, we can spend time dividing a polygon near a shadow border that will be flooded by the light of a source treated afterwards and thus become invisible.

3 Criteria for Subdivision

The goal of adaptive subdivision is to compute the radiosity function with almost uniform accuracy on all surfaces in the scene. A subdivision criterion should provide an estimate of the accuracy achieved within a given zone (polygon). When using sample points (vertices) and interpolation between these points, the error has two origins:

1. values at vertices are not computed exactly

2. values at other points are interpolated from the vertices' values

In this section, we focus on the second point and try to answer the following question: is the interpolation close enough to the actual intensity function? In order to isolate this particular problem and to get rid of the first reason for errors, we will work on very precise approximations of the values at the vertices, even though the method used to compute them might be unpractical in a real environment.

The soundness of the approximation obviously depends on the method used to interpolate. Although quadratic interpolation is available in hardware, the usual method for fast display on current graphics workstations is linear interpolation (Gouraud shading). It requires no more information than intensity values for the vertices. Testing if the linear interpolation is a good approximation amounts to checking whether the intensity function on the polygon is "far from linear" or not. Naturally this cannot be done knowing only radiosity values for the vertices. Gradients are the relevant information for this test.

The decision whether a patch needs subdivision or not is made by computing a function using information on the patch. Depending on whether the value obtained is below or above a given threshold, the patch is accepted or rejected. For the criterion to be useful, this function should be related to the error due to interpolation. If the unnecessary subdivision of some patches is bearable (although it would be better to avoid it), the criterion should ensure that the inaccuracy of accepted patches has an upper bound. Furthermore, the bound should be a monotonous function of the threshold.

Even if simpler schemes (involving only the difference of radiosity between the vertices) have been used with success (see [Cohen et al., 1986] for example), it seems that gradients could improve criteria for subdivision. To test this idea, we studied the relationship between three criterion functions and a reference error. The reference error is heavily sampled. In short, we integrate numerically the L_2 norm of the difference between the bilinear interpolation and the real function. The three criterion functions are:

- the difference of radiosity between vertices.
- the difference of radiosity gradients between vertices. At each vertex, two directional gradients are computed. For each direction, the difference between vertices is computed. The largest value is retained.
- the integral of the squared difference between the bilinear interpolation and a higher order polynomial interpolation $P(x, y)$. $P(x, y)$ is determined taking all information into account. 12 conditions are imposed: the value of $P(x_i, y_i)$ and two partial derivatives $\partial P/\partial x(x_i, y_i)$ and $\partial P/\partial y(x_i, y_i)$ are fixed at each vertex (x_i, y_i) $i = 1 \cdots 4$. On a rectangular domain, the integral is computed symbolically as a function of the radiosities and gradients at each vertex. The resulting expression can then be evaluated for different patches.

For these tests, high accuracy could be achieved for reasonable scenes using Wallace's method and heavily oversampling the source. We prefer to restrict ourselves to very simple test scenes where the receptor is illuminated by a square source and a shadow is cast by a halfplane. In this context, the contour integration technique is practical. Furthermore, form factors obtained with contour integration are accurate enough to allow the computation of gradients using finite differences. The issue of using a weaker estimate of the form factor and still being able to compute gradients is important and will be addressed in section 4.

The experiment is the following (see Figure 1). The source S is a square polygon. At a given distance from the source along a line Δ, a set of patches of a given size are drawn at random . Their center is taken from a disc in the plane perpendicular to Δ and they can have any orientation provided that they see the source. A half plane Π perpendicular to Δ casts a shadow on the patches. The plane can rotate around the line (angle θ). The parameters that can be changed are:

- the position of Π along Δ (in order to change the width of the penumbra zone).
- the size of the patches. The results presented are for square patches and do not change significantly when patches have reasonable aspect ratios, such as those obtained when using a binary subdivision scheme.

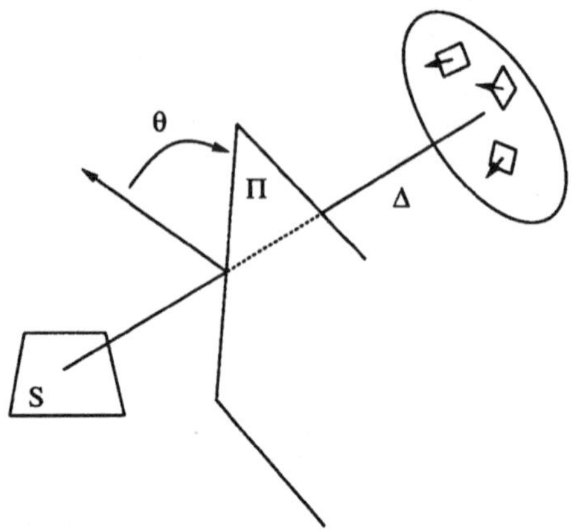

Fig. 1. Geometry of the experiments

- the direction of Δ with respect to the normal at the source. Again the results are for Δ aligned with the normal because there is no qualitative change for different directions.

For each experiment, Π takes several angles θ and for each angle a collection of patches is drawn. For each patch, the three criterion functions as well as the reference error are computed.

The results are presented in the form of clouds of points (Figures 2-5). Each point is a sample patch. The reference error is shown on the X axis and one of the three criterion function is shown on the Y axis. Ideally, there should be a linear relationship between the criterion function and the reference error. In this case, putting a threshold on either of them is equivalent. In the figures, points significantly above the diagonal represent patches for which the criterion function overestimates the actual error. They might be subdivided even though the approximation they provide is good enough. On the contrary, patches whose actual error is underestimated will correspond to points lying under the diagonal. They might escape necessary subdivision. Clearly the first case should be avoided because it leads to useless work but the second case must not happen because it produces erroneous results.

Figure 2 shows the results when no shadow is cast. In this case, the difference of radiosity between the vertices is clearly not related to the actual error. Patches can have low difference and high error as well as the opposite. Difference of gradients is not a very good approximation either. The quartic polynomial realizes a good fit of the true function and provides the most accurate error estimate.

In Figure 3, Π is set close to the source creating a wide zone of penumbra (around ten units). Patches are comparatively small (side of length 1). In this case again, the real radiosity function is nearly planar but not constant. For this reason, high differences of the radiosities are present even though a bilinear interpolation is very accurate. Other experiments show that these results do not vary significantly as a function of the size of the patch provided that the patch lies entirely in the penumbra. This is because the variation of the shadowing factor dominates the total variation the form factor making the penumbra zone quite uniform.

When the patch and the penumbra have comparable sizes (see Figure 4), the quality of the cubic interpolation lowers because the polynomial cannot keep up with the rate of variation of the real function. The difference of the gradients gives roughly the same results. The shape of the cloud of points for the difference of the radiosities starts to change. Patches cannot be totally lit or totally in the penumbra but cover both zones. A relationship with the actual error appears because the derivative of the form factor is

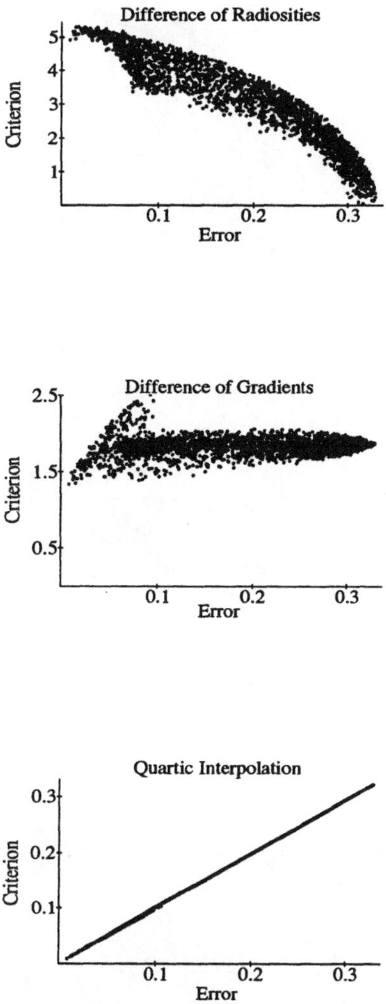

Fig 2. Results without shadow and with patches of size 0.5*

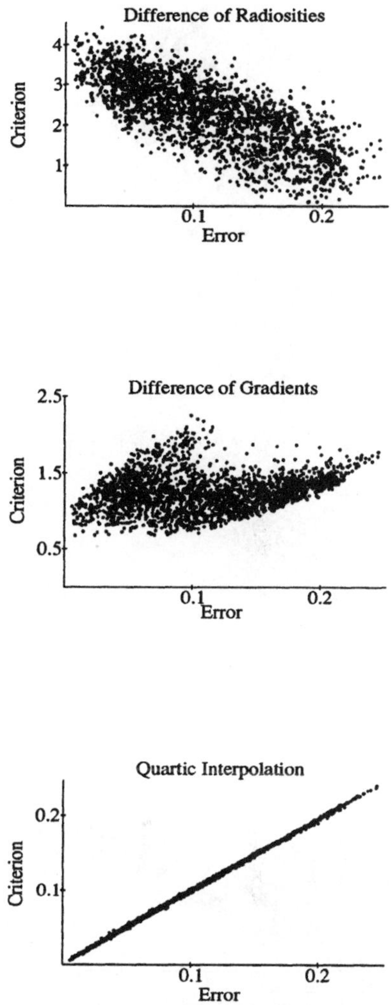

Fig 3. Results with a wide penumbra zone and patches of size 1*

*Values on both axes have been multiplied by 10^3.

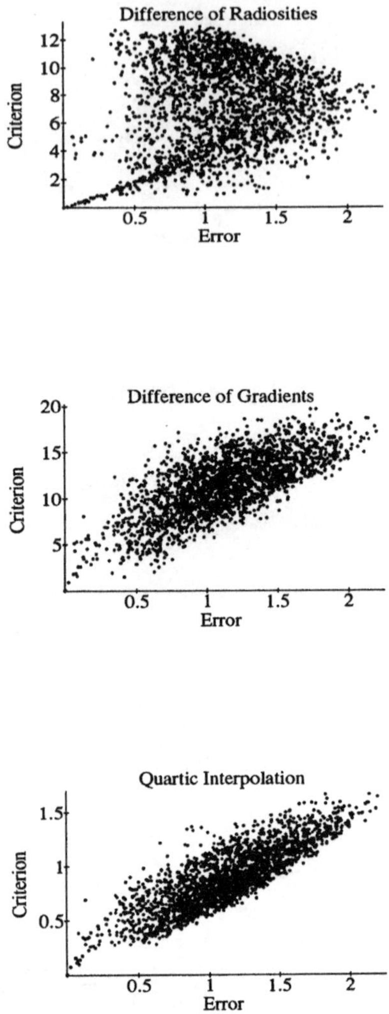

Fig 4. Results with a medium penumbra zone and patches of size 1*

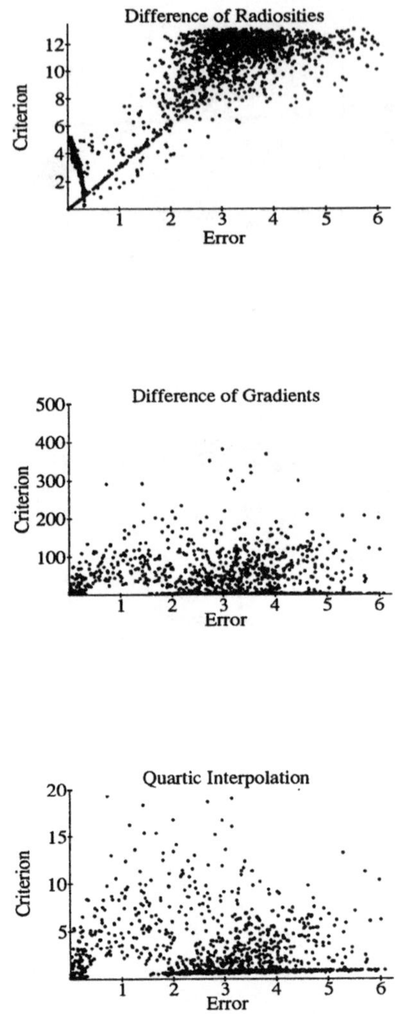

Fig 5. Results with a sharp shadow and patches of size 1*

---

*Values on both axes have been multiplied by 10^3.

not the same in the two zones. There is a difference in the radiosities as in the previous case but now, the linear approximation is not good. Compared to Figure 3, the actual error is now nearly ten times higher for patches of the same size.

Finally Figure 5 shows the result when the shadow is very sharp. The gradient based criteria are incorrect because gradients are small (compared to Figure 4 for example) for all vertices of the patch. Some lie in the lit zone and some in the shadow but the steep change of radiosity inbetween remains undetected. The cubic interpolation criterion does a little better because it takes into account both gradients and radiosities. However, difference of the radiosities is much better in this case.

To sum up, it appears that the use of gradients can potentially lead to better subdivisions. The only case where gradients fail to give a good criterion is when the shadow is sharp. Yet a comparison between gradients and the difference between the vertices' radiosities should spot these cases easily and allow a switch to another criterion. Overall, the cubic interpolation does not justify its higher complexity. A criterion based on the difference of gradients with the possibility of switching to the difference of radiosities in some cases seems promising.

4 Gradient Computations

In the previous section, the contour integration technique allowed us to estimate radiosity gradients with good precision. On actual scenes however, it requires too much computations to be practical. To be able to apply the new criteria, a method to evaluate gradients must be devised, using the ray tracing method for form factor computation.

The radiosity function is known by point sampling so finite differences can be used to compute its derivatives. It consists in taking the difference in radiosity at two neighboring points and gives only an approximation with a first error term of the order of h^2, where h is the distance between the points. For this reason, the two samples cannot be chosen too far apart. They cannot be too close either, because computed radiosity values have limited precision. When points are too close, the difference between their radiosities is of the same order as the error made at each point and obviously becomes meaningless.

To avoid doing more form factor computations, finite differences could be applied between the vertices of the mesh. However, except for possibly the finest subdivision levels, typical distances between neighboring vertices are too large. Another simple solution would be to use new samples created for this purpose around the vertices and compute their radiosities. In this case though, the amount of computation is increased threefold or fivefold depending on the finite differences scheme chosen (asymmetrical or symmetrical). Furthermore, the choice of the new points must take into account the spacing of sample points on the source and the distance of the obstacles to be able to detect variations in the radiosity.

To evaluate the form factor, the source is sampled adaptively and a set of roughly equivalent contributions is obtained. Each contribution is an elementary form factor and is accumulated in the final value if its associated ray is not intercepted. We use the elementary form factor described in [Wallace et al., 1989]:

$$dF_{A_2,dA_1} = dA_1 \cos\theta_1 \cos\theta_2 / (\pi r^2 + A_2) \tag{1}$$

where θ_1 is the angle between the normal $\vec{n_1}$ at dA_1 and the direction to the source, θ_2 is the angle between the normal $\vec{n_2}$ at the source and the direction to dA_1, $\vec{r}$ is the vector from dA_1 to the source and A_2 is the area of the source (see Figure 6). We will also use the notation $F(P)$ for dF_{A_2,dA_1} where P is the point on the receptor. We notice that the variation of the form factor on the receiving surface has two origins: the change of the cosines and distance terms in $F(P)$ and the change of the shadowing factor. The rays' only purpose is to test the shadowing factor. The variation of $F(P)$ is clearly unrelated. It can be estimated without sending new rays by directly computing the gradient in direction $\vec{v}$:

$$\Delta_{\vec{v}} F(P) = \lim_{h \to 0} \frac{F(P + h\vec{v}) - F(P)}{h} \tag{2}$$

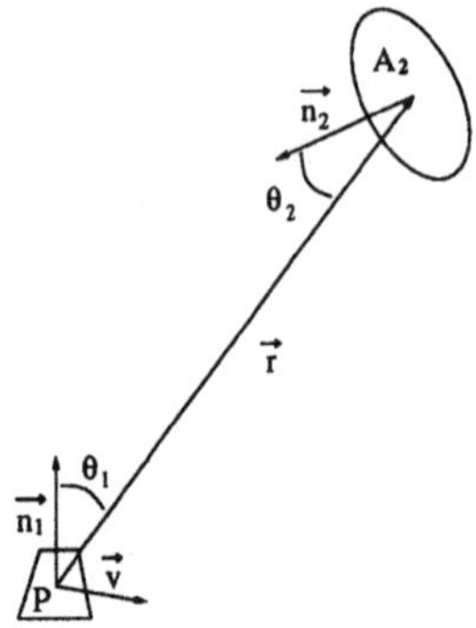

Fig. 6. Geometry of the elementary form factor

If $\vec{v} \cdot \vec{n_1} = 0$ (displacement in the tangent plane), the limit is given by:

$$\Delta_{\vec{v}} F(P) = \frac{\cos\theta_1 (r(\pi r^2 + A_2)\vec{v} \cdot \vec{n_2} - 2\cos\theta_2 (2\pi r^2 + A_2)\vec{v} \cdot \vec{r})}{r^2(\pi r^2 + A_2)^2} \tag{3}$$

and can be evaluated reusing a lot of intermediate results from the computation of $F(P)$.

If we can guess that a sample on the source remains totally visible when viewed from points in a neighborhood of P, formula 3 can be used to compute its gradient. When the sample remains totally invisible, the gradient is zero. It is therefore interesting to detect the samples whose shadowing factor is not likely to change in the neighborhood of the point. Using the subdivision of the source, it is simple to find samples having at least one neighbor with a different shadowing factor. For example, a tree structure explicitly coding the links between neighbors can be constructed once at the beginning. Each source is mapped in turn on this same structure which is then used to look for differences between adjacent samples. When no difference is detected, formula 3 is applied. In zones outside the penumbra (either in the light or in the shadow), all the samples are in this case. In the penumbra, some require more work because their shadowing factor is likely to change. But the benefit of this approach is to greatly reduce the number of samples needing further attention, making possible the use of a more complex procedure on them.

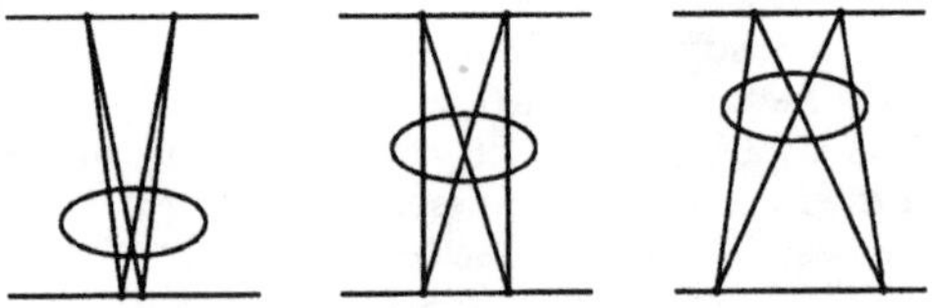

Fig. 7. Spacing of the rays

A method to efficiently choose the position of the new rays in order to detect changes in the shadowing factor is under investigation. The basic idea is the following: depending on the relative spacing of the sample points on the source and on the receptor, the rays show an almost uniform spacing at a certain distance (see Figure 7 for a similar situation in 2D). New rays are sent because an obstacle was hit so the distance to this obstacle is known and the spacing can be chosen accordingly. We can also notice that the new rays should fully benefit from last hit caching optimizations.

5 Subdivision Methods

This section describes a structure developed to store radiosity values on a subdivided polygon and primitives to refine the subdivision. The surface on which a shadow is cast is a four sided polygon but the method could also be applied to a spline patch by working in parameter space. We subdivide it into sub-polygons, the sample points being the vertices. We use a tree (bintree or quadtree) to record the recursive structure of the subdivision. In the structure, we choose not to duplicate vertices, i.e. there is only one memory representation of the vertex at a given location on the surface. This choice is necessary to avoid redundant calculation and the cost of the search operations used results in net savings in form factor computation. Although for pure Lambertian surfaces, the relevant information at a point is a single number, more complex global illumination models can greatly increase the amount of information stored in a vertex. Another good point of the uniqueness of vertices is therefore better memory utilization.

To avoid duplicating vertices, one must be able to find if there is already a vertex at some position in the tree structure. The search is done only on the topology of the mesh using adjacency relations between polygons. It consists in looking for the neighboring element in a direction to check whether it is more subdivided or not. For a quadtree, the neighbor searching algorithm is well known. For the bintree, on the other hand, algorithms need some modifications to run. In a quadtree, the neighbor finding algorithm consists in going up in the tree to reach (if possible) the smallest common ancestor of the polygon and its neighbor. As we go up, we record the path and we then go down following a symmetrical path. For the bintree however, there are two possible directions for the subdivision. Along the paths to a polygon and its neighbor, these directions can appear in different orders. But we notice that the only relevant moves are made in the direction orthogonal to the direction of the search so we store only these. When going down, if the subdivision is in the direction of the search, we go toward the original polygon and in the other case, we replay our recording. In this way, we are able to find a neighbor in the bintree (see Figure 8).

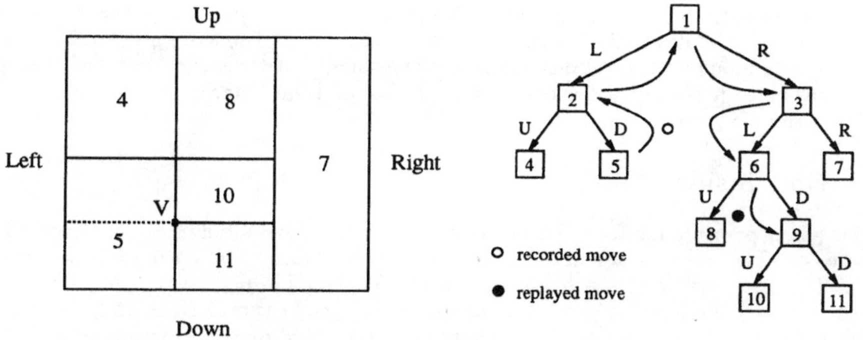

Fig 8. Before splitting polygon 5 along the dotted edge, we must test whether vertex V is already in the database or not. We look for the right neighbor of 5. Going up to 1 (smalles common ancestor), we record only the transversal moves (i.e. along the Up-Down axis). here we get (D). We then go down. To move from 1 to 3 we go in the subtree that does not contain 5; the move from 3 to 6 is automatic (we always go toward 5 in the Left-Right axis) and the move from 6 to 9 follows the recording. 9 is split along the Left-Right axis thus V already exists

The standard bintree or quadtree allows all the computation of intensities but problems appear when displaying the polygons. When one polygon is less subdivided than its neighbor, it has extra vertices on its edge. To avoid discontinuity of radiosity in the display these polygons must receive special treatment (triangulation for example). The identification of these polygons and of the vertices on edges can be done using the neighbor finding algorithm but can be optimized with an enrichment of the structure. Pointers are added from the vertices to their neighboring polygons. Using these pointers, it is easy to follow an edge reporting all vertices lying on it (see Figure 9). Besides increased memory cost, this requires update algorithms to maintain the coherence of the structure when a polygon is split. The general benefit is easier navigation in the graph of the vertices.

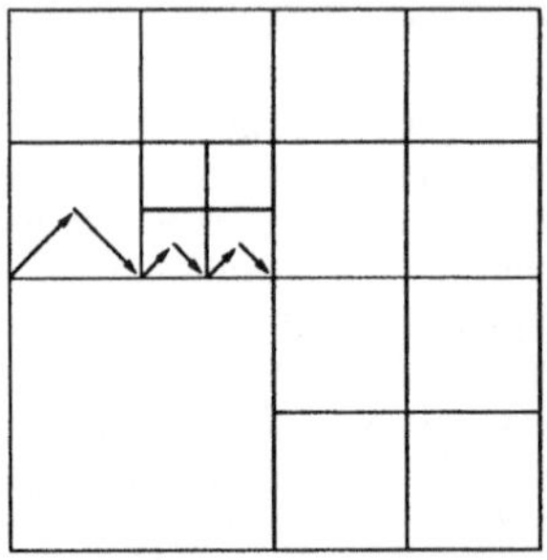

Fig. 9. Navigation on the edge

The bintree structure was developped to provide more balanced subdivision. By choosing at each step the right cut, the aspect ratios of the polygons are never bigger than two. Binary subdivision is also better than quaternary subdivision if one wants to experiment with non axis aligned cuts. In this case, quadtree performs badly because when a polygon with a high aspect ratio is created, it can never be eliminated. If the polygon is then subdivided, one of his subpolygons will have at least as high an aspect ratio. When cutting, it is also easier to choose one direction instead of two.

In short, the tree structure was chosen for its changeability. At each shooting step, it allows to start from the original polygon and to refine down to the appropriate level. It describes only the topology of the mesh and not the actual position of the vertices on the polygon. The drawback is memory cost because we need to store more information than the sample points positions and intensity values. But the structure is rich enough to allow us to perform various experiments on subdivision criteria and the positioning of new sample points.

6 Conclusion

This paper presented different problems related to adaptive subdivision for progressive radiosity. The use of gradients to improve the subdivision criterion is proposed along with the first part of an algorithm to compute gradients in real scenes. A binary tree structure was developed to provide an adaptive storage of the radiosities on the surfaces. These different techniques need to be gathered and tested on real environments.

Acknowledgements

We would like to thank Susan Spach for her comments on the paper and Frédéric Asensio for useful discussions.

References

Campbell, III, A. and Fussell, D. S. (1990). Adaptive mesh generation for global diffuse illumination. *Computer Graphics (SIGGRAPH '90 Proceedings)*, 24(4):155–164.

Cohen, M., Greenberg, D. P., Immel, D. S., and Brock, P. J. (1986). An efficient radiosity approach for realistic image synthesis. *IEEE Computer Graphics and Applications*, 6(3):26–35.

Hanrahan, P. and Salzman, D. (1990). A rapid hierarchical radiosity algorithm for unoccluded environments. In *Proceedings Eurographics Workshop on Photosimulation, Realism and Physics in Computer Graphics*, pages 151–171, Rennes, France.

Heckbert, P. (1990). Adaptive radiosity textures for bidirectional ray tracing. *Computer Graphics (SIGGRAPH '90 Proceedings)*, 24(4):145–154.

Wallace, J. R., Elmquist, K. A., and Haines, E. A. (1989). A ray tracing algorithm for progressive radiosity. *Computer Graphics (SIGGRAPH '89 Proceedings)*, 23(3):315–324.

Radiosity in Computer Graphics: A Proposed Alternative to the Hemi-cube Algorithm

Neil Gatenby, Terry Hewitt

Abstract

Radiosity is widely acknowledged as being the global illumination model responsible for the most realistic synthetic pictures yet produced [13]. This paper introduces an alternative to the hemi-cube algorithm which utilizes a unit hemisphere as its view volume and, as far as the authors are aware, is unique in that it:

1. Uses a discretization of the hemisphere which is new to radiosity and which splits the hemisphere's surface into a number of (near) equi-area regions, thereby simplifying the calculation of delta form factors.
2. Is the first such algorithm which, whilst utilizing a view volume (as opposed to a view screen) allows the user to trade off as much form factor accuracy as they wish, in exchange for a corresponding increase in form factor calculation speed.

1 Numerical Algorithms for Radiosity

Thermal engineers [9] have shown that if an environment under investigation is split into a number of finite regions (patches), then the fraction of energy leaving an ideal Lambertian patch i that impinges directly on a similar patch j is given by the *Form Factor*:

$$F_{ij} = \frac{1}{A_i} \int_{A_i} \int_{A_j} \frac{\cos\theta_i \cos\theta_j}{\pi r^2} H_{ij} dA_j dA_i \tag{1}$$

Where:

- A_i and A_j are the surface areas of the ith and j^{th} patches, respectively.
- dA_i and dA_j are elemental regions of the ith and j^{th} patches, respectively.
- H_{ij} is a term which accounts for occlusion between the two patches (possibly by themselves): H_{ij} takes the value 1 if a straight line joining dA_i to dA_j passes through no part of any patch in the enclosure, and 0 otherwise.
- r is the distance between dA_i and dA_j.
- θ_i is the angle subtended at dA_i by the normal to the ith patch at that point, and the straight line joining dA_i to dA_j. A similar definition holds for θ_j.

The most computationally-intensive process, when attempting to render an image using radiosity methods, is the calculation of these form factors for every pair of patches in the enclosure being modelled. Whilst Cohen & Greenberg's hemi-cube algorithm [3] *is* efficient – in that it can calculate the form factor between two partially occluded surfaces in interactive time – the fact that a full matrix radiosity solution will spend $O(N^2)$ time applying the algorithm suggests this as an obvious place to start looking for improvements.

Other researchers [1][6][7][10][11] have already proposed several alternatives to the basic algorithm. Most of these alternative methods [6][7][10][11] are aimed at improving the speed, rather than the accuracy, of the algorithm, whilst others [1][11] are concerned with the inaccuracies that exist due to the assumptions made when deriving the algorithm.

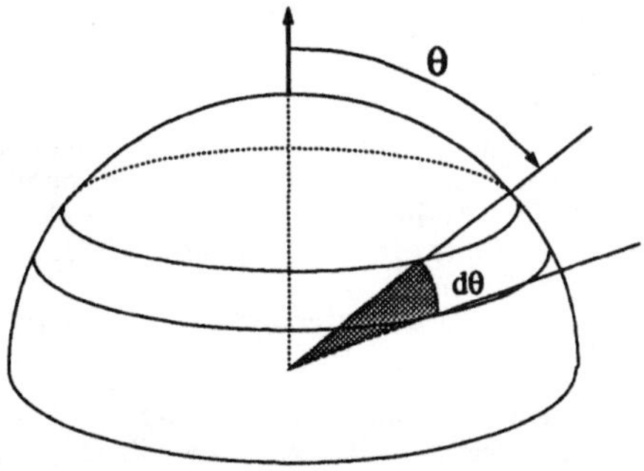

Fig 1. A small conic solid angle

One assumption, common to all purely numerical form factor algorithms, and first introduced in [3], is:

$$F_{ij} \approx \left(\int_{A_j} \frac{\cos\theta_i \cos\theta_j}{\pi r^2} H_{ij} dA_j \right) \frac{1}{A_i} \int_{A_i} dA_i$$
$$= \int_{A_j} \frac{\cos\theta_i \cos\theta_j}{\pi r^2} dA_j \qquad (2)$$

Which not only assumes that the patches are sufficiently far apart for the inner integral to be approximated as constant over the domain of the outer integral, but also that the two patches are not *partially* occluded.

1.1 Ray Traced Form Factors

Whereas the full matrix hemi-cube algorithm [3] uses projection *down* onto the hemi-cube's faces to calculate form factors, other researchers [5][6][10] have utilized ray tracing (ray casting) *out* from their view volume (view screen) to obtain the same results.

One noticeable advantage of utilizing ray tracing is that the directions can be jittered (randomly perturbated) in an effort to convert aliasing (which will effect form factor accuracy) into noise. Jittering allows the implementor to cast over a smaller set of directions than would be needed by a hemi-cube returning equally accurate results.

Directional Importance

Malley [6] argues that it is better, when ray casting, to concentrate one's efforts on those directions which give a greater contribution to the total. Exactly the same logic is applied here and has already been applied by others [5][7][10][11]. Whereas Malley [6] chose to pay less attention to those directions which contributed little to the whole, other researchers [5][7][10] have ignored these directions (near $\theta = \pi/2$ – see figure 1) altogether.

Sillion and Puech [10] have shown that the fraction of energy leaving the centre of a patch that goes unaccounted for when one ignores radiation leaving through a small conic solid angle such as that shown in figure 1, is given by:

$$\frac{dB}{B} = \sin 2\theta d\theta \qquad (3)$$

This is incorporated into the algorithm presented here – allowing what many applications would regard as a negligible (user-defined) loss of form-factor accuracy, in exchange for improved calculation speed. This 'Sillion-type' approach [10] is discussed in greater detail in section 2.2 of this paper.

2 A New Hemisphere Algorithm

2.1 Motives for Change

When Cohen & Greenberg [3] introduced their hemi-cube algorithm – one of the first numerical algorithms which solved equation 2 – they rejected the idea of using a hemisphere as view volume, stating: "Difficulty in creating equal sized elements on a sphere as well as creating a set of linear coordinates to uniquely describe locations on its surface makes this approach impractical" [3].

The motives behind seeking an alternative for Cohen & Greenberg's hemi-cube algorithm [3] were twofold:

- Firstly: Dissatisfaction with the lack of radial of symmetry in the hemi-cube method; symmetry which is so obviously present when calculating the form factor between an elemental area and a patch (see equation 2).
- Secondly: Repeated use of the Sutherland-Hodgman re-entrant polygon clipping algorithm [12]. Whilst the Sutherland-Hodgman clipper *is* efficient, it is not the sort of computational task one regards as desirable to apply $5M$ times (for an M element enclosure) when speed is an important factor.

2.2 A Discretization of the Unit Hemisphere

Having decided to utilize a unit hemisphere as a view volume with which to calculate the form factors for an enclosure, it was desirable to discretize its surface into regions of equal area. Whilst, strictly speaking, this was not achieved, a discretization *was* found which results in regions whose areas have a very small variance (see table 1).

The reason it is desirable to have small regions of equal area on the surface of the hemisphere (analogous to Cohen & Greenberg's hemi-cube *pixels*), is to simplify form factor calculation:

The approximation made in equation 2 – that the form factor double integral can be approximated as a single integral – is also made in the method proposed here. Given that this is the case, the form factor F_{ij} between two patches i and j can be found by constructing an imaginary unit hemisphere around the centre of patch i, with the normal to patch i passing through the apex of the hemisphere. If it was known which of equi-area surface regions were covered by the projection of patch j onto this hemisphere, then F_{ij} could be found by noting:

$$F_{ij} \approx \sum_{q=1}^{R} \Delta F_q \tag{4}$$

Where ΔF_q is the form factor between a differential area element at the centre of patch i and the qth surface region on the hemisphere covered by the projection of patch j onto its surface. This is exactly the approach taken in the hemi-cube algorithm [3].

Splitting the Hemisphere

Other researchers [4][11] have already discussed various ways in which the surface of a sphere can be discretized:

The discretization utilized by Spencer [11] – the 'traditional' spherical polar split – was considered as a possible candidate (Spencer's work was not known to the authors when the algorithm proposed here was devised) and rejected immediately because of the wildly varying areas of the surface elements. Whilst equi-area regions can be found by using unevenly spaced values of Φ in the discretization, this leads to a concentration on directions which contribute *least* to the form factor.

Foley *et al* [4] utilize a variant on a method which they reference as having been originally proposed by Renka [8], in which the surface of a unit sphere is approximated by a triangular mesh of points based on the regular tetrahedron – shown in figure 2(i) – whose

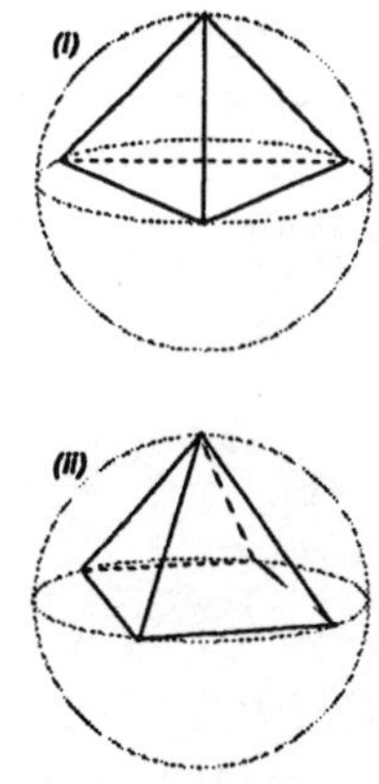

Fig 2. Two different tetrahedrons

vertices (on the unit sphere) are:

$$P_1 = (0,0,1)$$
$$P_2 = (\tfrac{\sqrt{2}}{3}, -\sqrt{\tfrac{2}{3}}, -\tfrac{1}{3})$$
$$P_3 = (\tfrac{\sqrt{2}}{3}, \sqrt{\tfrac{2}{3}}, -\tfrac{1}{3})$$
$$P_4 = (-\tfrac{2\sqrt{2}}{3}, 0, -\tfrac{1}{3})$$

This is one of the Platonic solids – and therefore not only are all its faces of equal area, but all its vertices lie in the same sphere (in this case the unit sphere centred at the origin) – Blinn [2] recommends the Platonic solids as being an excellent place to start looking for a regular discretization of a sphere.

Renka's method [8], as described by Foley *et al*, is to find the midpoint of each of the geodesic arcs that correspond to the tetrahedron's edges and to then join these midpoints to form four new spherical triangles. This process repeats until, after k such divisions, the original spherical triangle has been split into 4^k such triangles. This allows little control over the number of triangles allowed in the discretization – if one requires a triangle to be split into 3,000 triangles, then the options are either 1,024 or 4,096. It was for this reason that Foley *et al* rejected this method, but as the variance in the triangles' areas is smaller in Renka's method than in the method proposed by Foley *et al*, it is Renka's method that has been adopted for this algorithm. As far as the authors are aware, this is the first time such a discretization has been applied to form factor calculation.

Since it is the unit hemisphere, and not the unit sphere, that needs to be divided when dealing with opaque patches, the discretization used is based on one half of the platonic octahedron (figure 2(ii))whose vertices lie in the unit sphere centred at the origin[1] – the pertinent vertices being:

$$P_1 = (0,0,1)$$
$$P_2 = (1,0,0)$$
$$P_3 = (0,1,0)$$
$$P_4 = (-1,0,0)$$
$$P_5 = (0,-1,0)$$

After four and five Renka-type divisions, the four spherical triangles that form the original discretization of the hemisphere have been split into a total of 1024 and 4096 spherical triangles, respectively. The means and variances for these different distributions are shown in table 1.

---

[1]Originally, a tetrahedron (whose base lay in the plane $z = 0$) was used as a starting point for the discretization; but the octahedron results in spherical triangles whose areas have a smaller variance.

Table 1. Means and variances for different discretizations

Triangles	Mean(Areas)	Variance(Areas)
1024	6.13586 x 10^{-3}	2.21829 x 10^{-6}
4096	1.53501 x 10^{-3}	1.38123 x 10^{-7}

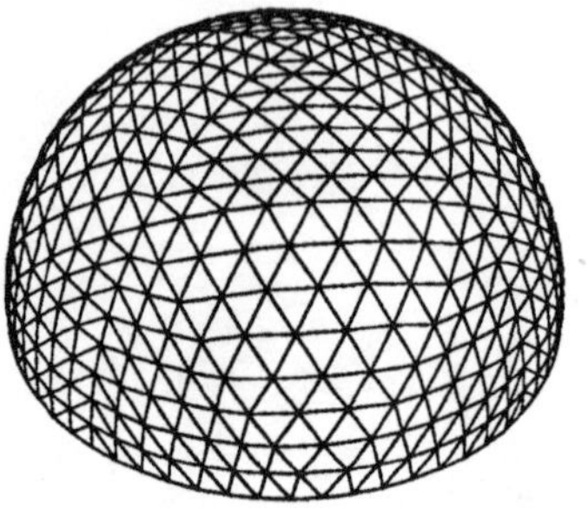

Fig 3. The discretization of the hemisphere

It can be seen from figure 3 that when each spherical triangle is split into four, the small triangles form themselves into 'rows' or 'rings' around the axis of symmetry of the hemisphere. Whilst these rows do not lie along convenient lines of latitude, they *can* be utilized in the algorithm, by ensuring the set of directions formed by the discretization is stored and can be easily accessed as 'rows' of directions. If this is done, then a 'Sillion-type' [10] algorithm (which ignores directions that only contribute a small amount to the form factor) can be instantly employed by simply ignoring the bottom k rows of directions – where k is some implementation-dependent small integer. The form factors obtained using this approach are then normalized so that each row of the form factor matrix sums to 1.

Calculation of Form Factors

In Cohen & Greenberg's hemi-cube method [3], a different delta form factor is associated with every pixel (symmetry aside). This means that for a 200 x 200 hemi-cube – i.e., one with 120,000 pixels – one must calculate and store 25,000 different delta form factors. In the method proposed here, as every surface element is approximated as being the same area, ΔA, the delta form factor associated with such a surface element (centre, on the unit hemisphere, given by the coordinates (x, y, z) – see figure 4) is given by:

$$\Delta F_{x,y,z} = \frac{\cos\theta_i \cos\theta_j}{\pi r^2}\Delta A \tag{5}$$

But, as can be seen from figure 4, the angle 'θ_j' is always zero, the distance 'r' is always 1, and '$\cos\theta_i$' is simply the z coordinate of the surface element's centre. So equation 5 simplifies to:

$$\Delta F_{x,y,z} = \left(\frac{z}{\pi}\right)\Delta A \tag{6}$$

Suppose the hemisphere's surface has been split into D such equi-area surface elements, then $\Delta A = 2\pi/D$ – so equation 6 simplifies further:

$$\Delta F_{x,y,z} = 2\left(\frac{z}{D}\right) \tag{7}$$

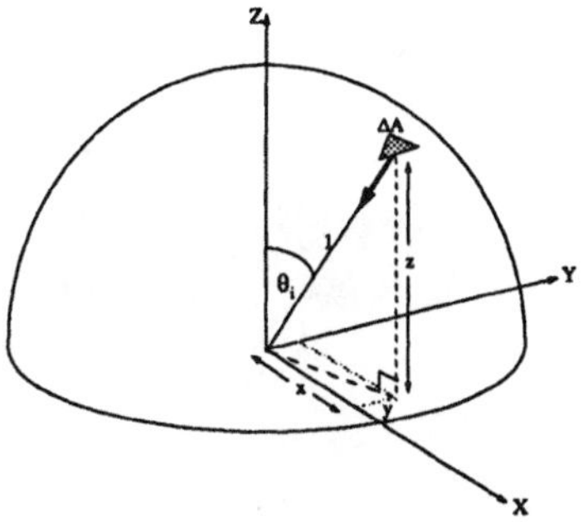

Fig 4. Delta form factors for surface elements on the unit hemisphere

Which is certainly easier to work out than the hemi-cube's delta form factors and, more importantly, is stored at negligible cost – no look-up table is required. Since the unit vector given by $(x,\ y,\ z)$ is already known, the delta form factor for the surface element associated with this direction is found simply by multiplying the (normalized) direction's z-component by $\frac{2}{D}$.

Having established the delta form factor associated with each direction, it remains to calculate the form factor for every patch in the enclosure relative to the base patch (about which the hemisphere has been constructed). An item buffer is used to achieve this, exactly as used by Cohen & Greenberg [3] – every direction has associated with it a pointer to a patch, and the distance to that patch – and these are updated if and only if a closer patch is found along that line of sight.

Two different methods of comparing every patch with every direction have been implemented. Briefly:

- One method simply goes through the current set of directions – which may or may not be a half-space, depending on whether or not a 'Sillion-type' approach is being employed – and compares the current direction with every patch's bounding sphere. If the current direction passes through this sphere then a more careful test, developed by the authors, is made. This test establishes whether or not the direction passes through the patch and, if so, how far away the patch is. The item buffer is updated (or not) accordingly.
- The second method utilizes the geometry of the discretization more than the first: For every patch, a bit-map is found that indicates which of the four quadrants the patch lies in. Also, '`top_row`' and '`bottom_row`' values are found that limit which rows are tested for the patch's presence. Only the quadrants that could possibly contain the patch (determined by the bit-map), and only the rows in those quadrants determined by the range ['`bottom_row`'...'`top_row`'] are tested for the patch's presence. The test to determine whether or not a patch can be 'seen' by a particular direction is exactly as for the first method (see figure 5).

Regardless of which method is used, the result is an item buffer that indicates what fraction of the energy leaving the centre of the base patch reaches which patches in the enclosure. This buffer can be traversed and a row of the form factor matrix calculated by simply applying equations 4 and 7.

3 Conclusions

Images produced using the Manchester hemisphere algorithm illustrate that it certainly works (see plates in section 4), but further research needs still to be done to confirm its accuracy:

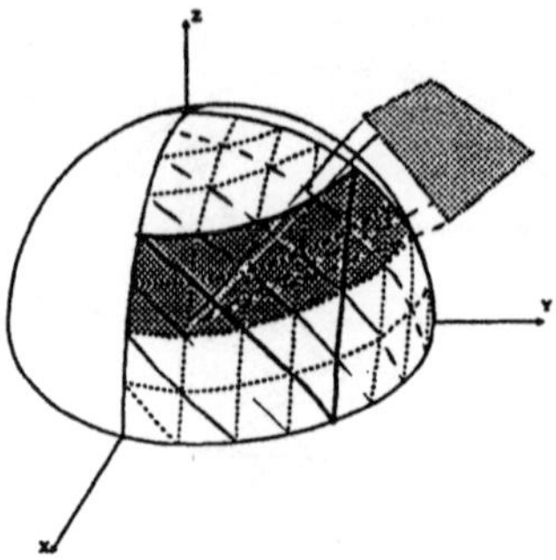

Fig 5. Calculation of form factors using triangular discretization of hemisphere

To date, because of time restrictions, no comparison of the form factors found using the new algorithm has been made with those found analytically. The authors suspect that, whilst the results will clearly be invalid when the proximity assumption is violated, they may also vary as a function of *target patch/hemisphere* orientation, due to the non-zero variance of the distribution of the spherical triangle's areas. This 'not-quite-uniform' distribution will yield areas of the hemisphere where triangles (and their corresponding directions) are clumped together, and others where they are sparse. It is felt that jittering the directions will overcome this problem, should it arise.

Similarly, a radiosity program which implements the hemi-cube algorithm has not yet been written, so accuracy and speed comparisons between the two algorithms have not been made.

A new alternative to the hemi-cube algorithm is presented which utilizes a hemisphere as its view volume, and which allows the user to trade form factor accuracy for speed.

4 Examples

The implemented program runs on a VAXstation 3540, and has been written in Pascal. The images were rendered using the DEC PHIGS PLUS primitive *Fill Area Set 3 with Data*.

Fig 6. A simply furnished room, daytime

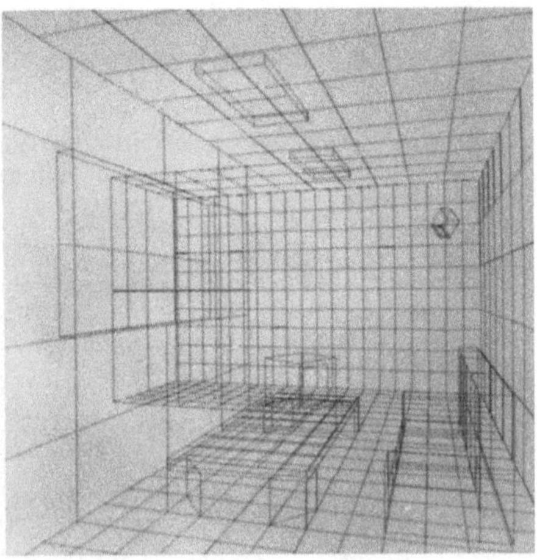

Fig 7. Patch/element discretization for figure 6

Figure 6 shows a simply furnished room. The scene is made up from 403 patches which have been split into a further 640 elements. Neither the strip lights nor the spot lights are 'on' in this instance, but both windows (one out of view) are acting as light sources.

Figure 7 shows the patch/element discretization for the room shown in figure 6.

Acknowledgements

The authors would like to thank the staff in the Computer Graphics Unit at the University of Manchester for their help and advice, particularly when discussing the algorithm's implementation.

Neil Gatenby would like to thank the Science and Engineering Research Council for their financial support.

References

[1] Daniel R. Baum, Holly E. Rushmeier, and James M. Winget. Improving Radiosity Solutions through the Use of Analytically Determined Form Factors. *ACM SIGGRAPH Computer Graphics*, 23(3):325–334, July 1989.

[2] James F. Blinn. Platonic Solids. *IEEE Computer Graphics & Applications*, 7(11):62–66, November 1987.

[3] Michael F. Cohen and Donald P. Greenberg. The Hemi-cube: A Radiosity Solution for Complex Environments. *ACM SIGGRAPH Computer Graphics*, 19(3):31–40, July 1985.

[4] Thomas A. Foley, David A. Lane, Gregory M. Nielson, and Ramamani Ramaraj. Visualising Functions over a Sphere. *IEEE Computer Graphics & Applications*, 10(1):32–40, January 1990.

[5] Neil Gatenby. Radiosity in Computer Graphics: A Proposed Alternative to the Hemicube Algorithm. Master's thesis, University of Manchester, 1991.

[6] Thomas J. V. Malley. A Shading Method for Computer Generated Images. Master's thesis, University of Utah, 1988.

[7] Rodney J. Recker, David W. George, and Donald P. Greenberg. Acceleration Techniques for Progressive Refinement Radiosity. *ACM SIGGRAPH Computer Graphics*, 24(2):59–66, March 1990.

[8] R. J. Renka. Interpolation of Data on the Surface of a Sphere. *ACM Transactions on Mathematical Software*, 10(4):417–436, December 1984.

[9] Robert Siegel and John R. Howell. *Thermal Radiation Heat Transfer*. McGraw-Hill Book Company, 1972.

[10] François Sillion and Claude Puech. A General Two-Pass Method Integrating Specular and Diffuse Reflection. *ACM SIGGRAPH Computer Graphics*, 23(3):335–344, July 1989.

[11] Stephen N. Spencer. The Hemisphere Radiosity Method: A tale of Two Algorithms. In K. Bouatouch and C. Bouville, editors, *Photorealism in Computer Graphics*, pages 127–135. Springer, 1992.

[12] I. E. Sutherland and G. W. Hodgman. Reentrant Polygon Clipping. *Communications of the ACM*, 17(1):32–42, January 1974.

[13] Alan H. Watt. *Fundamentals of Three-Dimensional Computer Graphics*. Addison-Wesley, 1989.

Physics-Based Lighting Models: Implementation Issues

Pierre Tellier and Kadi Bouatouch

Abstract

This paper provides tools to implement a physics-based lighting model. This latter uses a physics-based reflection model, spectral distribution of light powers and does not make any assumption on the specular behaviour of materials. The scenes are discretized into points instead of patches. A data structure named visibility graph is built to efficiently evaluate the visibility between two sample points. This global model is implemented according to a two-pass method.

1 Introduction

Photosimulation consists in producing highly realistic images. The realistic aspect of materials can be simulated only with the use of physics-based reflection and refraction models. Such models have been introduced by Cook and Torrance [7], from the works of Torrance, and Beckmann. To accurately evaluate the illumination of synthesized scenes a global model is required. The implementation of this global model can be performed according to two approaches: one-pass methods [13, 11, 14, 10] or two-pass methods [15, 16]. The one-pass methods perform all the illumination computations independently of the view point, allowing then a fast rendering of the same scene from different view points. However, these methods need a large amount of memory to store data. Another drawback is the aliasing defects due to sharp variations of specular reflections. To avoid these defects, a very dense sampling of the scene is indispensable, which would significantly increase the data to be stored. For these reasons, we have opted for a two-pass method.

As pointed out hereafter, in our method, the scene is dicretized into points instead of small patches. Moreover, our global model uses a physics-based reflection model, spectral distribution of the radiances (light power per unit surface and unit solid angle) and non perfectly specular materials.

After an introduction of our physics-based lighting model, we describe the different implementation steps of our method, then we give and discuss some experimental results.

2 The Global Model

A global illumination model must take into account all the reflections and refractions within the scene. To describe the mechanism of light transport we use the model introduced by Bouville and Bouatouch [2, 3]. This model consists of a set of equations which express (in terms of *radiance* since it is the quantity the eye is sensitive to) the radiance

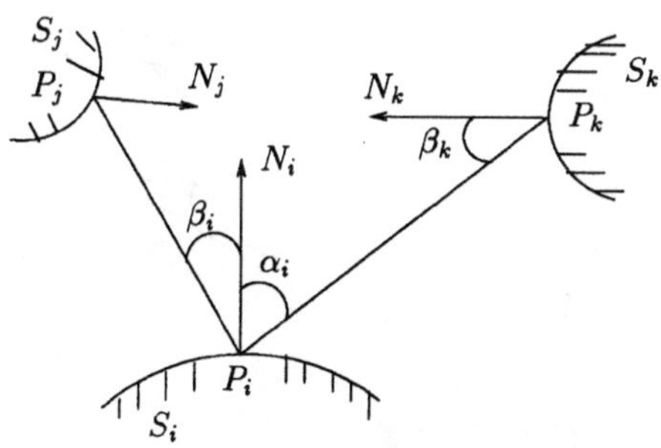

Fig. 1: Geometry of light transport mechanism

of a point P_i in the direction of P_j when illuminated by all the surfaces S_k (see figure 1):

$$L_\lambda(P_i, P_j) = h(P_i, P_j) * \left[L_\lambda^E(P_i) + \sum_k \int_{S_k} R_\lambda(P_k, P_i, P_j) L_\lambda(P_k, P_i) G(P_k, P_i) dS_k \right] \quad (1)$$

where

- $L_\lambda(P_i, P_j)$ is the radiance of P_i as seen from P_j,
- $h(P_i, P_j)$ is the visibility function,
- $L_\lambda^E(P_i)$ is the self-emitted radiance,
- $R_\lambda(P_k, P_i, P_j) = s_i R_\lambda^s(P_k, P_i, P_j) + d_i R_\lambda^d(P_i)$ is the bidirectionnal reflectance [7],
- $L_\lambda(P_k, P_i)$ is the radiance of P_k as seen from P_i,
- $G(P_k, P_i) = \frac{\cos\alpha_i \cos\beta_k}{||P_k P_i||^2}$ is a purely geometrical term,
- λ is a given wavelength.

The system made up of these equations is called *system of light energy balance equations.* Solving this system will provide the global radiance at each point of the scene.

3 The Method

3.1 Discretization into Points

3.1.1 Motivation

Any method used to implement a global illumination model requires a discretization of the scene either into patches or points. In our method, we have chosen to discretize the scene into points for the following reasons:

- we have to evaluate the radiance for each point of the scene in one direction (radiance is directional),
- to evaluate the radiance of a point, it is not necessary to consider all the points of the scene. Indeed, since the radiance is an integral, we can evaluate it by taking a certain number of samples of the variables of the integrand (Gauss or Monte Carlo methods),

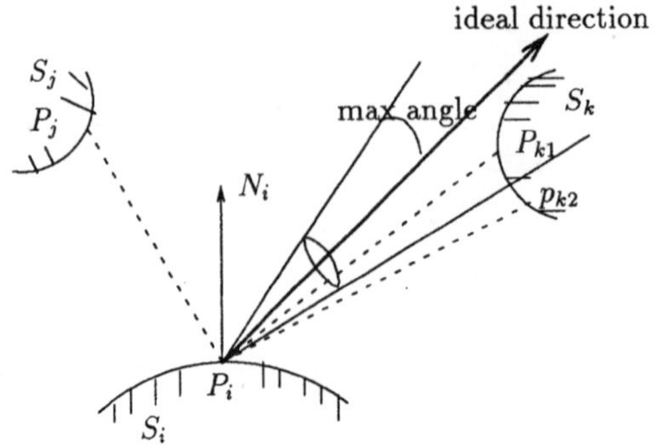

Fig. 2: The reflection cone

- visibility between two points can be easily computed by ray tracing,
- several kind of surfaces can be sampled into points: polygons, spheres, cylinders, cones, parametric surfaces etc.. Subdivision into patches can then be avoided, which allows the use of different kinds of geometric models.

Since the scene is discretized into points, only point-to-point light contributions have to be evaluated. In order to be able to discretize equation (1), with each point is associated a small surface element δS_k. Equation (1) becomes then:

$$L_{ij} \approx h_{ij} \left[L_i^E + \sum_{k=1}^{N} R_{kij} G_{ki} L_{ki} \delta S_k \right] \tag{2}$$

where N is the number of sample points in the environment. Using this point sampling and separating the diffuse and specular reflections, we obtain:

$$L_{ij} = h_{ij} \left[L_i^E + L_i^d + L_{ij}^s \right] \tag{3}$$

$$L_i^d = \sum_{k=1}^{N} d_i R_i^d G_{ki} L_{ki} \delta S_k = \sum_{k=1}^{N} \mathcal{D}(L_{ki}) \tag{4}$$

$$L_{ij}^s = \sum_{k=1}^{n} s_i R_{kij}^s G_{ki} L_{ki} \delta S_k = \sum_{k=1}^{n} \mathcal{S}(L_{ki}) \tag{5}$$

where L_i^D is the diffuse component of the reflected light, L_i^S the specular component and n is the number of points included in the *reflection cone* which bounds the specular component R^s of R. The operators $\mathcal{D}$ and $\mathcal{S}$ are the global diffuse and global specular operators respectively.

Such reflection cones are defined by an axis which is the ideal reflection direction, and by an angle which depends on the physical properties of the materials (see figure 2). Actually, this angle depends on the Beckmann distribution of micro-facets and is precomputed for different values of the parameter m of the Beckmann function. If needed, the value of the angle is interpolated using precomputed values. The diffuse component is evaluated by adding the contribution of all points in the scene to a given point, while the specular component takes into account only the points whose contribution will be significant, i.e. the points included in the reflection cone. The points within to the reflection cone are found as follows. First, we find all boxes which intersect the cone. Next only the points included in these boxes have to be checked for intersection. To efficiently reduce the number of sampled directions, the specular directions could be determined using the Gauss method.

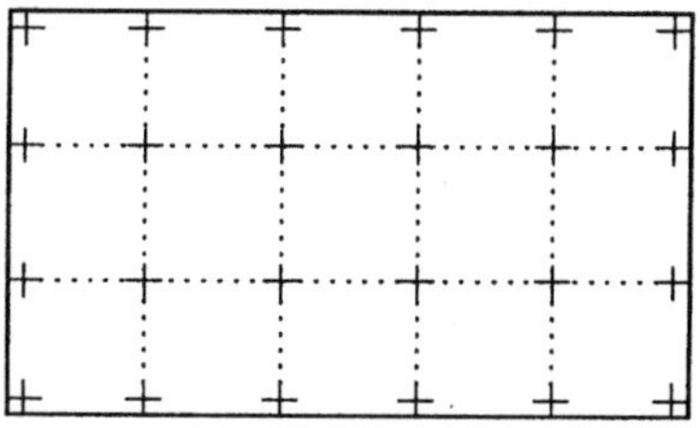

Fig. 3: Sampling a polygon into points

3.1.2 The Discretization Method

Each surface is sampled into points, according to a method nearly similar to the one described in [12] (see figure 3). This method differs from Nishita's in the way the sample points are arranged. Indeed, in our method, there are no sample points on the surface boundaries, which allows to avoid unrealistic shadows along edges shared by two surfaces. Moreover, the number of samples is arbitrarly chosen by the user. Other sampling strategies are being investigated. Note that with each sample point is associated a surface area δS determined as in [12].

3.2 First Pass

3.2.1 Extended Form Factors

Since a reflected light light can be separated into diffuse and specular components [2], the two-pass method consists of two main steps: the first one is a view independent computation of the *Global Diffuse Radiance* while the second evaluates the view dependent *Global Specular Component.*

Let us rewrite equation (1):

$$\begin{aligned} L_{ij} &= h_{ij}\left[L_i^E + \sum_{k=1}^{N}(d_i R_i^d + s_i R_{kij}^s) G_{ki} L_{ki}\delta S_k\right] && (6) \\ &= h_{ij}\left[L_i^E + \sum_{k=1}^{N} d_i R_i^d G_{ki} L_{ki}\delta S_k + \sum_{k=1}^{n} s_i R_{kij}^s G_{ki} L_{ki}\delta S_k\right] && (7) \\ &= h_{ij}\left[L_i^E + \sum_{k=1}^{N}\mathcal{D}(L_{ki}) + \sum_{k=1}^{n}\mathcal{S}(L_{ki})\right] && (8) \end{aligned}$$

where $\mathcal{D}$ and $\mathcal{S}$ are global diffuse and global specular operators respectively. The expression of the global diffuse radiance L_i^d of a point is then given by:

$$L_i^d = L_i^E + \sum_{k=1}^{N} d_i R_i^d G_{ki} L_{ki} dS = L_i^E + \sum_{k=1}^{N}\mathcal{D}(L_{ki}). \tag{9}$$

The expression of L_i^d, as function of L_j^d is then:

$$L_i^d = L_i^E + \sum_{j=1}^{N}(\mathcal{D} + \mathcal{D}(\sum_{k=1}^{n}\mathcal{S}) + \mathcal{D}(\sum_{l=1}^{n}\mathcal{S}(\sum_{m=1}^{n}\mathcal{S})) + \ldots)L_j^d. \tag{10}$$

The term $[\mathcal{D} + \mathcal{D}(\sum_k \mathcal{S}) + \mathcal{D}(\sum_l \mathcal{S}(\sum_m \mathcal{S})) + \ldots]$ is called *Extended Form Factor* and is named EFF_{ji} by analogy with radiosity.

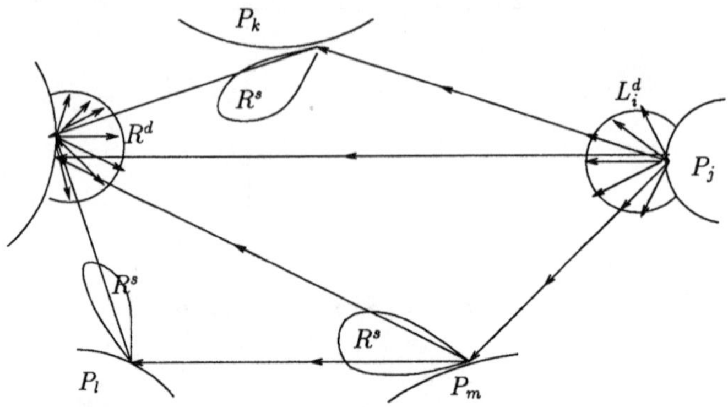

Fig. 4: Extended form factor FFE_{ji}

It is the proportion of diffuse light emitted by the surface S_j that contributes to the global diffuse radiance of the point P_i (see figure 4). Using these terms we obtain a system of N equations of N unknowns:

$$L_i^d = L_i^E + \sum_j \left[EFF_{ji} * L_j^d \right] \tag{11}$$

This system can be solved (for each wavelength) by using the same methods as for radiosity (complete or iterative solution [5]).

The global model implementations performed by Wallace [16] and Sillion [15] use this decomposition but both neglect the effects of non purely specular materials, use an empirical reflectance model and deal only with the three RGB components.

3.2.2 Algorithm

The following algorithm computes the contribution of one given point to all its environment. This algorithm utilizes a shooting process since it allows a progressive solution.

```
/* statements */

spectrum EFF_λ[N][N] /* matrix of spectral extended form factors */
/* N is the number of sample points */

spectrum spec1 = {1.0, ..., 1.0} /* spectrum initialized to 1.0 */


EvaluateEFF(j) /* computes the j^th column of extended form factors */

/* j:  emitting point */
{
     for all points P_i {
      compute FF_ji = h_ji G_ji δS_j      /* geometrical term */
      EFF_λ[j][i] = EFF_λ[j][i] + FF_ji * d_i R^d_λi     /* for each spectrum sample */
      if (s_i ≠ 0) {     /* specular surface */
       GlobSpecOp(j, j, i, spec1, 1)
      }
     }
}
```

The procedure **EvaluateEFF** evaluates the direct contribution of the global diffuse radiance of P_j to the global diffuse radiance of all the points of the scene. All points being illuminated by P_j and belonging to a non perfectly diffuse surface emit the specularly reflected light to the environment. This is made possible thanks to the procedure **GlobSpecOp**.

```
GlobSpecOp(j, l, i, ΔEFF_λ, lg) /* Global Specular Operator */

/* j:  emitting point */
/* i:  last point met */
/* l:  previous point in light path */
/* ΔEFF_λ:  cumulated contribution of P_j to P_i via P_l */
/* lg:  length of path from P_j to P_i */

{
/* light is emitted by P_l and reflected by P_i towards points P_k*/

    compute FF_li = h_li G_li δS_l       /* geometrical term */
    for all points P_k ∈ reflection cone {
     compute FF_ik = h_ik G_ik δS_i      /* geometrical term */
     ΔEFF_λ = ΔEFF_λ * FF_li * s_i R^S_λlik /* for each spectrum sample */
     EFF_λ[j][k] = EFF_λ[j][k] + ΔEFF_λ * FF_ik * d_k R^d_λk
     if ((lg < lgmax) and (ΔEFF_λ > EFFMIN_λ) and (s_k ≠ 0)) {
      GlobSpecOp(j, i, k, ΔEFF_λ,lg+1)
     }
    }
}
```

3.2.3 The Visibility Graph

Since the scene is sampled into points, it seems worthwhile to build a visibility graph giving a boolean visibility information for all pairs of points, instead of building a valuated graph whose complexity is $O(N^3)$ for each wavelength as done by Buckalew [4].

This visibility graph is very attractive but requires a large memory. To cope with this problem of memory size, we propose the following visibility strategy. The scene is subdivided into a 3D regular grid whose elements are *voxels*. Then the visibility graph is built giving a visibility information between each pair of voxels in place of sample points. The data structure to store the visibility information between two voxels is :

```
typedef struct
{
boolean visible;
graph *subgraph;
}link;
```

The field subgraph is a pointer to a boolean matrix storing visibility information between the points of the two voxels. Indeed, let Vox1 and Vox2 be two voxels. Three cases can be considered:

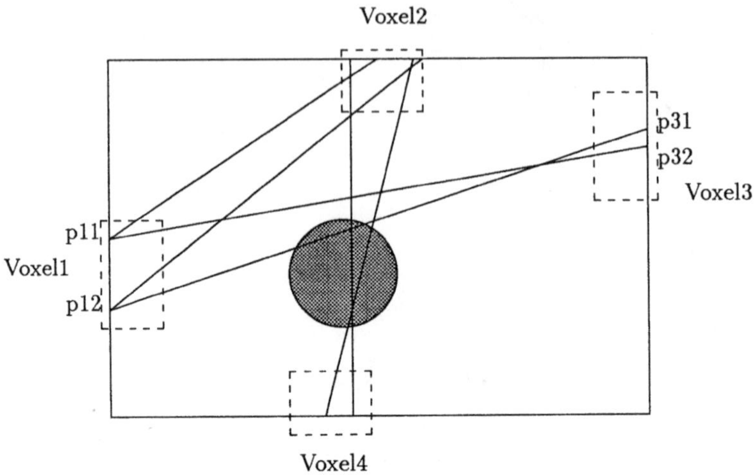

Voxel1 and Voxel2 are fully visible

Voxel2 and Voxel4 are fully hidden

Voxel1 and Voxel4 are partially hidden:

the subgraph (p11,p31),(p11,p32),(p12,p31),(p12,p32) is stored

Fig. 5: Voxel-to-voxel visibility information

1. all the points of Vox1 are visible from all the points of Vox2.
```
link(Vox1,Vox2).visible = true;
link(Vox1,Vox2).subgraph = null;
```
2. no points of Vox1 are visible from any point of Vox2.
```
link(Vox1,Vox2).visible = false;
link(Vox1,Vox2).subgraph = null;
```
3. the two voxels are partially hidden (for example Voxel1 and Voxel3 of figure 5). In such a case we decide to store the visibility information between all the pairs of points included in the two voxels.
```
link(Vox1,Vox2).visible = false;
link(Vox1,Vox2).subgraph = subgraph(Vox1,Vox2);
```

In most scenes we have treated in our experiments, this strategy appeared rather efficient but we must keep in mind that the real memory complexity of this graph is always $O(N^2)$.

If there is not enough memory available to store this graph, one method consists in reducing the size of the data structure `link`, by replacing the field `subgraph` by a pointer to the object lying between the two voxels as shown in figure 5. In this case, the visibility computation is performed according to the *light buffer* method [8].

During the evaluation of the visibility graph, the distance between each pair of points is computed. If the distance between two points is small compared to their associated surface element (important solid angle), the corresponding surface elements are locally subdivided. We obtain then new sample points with smaller associated surface elements. Thereby, the solid angles between these two points become smaller, which makes the computation more accurate.

number of voxels	1728
number of empty voxels	946
number of pairs of points	7 783 485
number of pairs of visible voxels	117 336
number of pairs of hidden voxels	1 231 367
number of pairs of partially hidden vovels	143 425
number of pairs of points stored	5 889 008
graph computation time	4200 sec.

Table 1: Cost of the visibility graph

Note that this graph can be easily updated in case of small movements of some objects of the scene.

3.3 Second Pass

In the second pass the global specular operator is evaluated by means of a distributed ray tracing [6]. Note that this step does not entail the shooting of rays towards light sources (shadow rays), since the global diffuse component of the sample points (of the scene) is already available. Moreover, the intersection between a shot ray and the scene results in a point which may not be a sample point. However, the global diffuse component of the radiance at this intersection point can be interpolated by using the diffuse spectral radiances of the four closest sample points computed at the first pass as done in [1, 9].

4 Results

As our implementation is very young, our program is still being tested and not optimized. That is why we also express the cost of the evaluation of the matrix of extended form factors in terms of numbers of calls to the visibility function h_{ij}. The cost of our method is given, using or not the visibility graph. The results obtained are in favour of the use of this graph.

Our test scene is made up of 134 polygons. The materials of the objects of the scene are gold, sand, concrete, brown stone, blue and green enamel. All these materials are perfectly diffuse except gold whose parameters are: $s = 0.9$, $d = 0.1$ and $m = 0.1$ (Beckmann term). The scene includes two primary light sources which are normalized D6500 white sources. The sampling of the scene results in 3946 points, 196 of which lie on specular surfaces (gold). To emphasize the influence of specular materials on the global diffuse radiance, our test scene was processed according to three different ways, giving the three following images:

Figure 6 (on color plates): all the materials are assumed to be perfectly diffuse,

Figure 7 (on color plates): scene containing diffuse and specular materials, image resulting from the first pass,

Figure 8 (on color plates): scene containing diffuse and specular materials, final image obtained after the two passes.

Note that the light spots near the leg of the table are the global diffuse component due to the specular properties of this golden leg.

The memory cost (table 1) of the visibility graph seems very important. It could be reduced for most scenes with a more accurate subdivision of the scene (adaptive rather than regular).

The computation times (table 2) obtained with our current method should be considerabily reduced if the following remarks are accounted for:

	With Graph	Without Graph
Time (seconds)	13 228	40 741
Number of visibility computations	7 783 485	42 433 940

Table 2: Computation of the matrix of extended form factors

- the spatial subdivision performed for the construction of the visibility should be used to accelerate the visibility computations,
- the intersection procedure that fits in our method should only return a boolean visibility information between two points, it does not need to calculate the coordinates of the intersection points.
- the visibility of a set of points could be evaluated simultaneously (between two voxels),
- The number of specular directions could be reduced by making the angle defining the reflection cone dependent on the specular reflection R^s instead of the Beckmann distribution.

5 Conclusion

Unlike the models already implemented, our illumination model accounts for a physics-based reflection model, spectra instead of RGB components, spectral reflectance and transmittance of materials as well as color science. In our implementation, the scene is sampled into points instead of small patches. To prove that point sampling is correct, we have generated an image with that kind of sampling. It has been compared with the image of the same scene generated by a technique based on discretization into patches . The visual results seem very similar.

Even if the visibility graph occupies an important memory size, it significantly reduces the synthesis time, and in addition is well suited for an adaptive sampling which improves the precision of the computations of the unit solid angles. Note that these solid angles are evaluated from the areas δS of the surface elements associated with the sample points.

As our method cannot be analytically verified, we expect to compare our results with some others and with pictures of real scenes. Furthermore, we are extending our method to handle transparent materials.

References

[1] J. Arvo. Backward ray-tracing. In *Course of the SIGGRAPH'86. Developments in Ray-Tracing*, 1986.

[2] C. Bouville and K. Bouatouch. A unified approach to global illumination models. In *PIXIM'89 Conference*, pages 250–263, September 1989.

[3] C. Bouville, K. Bouatouch, P. Tellier, and X. Pueyo. Theoretical analysis of global illumination models. In *Eurographics Workshop on Photosimulation, Realism and Physics in Computer Graphics*, pages 53–66, Rennes, France, June 1990.

[4] C. Buckalew and D. Fussel. Illumination networks: fast realistic rendering with general reflectance functions. In *SIGGRAPH'89 Conference Proceedings*, pages 89–98, July 1989.

[5] Michael F. Cohen, Shenchang E. Chen, John R. Wallace, and Donald P. Greenberg. A progessive refinement approach to fast radiosity image generation. In *SIGGRAPH'88 Conference Proceedings*, pages 75–84, august 1988.

[6] R. L. Cook, T. Porter, and L. Carpenter. Distributed ray tracing. In *SIGGRAPH'84 Conference Proceedings*, 1984.

[7] R.L. Cook and K.E. Torrance. A reflectance model for computer graphics. *ACM transactions on graphics*, 1(1):7–24, January 1982.

[8] E.A. Haines and D.P. Greenberg. The light buffer: a shadow testing accelerator. *IEEE Computer Graphics and Applications*, 6(9):6–16, September 1986.

[9] P.S. Heckbert. Adaptative radiosity textures for bidirectional ray tracing. In *SIGGRAPH'90 Conference Proceedings*, pages 145–154, August 1990.

[10] D. S. Immel, M. F. Cohen, and D. P. Greenberg. A radiosity method for non-diffused environments. *SIGGRAPH'86 Conference Proceedings*, 20(4):133–142, July 1986.

[11] S.P. Mudur and S.N. Pattanaik. Multidimensional illumination functions for visualization of complex 3d environment. *Visualization and Computer Animation*, 1(2):49–58, 1990.

[12] T. Nishita and E. Nakamae. Continuous tone representation of three-dimensional objects taking into account of shadows and interreflection. In *SIGGRAPH'85 Conference Proceedings*, pages 23–30, July 1985.

[13] B. Le Saec and C. Schlick. A progressive ray-tracing-based radiosity with general reflectance functions. In *Eurographics Workshop on Photosimulation, Realism and Physics in Computer Graphics*, pages 103–116, Rennes, France, June 1990.

[14] M. Shao, Q. Peng, and Y. Liang. A new radiosity approach by procedural refinements for realistic image synthesis. In *SIGGRAPH'88 Conference Proceedings*, pages 93–101, ACM, August 1988.

[15] F. Sillion and C. Puech. A general two-pass method integrating specular and diffuse reflection. In *SIGGRAPH'89 Conference Proceedings*, pages 335–344, July 1989.

[16] J. R. Wallace, K. A. Elmquist, and E. A. Haines. A ray tracing algorithm for progressive radiosity. In *SIGGRAPH'89 Conference Proceedings*, pages 315–324, July 1989.

Shaft Culling for Efficient Ray-Cast Radiosity

Eric A. Haines, John R. Wallace

Abstract

In radiosity algorithms, much time is spent computing visibility between two surfaces. One approach to approximating this visibility is to use ray casting methods. A new algorithm is presented which takes advantage of object coherency when using ray casting for radiosity. An efficient method is presented to form a volume between the emitter and receiver, and then generate a candidate list of items partially or wholly within the volume. Using this list, ray casting is performed to determine the amount of visibility between surfaces. Statistics are presented showing the decrease in overall computation time compared to a traditional ray casting technique.

1 Introduction

Within radiosity implementations the most time consuming procedure is determining the visibility between surfaces. Determining the amount of light that one surface transfers to another is made difficult by other objects which can obscure some of this light. An exact solution of this problem is usually not attempted; rather, some point sampling of the two surfaces is performed and the visibility approximated.

In [Wallace et al., 1989] the authors presented a procedure of ray casting to perform progressive radiosity computations. Ray casting between surface vertices and samples on the emitter was used to determine visibility. The number of sample points on the emitter could vary (i.e. be anything between 1 and 25, typically 4), and the surfaces could be tessellated to generate more samples. The amount of shadowing for each sample point on each surface was computed using a ray caster with a hierarchical bounding volume efficiency scheme.

Ray tracing efficiency schemes have exploited a number of forms of coherence to decrease the amount of time spent testing rays against the environment. One set of rays which have a large degree of coherence are those starting at the eye; such rays have a common origin and a predictable set of directions. This coherence has been used in a variety of ways, such as the item buffer [Weghorst et al., 1984] and hybrid hidden-surface/ray-tracing algorithms [Nakamae et al., 1989; Salesin & Stolfi, 1989]. However, not as much is known in advance about all other rays, i.e. those for shadow testing, reflection, and refraction.

In shadow testing we know the rays will end at some light source. This fact has been used in efficiency schemes such as the light buffer [Haines & Greenberg, 1986]. In the light buffer algorithm, a preprocess computes lists of candidates for all directions from the light. The appropriate list is retrieved for a given direction from the light and used for ray tracing. The lists are much shorter to test than the full environment. However, because there is no knowledge of where the shadow test rays originate, much effort is wasted in creating visibility lists for volumes of space which are never accessed. Also, the light buffer has a problem of creating candidate lists at a uniform resolution, that is, space is partitioned into volumes of approximately the same size. In practice there are volumes with a large degree of complexity that could use further subdivision, while other volumes contain few objects. The light buffer suffers from not knowing in advance which directions rays will come towards each light source.

Ray classification [Arvo & Kirk 1987] uses 5D volumes to exploit ray location and direction coherency. However, when forming a 5D volume it is not known how many times this volume will be accessed, so a caching scheme is used to save candidate sets. One problem is that the cost of building the volume may not be offset by the savings from using it. Caching schemes usually result in the candidate list for a particular volume being regenerated a number of times.

In [Marks et al., 1990] are explorations of ray casting efficiency for radiosity visibility computations. For environments made entirely of quadrilaterals, they create a decahedral volume between the two test patches. If no objects are found to be inside this volume, the patches must be fully visible; otherwise, the larger patch is subdivided via a quadtree and the intervisibility between sub-patches is checked. At some level standard ray casting would finally be performed as needed. They found that this technique provided acceleration for scenes with much intervisibility coherence, but in others cost a little more time.

In this paper we present a method of ray casting that takes advantage of similar forms of coherence. One source of coherence is that in radiosity algorithms a single object (e.g. a light emitter, such as a fluorescent ceiling panel) is tested for visibility with a large number of other objects. Note that this is true for both progressive radiosity [Cohen et al., 1988] and full matrix solution radiosity [Cohen & Greenberg, 1985]. Another form of coherence is that rays between two objects are limited to a particular volume of space. We also know in advance approximately how many rays will be cast between these two objects. These properties of radiosity algorithms provide considerable opportunities for increasing efficiency.

2 Algorithm

We define an *object* to be some single renderable primitive, e.g. a polygon, a torus, etc. An *item* is defined as an object or a bounding volume containing one or more other items. Some examples of items include: a spline surface, a bounding sphere containing three objects, and a root bounding box of a bounding volume hierarchy.

At the start we are given a set of objects, and have built a bounding volume hierarchy by some method [Goldsmith & Salmon, 1987; Kay & Kajiya, 1986]. We wish to find some set of items which can potentially block the visibility between the emitter and the receiver (we call these two items *reference items*). This set of items will be called the *candidate list*. Each time we wish to determine the visibility of some sample point to some other sample point, we use a ray caster to test all the items on the candidate list until an intersection is found or the list is exhausted.

One trivial solution would be to form a candidate list of all objects in the scene. This is equivalent to what a ray tracer without any efficiency scheme does - all the objects are tested against the shadow ray. Another trivial solution would be to give the root node of the bounding volume hierarchy as the only candidate on the list. Standard traversal of the bounding volume hierarchy then ensues for shadow testing.

Imagine we define a volume in space which contains the two reference items and all points between them. That is, any ray starting on one reference item and ending at the other is fully contained inside this volume we have defined. We can form a better candidate list by comparing this volume with the various items in the hierarchy; if an item is outside of the volume, it cannot be hit by any rays between the two reference items and so cannot occlude visibility between them. In fact, we could find that there are no objects at all between the reference items, in which case we know that the two reference items are fully visible to one another without any ray casting being necessary.

An important consideration is the amount of time spent creating this candidate list versus the time saved by using it. In radiosity algorithms we typically will sample each emitter a number of places on its surface, and each object receiving energy will typically have a set of points (e.g. a mesh) on its surface from which we wish to determine the emitter's visibility. For example, if we sample our light source 4 times per test point, and have an object with a 5 x 5 grid of test points, there are potentially 4 x 5 x 5 rays we need to cast between the surface and the light. The fact that we can know in advance about how many tests we will be doing means that we can much better approximate how much time to put into forming a candidate list.

This fact points toward a simple test to determine whether to create a candidate list. First, if either or both items are polygons, culling tests can be performed: if the emitter faces away from the receiver, or vice versa, then no energy is transferred. Else, find the approximate number of potential rays between the two reference items. If it is greater than some preset value, then form the candidate list and use this. If less, then don't bother, as the time saved per ray may not add up to be more than the time lost forming the candidate list. This value is best determined empirically for the system used; we think 10 rays is a reasonable break-even point. In fact, in most cases we test four points on the emitter against each object, so even an untessellated triangular receiver will generate 4 x 3 potential rays. This means that in almost all cases we will want to form a candidate list, so can dispense with this test altogether since we will almost always pass it.

The candidate list algorithm itself consists of three parts: forming the testing volume, creating the candidate list by using this volume, and accessing the candidate list for visibility determination between samples.

2.1 Forming the Testing Structure

The testing structure is formed by doing the following:

1. Obtain the bounding boxes for the reference items.
2. Compute the extent bounding box containing both reference items.
3. Create the plane set between the two reference items' boxes.

The first step is to form a bounding box for each reference item. Since we are using a bounding volume hierarchy, these boxes will be available or easily derivable. Each box is axis aligned and so consists of a minimum and maximum corner. These will be referred to by *lo.x*, *lo.y*, *lo.z* for the minimum corner, *hi.x*, *hi.y*, *hi.z* for the maximum.

In the second step an *extent box* is formed which contains both reference items. The extent box is formed by comparing the low and high corners of the reference boxes and taking the minimum and maximum, respectively. In this step we also wish to identify all edges of this box which are not a part of either reference box; call these *culled* edges.

The third step is to form a set of planes which connect the two reference boxes. The idea is to connect the edges of the reference boxes so that a minimal bounding volume is formed. Another way of thinking of this process is that each plane added whittles away the space between the reference boxes and one of the culled extent box edges identified in step two. This volume will be referred to as a *shaft*. These three steps are shown in Figure 1.

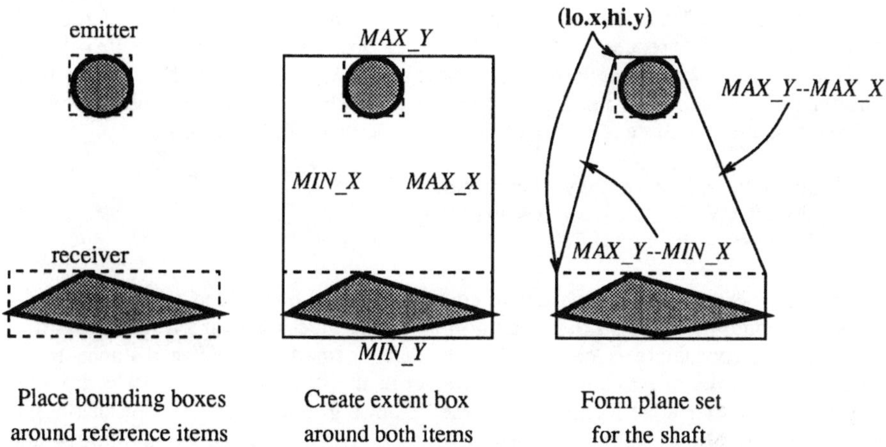

Fig. 1. Shaft formation process

When forming the extent box we also keep track of which coordinates of which box become the minimum or maximum. These lists will be used to identify culled edges. Two lists, one for each reference item, are needed. Each list can have a maximum of six entries, and each entry records a coordinate type (minimum or maximum) and direction (*X*, *Y*, or *Z*), e.g. *MIN_Z*. The minimums of each reference box are compared: whichever is smaller is noted on the list for that reference item. If the minimums are exactly equal, nothing is put on either list. Likewise, the maximums are compared and the higher coordinate type is stored.

We will walk through what lists are formed for Figure 1. We first compare the lower extents (i.e. *lo.x* and *lo.y*) of the emitter and receiver. The receiver's lower extent is less than the emitter's in both cases, so the receiver list has *MIN_X* and *MIN_Y* placed on it. The higher extents of the two boxes are now compared, and since the emitter's *hi.y* is greater than the receiver's, *MAX_Y* is placed on the emitter's list. Similarly, *MAX_X* is added to the receiver's list. The *Z* values are not shown for this 2D example, but would normally be tested and listed.

The third step is to generate all the planes which connect the edges of the two reference boxes. This turns out to be analogous to forming all combinations of elements on one reference list with those on the other, discarding combinations with matching directions. Each combination of elements corresponds to a plane connecting the edges of the reference boxes.

For example, in Figure 1 the *MAX_Y* of the emitter can be combined with the *MIN_X*, *MIN_Y*, and *MAX_X* of the receiver. *MAX_Y--MIN_Y* is discarded as a combination since the directions match. *MAX_Y--MIN_X* means that we need to form a plane between the edge on each box which has *hi.y* and *lo.x* constant for that box. In other words, two coordinates of the edge are fixed, and so the third, unmentioned, direction (in this case, *Z*) varies. Explicitly, we wish to connect the edges formed by the two vertices (*lo.x,hi.y,lo.z*) and (*lo.x,hi.y,hi.z*) on each reference box with a plane. Another way to think of this plane is that it cuts away the space in the extent box associated with the culled edge *MAX_Y--MIN_X*.

Forming such a plane is straightforward. The plane's normal has three coordinates. The "unmentioned" coordinate is always 0. The other two coordinates of the normal consist of the differences between the other axis' specified coordinates for each reference box. Carrying on with our example of *MAX_Y--MIN_X* (which means that the emitter's *hi.y* value is the maximum *Y* of the extent box, paired with the receiver's *lo.x* value being the minimum *X* of the extent box), the plane in this case is:

X.direction = Receiver.hi.y - Emitter.hi.y
Y.direction = Emitter.lo.x - Receiver.lo.x
Z.direction = 0

Each pair of extents has its own set of equations and its own culled edge associated with it. Care should be taken so that each normal formed faces outward from the shaft.

One last coordinate is needed for the plane formed: a relative distance from the origin. This is the negative of the dot product of the direction vector just computed and any point on either edge forming the plane. Knowing in advance that one of the plane's coordinates is always 0, this calculation can be streamlined to two multiplies and an addition. Our Figure 1 example yields:

P.distance = -(X.direction * lo.x + Y.direction * hi.y)

using the *lo.x* and *hi.y* of either reference box.

Each plane formed is stored in a list, which is called the *plane set*. Note that it is possible for no planes to be formed. For example, this occurs when one reference box is entirely inside the other box. A maximum of eight planes can be formed between any two boxes; see Figure 2 for some configurations where eight planes are formed between two reference boxes.

2.2 Creating the Candidate List

Now that the extent box and plane set have formed a shaft, we will use them to determine if items overlap it. To begin, the root item of the bounding volume hierarchy is compared to the shaft. Three possibilities can occur:

1. The item is entirely outside of the shaft.
2. The item is entirely inside the shaft.
3. Else, the item overlaps the shaft.

If the item is entirely outside of the shaft, then the item cannot obscure the visibility of the reference items and so can be ignored.

If entirely inside the shaft or overlapping, one strategy, called "always open", is to always open the item if it is a bounding volume and then recursively test its children against the shaft. The candidate list will then consist entirely of objects (no bounding volumes) inside or overlapping the shaft at the end of testing.

For all other strategies, if the item is fully inside the shaft, it is put on the candidate list. Bounding volumes save time in ray tracing whenever they are *missed* by rays; such a ray/bounding volume test is then cheaper than all the items inside the bounding

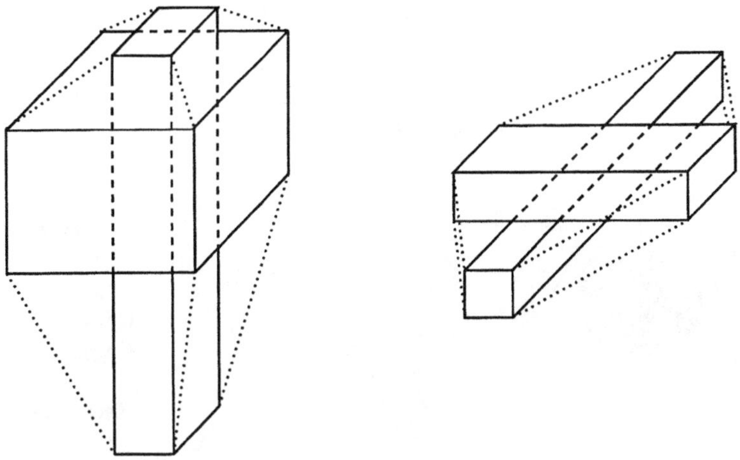

Fig. 2. Eight plane shafts

volume being tested and . . .sed. The philosophy here is that a bounding volume fully inside the shaft is more likely to be missed than an overlapping bounding volume, and so more likely to save time.

The overlap case is where things become interesting. If the overlapping item is an object, then it is put on the candidate list since it can obscure visibility. Many strategies are possible for dealing with an overlapping bounding volume. One is to always put these bounding volumes on the candidate list, on the theory that we want to minimize time spent forming the list, and the bounding volume is meant to help ray casting efficiency. In this case, overlapping items are treated the same as those inside the shaft. This is called the "keep closed" strategy.

The converse, called "overlap open", is to always open up overlapping bounding volumes and test all the children against the shaft, recursively. The idea here is that there are generally only a few objects between any two reference items, so opening the boxes now will save many intersection tests of them during ray casting.

However, some time can be wasted opening bounding boxes when unnecessary. For example, imagine a bounding box has four objects in it, three of which are barely in the shaft. Opening the box culls out the fourth object not in the shaft, but forces us to test the other three objects against each ray. If we had left the box intact, and the box overlapped the shaft only a little bit, each ray would be tested (and would most likely miss) against the box, thereby saving us testing the three objects.

The "ratio open" strategy addresses this situation by providing a ratio and testing to see if it is worth opening a bounding box. For example, we might give the ratio 0.4. If more than 40% of the items in a bounding box are found to be in or overlap the shaft, don't open the bounding box; rather, put it on the candidate list. It should be noted that, regardless of the ratio, if only one child of the bounding volume overlaps the shaft, then the bounding volume should probably be discarded. About the only reason to keep such a bounding volume around is if the child item is much more difficult to test for a miss than the bounding volume.

In Figure 3 a set of six objects (1-6) and two bounding volumes (A and B) are compared using the various strategies, with the candidate lists shown for each. Object 1 is eliminated by all strategies. Bounding box *A* is opened by the "always open" strategy, otherwise left closed. Bounding box *B* depends on the strategy; for example, the number of objects within or overlapping the shaft in the "ratio open" strategy is greater than 40%, so the bounding box is not opened.

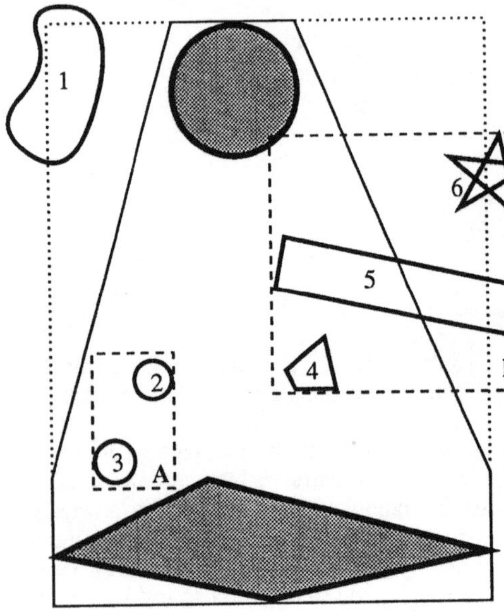

Always Open
Object 2
Object 3
Object 4
Object 5

Keep Closed
Bounding Box A
Bounding Box B

Overlap Open
Bounding Box A
Object 4
Object 5

Ratio Open (0.4)
Bounding Box A
Bounding Box B

Fig. 3. Shaft strategies

Further enhancements of these strategies are presented later. The performance of the various strategies is explored in the Statistics section.

2.3 Shaft Cull Testing

Regardless of which strategy is used for traversing the hierarchy and forming the candidate list, the algorithm which remains the same is the comparison of a shaft and a bounding volume. This test is called the *cull test*, and the bounding volume tested is called the *test volume*. The two bounding volume types we will consider are the sphere and the axis aligned box. The process for the sphere is presented first, and then a referencing scheme is described which makes the box about as fast to test.

The steps in the test are as follows:

1. See if the test volume is inside, outside, or overlapping the extent box.
2. If outside the extent box, testing is done. Else, check if the test volume overlaps either reference item.
3. If it overlaps, testing is done. Else, compare the test volume to the plane set.
4. If outside the plane set, testing is done. Else (depending on the strategy), either test if the item is fully inside the shaft or perform the "always open" strategy.

The first step is a simple comparison of the extent box with the test sphere. We use Arvo's algorithm for box/sphere comparison [Arvo, 90]. In this test each coordinate axis is treated separately. If the sphere's center coordinate for an axis is outside the box's extent along that axis, then take the difference between the sphere's coordinate and the closer extent value and square it. Sum and check these squared values (if any) for the axes; if greater than the radius squared, then the sphere is outside the box.

In the second step we do a similar test as the first step for the sphere and each of the reference item boxes. If the test sphere overlaps either reference item, then it must overlap the shaft and testing is done.

Testing the sphere against the plane set is straightforward. To simplify computations, each plane equation's normal should be normalized and the plane distance value scaled accordingly. The distance from the sphere's center to the plane is defined as:

T = dot_product(Plane.normal, Sphere.center) + Plane.distance

For each plane, the distance of the sphere's center from the plane is compared to the sphere's radius:

If T > Sphere.radius, then the sphere is outside the plane

If the sphere is found to be outside of any plane, then it must be outside the shaft. If the sphere survives all these tests, it is probably overlapping or inside the shaft. Note that a sphere can overlap the extent box and one or more planes and yet still be outside of the shaft. We call these *false positives*, as they are falsely identified as overlapping the shaft. Such false categorization does not causes any errors but can cost speed, as some bounding spheres which could actually be ignored may instead be placed on the candidate list and wastefully tested against rays. If a shaft is being formed for a particularly large set of rays, it may be worth fully checking for these false positives. Figure 4 shows a portion of a bounding sphere which is a false positive.

Depending on the strategy, we may want to determine whether the sphere is fully inside the shaft. If any of the sphere's extents are outside the shaft's extent box, then the sphere cannot be fully inside the shaft. Otherwise, the sphere is again tested against the plane set. The test is similar to the above:

If T > -Sphere.radius, then the sphere is fully inside the plane

Note that the comparison computed is simply 2 * Sphere.radius more than that calculated for the outside test, so the outside test's results can be used to quickly test this condition. If the sphere is found to be inside all planes in the plane set, then it is fully inside the shaft; else, as soon as it fails to be fully inside any plane, it is categorized as overlapping the shaft.

Boxes can be tested in a similar fashion to spheres. We do not need to normalize the plane set normals if only boxes are tested against them. In the sphere test the radius was used for comparison, so normalization was important. In the box test we will find the distance between various corner points and each plane and test the sign of these distances, so the scale of the distance is irrelevant.

An insight into the nature of testing any box against a plane is important in understanding the box and shaft test. First, we define the distance of a point from the

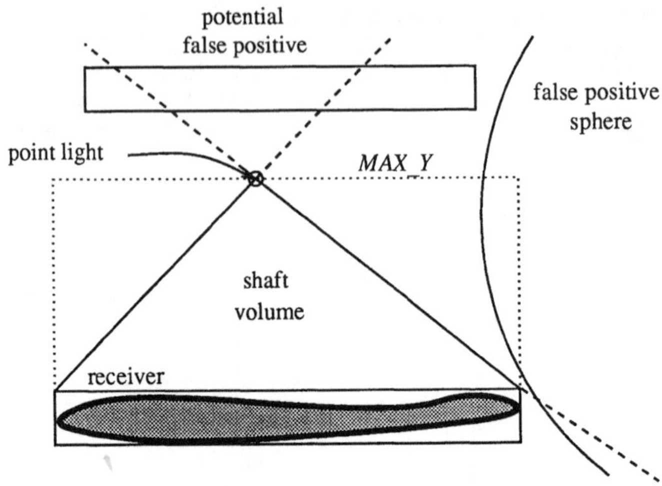

Fig. 4. False positives

plane as:

T = dot_product(Plane.normal, Point) + Plane.distance

The plane defines a half space. Similar to the sphere test, if $T > 0$ then the point is outside the half-space, if $T = 0$ it is on the plane itself, and $T < 0$ means inside. When referring to the distance from the half-space, note that we mean this signed distance, i.e. a point with $T = 2$ is termed to be farther away from the half-space than a point with $T = -9.3$. Points inside the half-space are considered to be closer to it.

Given a particular plane, there is at least one corner which is always going to be the farthest away from the half-space, and at least one that is always nearest. The farthest corner is a function of the normal of the plane: if the sign of a particular coordinate is negative, use the *lo* value of the box, if the sign is positive use the *hi*, if 0 either value can be used. The nearest corner is just the opposite of these conditions: *lo* for positive, etc.

For example, given a plane with a normal [-3 8 4], the farthest box corner is always [*lo.x,hi.y,hi.z*] and the nearest always [*hi.x,lo.y,lo.z*]. Each plane points into one octant of the *XYZ* reference frame, which corresponds to one corner of all boxes tested against it. A fast way to obtain the box's relevant corner during testing is to store a little extra data for each plane as it is formed. Store the offsets to the coordinates needed for the far corner and near corner with each plane, e.g. if *lo.x* = 0, *lo.y* = 1, etc, then the farthest box corner's offsets in our example would be [0,4,5] and the nearest [3,1,2]. Note that near corner offsets equal ((far offsets + 3) modulo 6). These offsets can then be used directly to obtain the desired corner, as opposed to wasting time with comparison tests to access it.

Using this observation, the outside/inside/overlap test for boxes is done similarly to the sphere test. First check the test box against the extent box: if there is no overlap, the box is definitely outside and testing ends. Else check the test box against the reference item boxes: if there is any overlap, the test box must overlap the shaft and testing is done. Else test each plane in the plane set against the box's nearest corner for that plane:

If dot_product(Plane.normal, Box.near_corner) + Plane.distance > 0,
box is outside the plane (and the shaft)

If the box survives all of these tests, then it must overlap or be inside the shaft. In Figure 3 object 1 overlaps the extent box, but is outside the *MAX_Y--MIN_X* plane so it is discarded. Bounding boxes *A* and *B* survive the plane set test.

From our earlier extent box testing we know if the test box was fully inside the extent box; if it is not then it must overlap the shaft. For example, in Figure 3 bounding box *B* is not fully in the extent box, so we know that it must overlap the shaft. Else, if needed, we can find the exact state of the box by testing planes in the plane set against the corresponding farthest corners:

If dot_product(Plane.normal, Box.far_corner) + Plane.distance > 0,
box overlaps shaft

If the box does not overlap any plane in the set, then it is fully inside the shaft. Bounding box *A* in Figure 3 needs to be tested in this fashion, as it was fully inside the extent box but it was not known if it was fully inside the plane set.

The box test uses a small number of operations, which is made possible by accessing only the relevant corner for testing. The traditional method for testing overlap is to perform a full culling test of the box against shaft's extent box and plane set and see if anything was left inside, which is much more than is needed to classify the test box. Using interval analysis a faster classification test can be devised, but this type of testing is conservative and inexact: such an algorithm errs on the side of classifying some boxes which are outside as being overlapping, for example.

2.4 Candidate List Access

For each combination of sample points on the receiver's surface and on the emitter's surface a ray is cast, if both surfaces face each other. The ray is tested against the candidate list, with bounding volumes hit being opened and their children tested. As soon as an opaque object is hit the shadow testing is over, with the receiver's point in shadow.

One implementation of this process is to copy the candidate list to a stack. Pop an item and test it for intersection, and push the children of any bounding volume intersected on the stack. A hit or an empty stack ends testing.

3 Enhancements

In this specific use of shafts to test rays traveling from an emitter to a receiver there are a number of important enhancements we can make to the basic shaft testing algorithm. We know that rays cast between these two reference items must originate in one box and end in the other. Therefore, if our test volume fully encompasses either reference box, then the item must *always* be tested for intersection by any ray cast. So, when testing overlap with the extent box, we can also check if the test volume encompasses either reference box. If so, and the item is a bounding volume, we should then always open the volume and test its children, since we know the bounding volume will always be hit. If the item is an object it should be placed on the candidate list. Box/box comparison is just extent overlap testing. Sphere/box comparison can be done by using Arvo's method [Arvo, 1990] Essentially, sum the squares of the distances between each axes' sphere center coordinate and most distant box extent coordinate; if less than the sphere's radius squared, the box is fully inside the sphere.

The realization that all volumes which contain a reference box must be hit, combined with the fact that in radiosity we compute the effect of one emitter against a large number of receivers, leads to another useful preprocess. In fact, this preprocess can be used to prune the hierarchy of useless bounding volumes and so increase the efficiency of a traditional ray caster. When we first begin testing the emitter with the set of receivers, we can traverse the bounding volume hierarchy once and find a candidate list (called the *emitter list*) of those items which do not enclose the emitter's box. We then can use this list when testing the hierarchy against any shaft, since all shafts for this step have the emitter as a reference box. Instead of starting at the root node and testing the whole hierarchy against each new shaft (and finding that the same items enclose the emitter's box and so must be opened again and again), we now test items on the emitter list against the shaft. For each shaft, only the receiver's box then needs to be tested for enclosure within the test item, versus the earlier method in which both reference items were so tested.

One other advantage to building an emitter list is that all the items placed on it can be categorized as either overlapping or outside the emitter. When it comes time to check items on this list against the shaft, we know that an item which overlaps the emitter cannot be categorized as outside.

One efficient way to form the emitter list is to start at the emitter and travel up the hierarchy. The parent of the emitter must contain the emitter, so can be opened. The other children of the parent are then tested to see if they enclose the emitter, and if not are placed on the emitter list, else they are opened in turn and their children tested. When all of the parent's other children have been categorized, we move to the parent of this parent (which again must contain the emitter) and repeat the process, ending when the root item's children are categorized. Note: the emitter itself may or may not be placed on the emitter list, depending on whether it can occlude itself.

If the emitter casts energy in a limited set of directions (e.g. a one-sided polygonal light will not illuminate anything behind it), then all items behind it can be eliminated once at the beginning of the step. Another technique is useful if surfaces have only one side, but none of these non-existent backfaces are visible to any frontface. For example, in a solids modelling system the insides (backfaces) of the objects can not be seen and so do not have to be rendered. In such cases, the light reference object can be used to cull away all polygons which face away from the light. This can be done by the same methods used to compare a plane and a bounding volume (the emitter, in this case).

Another idea is to mark within the hierarchy when various sons are outside the shaft. When creating the candidate list, the "ratio open" strategy finds which children of a bounding volume are outside the shaft. If the bounding box is put on the candidate list, mark the children of this box which are outside the shaft by using a mailbox method [Arnaldi et al., 1987], with a new identifier for each new shaft. Now when the ray caster is called, it also checks if a child is marked as outside; if so, the child is not tested for intersection.

One problem with using the receiver's bounding box is that this volume often slightly overlaps adjacent polygons which do not have any possibility of shadowing the receiver. Using a small tolerance to cull out such polygons can eliminate a few of these from the candidate list (though the shaft will still catch a few). If such adjacent, non-occluding polygons can be identified by other means, they can all be eliminated from testing.

An enhancement which is worth using is shadow coherency testing [Haines & Greenberg, 1986]. This method keeps track of what object (if any) the previous ray hit, and tests the current ray against this object. The theory is that whatever shadowed the last ray will probably shadow the current ray, since the current ray is usually similar to the previous ray. This technique is useful for all ray tracers and casters, and the shaft tester is no exception. In fact, when a new receiver is tested, we should not build a shaft between the reference items until the shadow coherency object is missed. We may find that the previous shadowing object occludes all rays between the new receiver and the light, too.

4 Other Uses

Though presented as an algorithm for speeding ray casting testing between two surfaces, the shaft culling technique is applicable to other computer graphics problems. For example, it is a fast way of determining the candidates in a 5D classification scheme ray tracer [Arvo & Kirk, 1987]. In their method, all vertices of the bounding polyhedron for an object are tested against each plane. The shaft method is much faster than their scheme when testing boxes for inclusion, as only one corner of any box needs to be tested against each plane.

We also find shaft culling to be a useful tool in minimizing the work done by other algorithms. For example, if we want to perform an analytic solution for point to patch visibility, shaft culling gives us a short list of candidates which to send to the analytic algorithm.

5 Statistics

Each object in the scene has its own bounding box, and the bounding boxes were clustered using Goldsmith and Salmon's hierarchy building algorithm, as it was found to yield faster intersection timings than Kay and Kajiya's scheme. The regular ray caster has been optimized over a number of years, so provided a tough standard of comparison.

A simple environment was used to compare the performances of the various strategies. The office scene in [Wallace et al., 1989] was used in two different forms: standard and premeshed. The standard scene was meshed by the radiosity software, then shafts were formed using the original polygons. This means that shafts generated for the walls and floor have a large number of rays generated (due to the internal mesh points in them), but the shafts are proportionally wider and so include more candidates for testing. The premeshed database was identical, except that the floor and walls were tessellated into polygons before input. This causes more shafts to be formed, with less rays per shaft, but makes for tighter shaft for those rays. Both scenes have 10 area light sources. The standard database has 906 polygons before meshing, the premeshed scene has 2563 polygons before further meshing. These two versions of the same scene were used to examine the effect of shaft width on the various strategies.

The code to test shafts had various inefficiencies due to the nature of the calling routines. For example, shafts were formed using the entire emitter, though the samples on the emitter could have been bounded by a smaller bounding box. Shafts were always formed, even if the shaft had only one uncached ray left to shoot.

The first testing done was to see the effect of shadow coherency, a proven algorithmic improvement. The standard and premeshed scenes have similar results. In the simplest test case of one sample per emitter and 20 steps of progressive radiosity, about 54% of the vertices were shadowed. Of these shadowed vertices, about 78% were found occluded by the shadow cache; that is, whatever had previously shadowed the last ray also shadowed this one. The total number of intersection tests for the standard scene was 3262K intersection tests of bounding boxes and polygons without and 2252K tests with shadow caching, a 31% decrease.

We then tested the effect of forming a candidate list for the emitter at the beginning of each step of progressive radiosity. We also culled away any bounding boxes which were entirely behind the area light source. This candidate list was then used for regular ray casting with shadow caching. For the simplest test case this provided a 17% decrease in number of intersection tests for the standard scene, 12% for the premeshed scene. Because the cost of this preprocess was so minimal and always saved time overall, it and shadow caching are used in all the tests that follow. The baseline ray caster used for comparison also uses both of these algorithmic improvements.

Two sets of tests were done on each of the databases for each strategy. We varied the number of samples on each emitter for these tests to see the effect on overall performance. Shadow caching quickly found the visibility of some of the rays cast. Since we are trying to understand the effect of the shaft, we did not include the rays which were shadow cached (each ray has one intersection) in the statistics. In the standard scene, this meant averages of 8.1 and 34.0 uncached rays per shaft; the premeshed scene has less samples per polygon on the average, so averages were lower at 3.9 and 18.5 rays/shaft.

Statistics for the low and high sampling rates for each scene and strategy are shown in Table 1. Tests were done on a Hewlett-Packard Series 9000 model 433 workstation with unoptimized, debuggable, profiled code. The "tests/ray" is the total of bounding box and object tests made per ray: recall that each object has its own bounding box. The times are the average costs in microseconds for shooting an uncached ray, with the cost of building and testing the shaft included. In general, the number of intersection tests per ray were similar at the two sampling rates, and so are presented as an average of these two.

TABLE 1. Statistics for Office Databases

Strategy	standard			premesh		
	tests/ray	low time	high time	test/ray	low time	high time
baseline	44.6	787	762	79.2	1487	1355
keep closed	28.4	642	534	47.8	1094	919
always open	132.9	2268	1785	27.5	1289	694
overlap open	66.7	1382	1008	25.8	1283	680
ratio open 0.2	22.1	670	482	27.1	1104	685
ratio open 0.4	21.9	681	480	26.7	1135	682
ratio open 0.6	22.1	679	478	25.0	1154	654
ratio open 0.8	24.1	753	524	21.1	1327	631

Oddly enough, the percentage of rays which hit occluders was lower at the higher sampling rate. This also had the peculiar effect of lowering the time cost per ray for the higher sampling rate by 3-9% for the baseline ray caster.

Which beam culling strategy to use is an interesting question, as there is no strategy which has dominance over the others. The "keep closed" strategy minimizes the time spent doing the shaft culling preprocess, as most items are quickly classified as outside the shaft or overlapping. Shaft culling eliminates those items which definitely have no chance of being hit, saving the costs of intersecting (and missing) these items repeatedly by rays in the shaft. Because this strategy does spend such a small amount of time building the candidate list and is so conservative (bounding volumes are not opened unless definitely hit), it always yields some speed up. Time saved ranged from 21% to 32%.

The "always open" strategy failed for the standard office scene and the low sampling rate premesh, but performed quite well for the high sampling rate premesh scene. We believe this is because of two factors. In the standard scene, the wall and floor shafts have many samples and shafts formed using them encompass much of the office. This strategy removes all the bounding boxes for items in these shafts. Many of these bounding boxes would have been more profitable to keep around and use for the standard scene. This leads to excessive intersection rates for the standard scene. The premesh scene has shafts with smaller volumes, and so less likelihood that bounding boxes would help much.

Opening all these boxes during the preprocess also costs time when forming the shaft's candidate list. For the premesh scene, it is not until this cost is amortized over a large number of rays that the time per ray drops to acceptable levels. However, when the shafts are mostly narrow (as in the premesh scene) this strategy can yield greater savings than the "keep closed" strategy.

The "overlap open" strategy is not much different than "always open". The preprocess cost was higher due to having to do extra tests, and some savings were garnered by keeping around bounding volumes which were fully in the shaft. Looking more closely at the performance of this strategy, we found that there were very few boxes with more than one item in them which were fully inside shafts: 0.71 boxes per standard scene shaft and 0.36 boxes per premeshed shaft. Such complex boxes contained an average of 33.1 objects as descendants for the standard scene and only 4.3 children for the premeshed scene. Such boxes, then, proved useful in improving standard scene rates, but were of little use in the premesh scene.

The "ratio open" strategy gave the best results in terms of number of intersection tests. At the best ratio, they cut the number of intersection tests performed by a factor of 2.0 for the standard scene and 3.8 for the premeshed scene. With the preprocess time included, speedup ratios of 1.6 and 2.2 were obtained for the maximum sampling rates. Note that for the standard scene in particular the lower ratios caused less boxes to be opened, so that less time was spent on shaft culling itself. Even though this led to slightly higher numbers of tests per ray, the overall cost was lower.

It is interesting to examine how the ratio is related to the database type. A low ratio corresponds to keeping bounding volumes closed, while a high ratio opens them more frequently. With the large, wide shafts in the standard scene, it seems to be profitable to keep bounding volumes closed, while the premesh scene's narrow, tight shafts benefit little from bounding boxes. However, the 0.8 ratio does perform better than the 1.0 ratio, which is exactly equivalent to the "overlap open" strategy, so it is still worth keeping around a bounding volume in which most or all of its children overlap the shaft.

The statistics shown for "ratio open" use the mailbox marking enhancement for boxes found outside the shaft (i.e. such boxes are tagged so that during ray casting they are not tested, even though in the hierarchy). This enhancement helps more for lower ratios, as more boxes could be outside the shaft but still a part of the hierarchy. For example, for the premeshed scene, tagging with a ratio of 0.2 saved 2.6 bounding box tests per ray, at 0.6 saved 1.4, while tagging at 0.8 saved less than 0.01.

Factoring in the speedup due to "ratio open" with the savings from emitter list preprocessing, we obtain a maximum speedup of 2.44 for the new algorithms presented when compared to a standard ray caster with shadow coherency.

6 Future Work

What follows are some areas of research which could be done. We present them in hopes of sparking interest in pursuing further research in this area.

One improvement suggested by the statistics is gathering clumps of rays for shaft formation. For example, the floor is a receiver that may have a thousand grid points on it which need to be tested, but since the floor is so large, a shaft between it and a light source may cull out very little of the whole environment. However, limiting shaft creation to single mesh elements may create efficient shafts which then do not get used by many rays. Our standard and premesh scenes reflect these extremes. A happy medium might be to collect a set of rays from a receiving object until a certain number is reached; at this point build a shaft around the set of rays and test each ray in turn. This method has the advantages of a high number of rays to amortize the shaft culling costs, while hopefully creating a set of smaller shafts than if the whole surface was tested using one shaft.

An important property of such beams is that each cluster can use the parent object's candidate list as a starting point. That is, each sub-shaft formed from a set of rays on the object must have a candidate list which is encompassed by the object's candidate list. Using the object's candidate list will give us a head start on forming a sub-shaft's list. In fact, if we had a large enough number of samples on the emitter, it might be worth forming a separate shaft for each individual sample point on the receiver (or vice versa).

Another use of this nesting of shafts idea is superstructuring. Imagine you have stored for each bounding volume the number of sample points on objects contained in the hierarchy below it. The number of points could be used to decide when to create a shaft: if within some given range (i.e. not too many and not too few), a shaft could be formed for the bounding volume containing receivers instead of a particular receiver. In this way all objects will have access to a candidate list of some sort, and this list will always be better than or equal to (i.e. have the same number of items or less) our emitter's original list. Other ideas are to use the number of samples per unit of surface area of the bounding volume (which should be related to the flux or density of rays between the reference items) to determine when to make a shaft.

Candidate list sorting is a simple idea worth exploring. Given a candidate list and a set of rays, there exists an optimal ordering of the list in order to minimize ray/item intersection time. What criteria to use to sort this list (and whether the list is even worth sorting) is a subject that has never been researched. One important variable to note is what purpose the ray has: the optimal shadow test list (where the first hit is desired) is almost certainly different that the optimal shading ray test list (where the closest object is desired). Such analysis could prove fruitful in other areas of graphics.

The basic shaft culling algorithm always forms the full plane set. However, some of these planes may whittle very little space away from the extent box's volume. A quick approximation of how much space a plane trims from the extent box volume could be used to determine whether it appears profitable to use that plane for testing. Sorting the planes by their approximate contribution is another speedup; we want to find if an object is outside the shaft as soon as possible, so using the plane which chops the most space first will minimize shaft culling time. Figure 2 shows some eight-plane shafts; all planes make a fair contribution to the shaft formed at the right, but the left shaft's top four planes chop off little volume of the extent box compared to the bottom four.

New strategies are an area open for research. The "ratio open" strategy is effective at high sampling rates, "keep closed" for lower rates; this suggests making strategy selection dependent upon the number of rays to be tested. Strategies based upon other factors (e.g. approximate solid angle or relative shaft volume) are also worth investigating.

The extension of shafts to other efficiency schemes is also of interest. As has been shown, 5D classification can immediately benefit. Grid and octree efficiency algorithms could be used with plane sets; however, the rules for finding items to test would have to be reworked. Boxes which are not axis aligned, slab sets [Kay & Kajiya, 1986], and ellipsoids are bounding volumes which can fit into shaft testing with some work. There may be some interesting insights to be gained by exploring how to test these efficiently.

The mathematical nature of shafts is an area of research in its own right. Bounding boxes which are tested against a shaft are always correctly categorized (though of course their contents may not be). We strongly believe that shaft formation yields no boxes which are falsely categorized (i.e. false positives). However, an arbitrary set of planes forming a polyhedron cannot be used to test bounding boxes without the risk of false positives. In Figure 4 is a case in which the *MAX_Y* plane seemingly adds nothing to the volume, since all the other planes fully demark the volume's bounds. However, a box above the point light is within the bottom three extent box planes and overlaps the two planes in the plane set - it is the *MAX_Y* plane which is necessary to correctly categorize the box as being outside the shaft. We have constructed an informal proof showing that shafts do not give false positives for box testing, but have not formally studied what are necessary and sufficient conditions to avoid miscategorization.

7 Summary

We have presented methods to improve the use of ray casting in determining visibility for radiosity. Taking advantage of object coherency, we form candidate lists for ray casting from a hierarchical bounding volume structure by a variety of means. One operation presented is the building and use of the volume between two items to cull out those items in the environment which cannot occlude visibility and to form an efficient list of items for intersection testing. A number of efficiencies are explained which make this shaft forming and testing algorithm more profitable to use. Some future research directions are identified.

Acknowledgements

Kells Elmquist first posed the problem of how to find the set of objects between two given objects. The algorithm presented was tested using the ARCore radiosity and ray tracing software developed by a large number of people at 3D/Eye and Hewlett-Packard. Thanks to Juhana Kouhia for careful proofreading of the draft and final paper.

References

Arnaldi, B., Priol, T. & Bouatouch, K. (1987) *The Visual Computer*, **3**,3 98-108, "A New Space Subdivision Method for Ray Tracing CSG Modelled Scenes."

Arvo, J. & Kirk, D. (1987) *Computer Graphics* (SIGGRAPH '87 Proceedings), **21**,4 55-64, "Fast Ray Tracing by Ray Classification."

Arvo, J. (1990) **in:** Glassner, A.S. (ed) *Graphics Gems*, Academic Press, London, 335-339, "A Simple Method for Box-Sphere Intersection Testing."

Cohen, M. & Greenberg, D.P. (1985) *Computer Graphics* (SIGGRAPH '85 Proceedings), **19**,3 31-40, "The Hemi-Cube: A Radiosity Solution for Complex Environments."

Cohen, M., Chen, S.E., Wallace, J.R. & Greenberg, D.P. (1988) *Computer Graphics* (SIGGRAPH '88 Proceedings), **22**,4 75-84, "A Progressive Refinement Approach to Fast Radiosity Image Generation."

Goldsmith, J. & Salmon, J. (1987) *IEEE Computer Graphics and Applications*, **7**,5 14-20, "Automatic Creation of Object Hierarchies for Ray Tracing."

Haines, E.A. & Greenberg, D.P. (1986) *IEEE Computer Graphics and Applications*, **6**,9 6-16, "The Light Buffer: A Ray Tracer Shadow Testing Accelerator."

Kay, T.L. & Kajiya, J.T. (1986) *Computer Graphics* (SIGGRAPH '86 Proceedings), **20**,4 269-278, "Ray Tracing Complex Scenes."

Marks, J., Walsh, R., Christensen, J. & Friedell, M. (1990) *Proceedings of Graphics Interface '90*, Canadian Information Processing Society, Toronto, Ontario, 17-30, "Image and Intervisibility Coherence in Rendering."

Nakamae, E., Ishizaki, T., Nishita, T. & Takita, S. (1989) *IEEE Computer Graphics and Applications*, **9**,2 21-29, "Compositing 3D Images with Antialiasing and Various Shading Effects."

Salesin, D. & Stolfi, J. (1989) *Proceedings of the PIXIM '89 Conference*, Hermes Editions, Paris, France, 451-466, "The ZZ-Buffer: a Simple and Efficient Rendering Algorithm with Reliable Antialiasing."

Wallace, J.R., Elmquist, K.A. & Haines, E.A. (1989) *Computer Graphics* (SIGGRAPH '89 Proceedings), **23**,3 315-324, "A Ray Tracing Algorithm for Progressive Radiosity."

Weghorst, H., Hooper, G. & Greenberg, D.P. (1984) *ACM Transactions on Graphics*, **3**,1 52-69, "Improved Computational Methods for Ray Tracing."

Progressive Refinement Radiosity on a Transputer Network

Martin Feda, Werner Purgathofer

Abstract

A parallel implementation of the progressive refinement algorithm for radiosity computation on a transputer network is described. Worker transputers compute the radiosity values of all patches using hemi-cubes, while rendering transputers generate images of the actual results. The scene is split up into parts of equal size and distributed among the workers to overcome the limitations of local memory. The workers are configured in a minimum pathlength network to reduce the communication overhead for loading non-local scene data and for updating patch radiosities.

1 Introduction

This paper presents a parallel implementation of the progressive refinement algorithm of the radiosity approach on a transputer network.

During the past two years, the progressive refinement radiosity algorithm [Cohen et al. 1988] has proved to be a powerful method for high-quality image generation. It eliminates the enormous memory requirements of the original full-matrix radiosity computation, and it provides an useful solution almost immediately, which is then successively improved. Images visualizing the actual radiosity values can be generated continuously.

Based on the progressive refinement algorithm several improvements of accuracy (e.g. [Baum et al. 1989] and [Wallace et al. 1989]) and calculation time (e.g. [Baum and Winget 1990] and [Recker et al. 1990]) were suggested. The most important method to speed up the radiosity calculation is the use of parallelism. While Baum and Winget describe an implementation on a multiprocessor graphics workstation calculating hemi-cubes by a graphics processor, Recker et al. use a cluster of workstations.

A much cheaper alternative are transputer networks. Transputers are characterized by a processor with its local memory and four fast serial interconnection links. The main problem of transputers is their limited local memory. In contrast to the workstations used by Recker et al., it is not possible to store the complete scene description at each transputer.

This paper focusses on strategies for data exchange between transputers to find the globally best shooting patch, to calculate form-factors and to update radiosities.

2 The Progressive Refinement Algorithm of the Radiosity Approach

The radiosity approach computes diffuse global illumination of surfaces by determining the dependence of light leaving one patch and arriving on another one. This dependence is described by a system of equations:

$$B_i := E_i + \rho_i * \sum_{j=1}^{n} F_{ij} B_j$$

B_i	radiosity of patch i	ρ_i	diffuse reflectivity of patch i
E_i	light emitted from patch i	F_{ij}	geometric form-factor

In the full-matrix method all form-factors are computed in a first step to determine all equations, which are then solved. Finally the image is rendered.

In contrast, the progressive refinement algorithm increments the radiosities of all patches by successively shooting light out from the patches which have the maximum undistributed energy at this moment. The initial values of the radiosities are set to the patch emissions E_i. For each shooting patch i all form-factors F_{ij} to all receiving patches j are calculated by a single hemi-cube placed on patch i:

For all j: $\quad B_j := B_j + B_i\,(\rho_j F_{ji}) \quad$ where $F_{ji} := F_{ij}\, A_i / A_j$

A_i area of patch i

This method eliminates the memory requirement of the form-factor matrix. Images can be generated continuously. A useful radiosity estimate is provided almost immediately.

3 System Overview

Our system consists of 17 INMOS T800 transputers with four 20 MBit/s links, 2MB local memory and an on-chip floating point unit. One link of the transputer at the root of the network is connected to a PC, which acts as host of the system. All program code and data must be loaded from the host onto the network, and the generated images are displayed on the host's screen.

There are two main tasks which must be performed to generate an image: the radiosity of all patches must be computed, and an image must be generated using these radiosity values.

In our system each transputer is assigned to only one of these two tasks. Transputers called "renderers" continuously generate images of the actual radiosity values, which are improved at the same time by the "workers". Thus far several configurations with up to 8 workers and up to 8 renderers were used. The root transputer called "master", which connects the network to the host, performs data input and output, data exchange between workers and renderers, and control tasks. Fig. 1 shows the structure of the complete network:

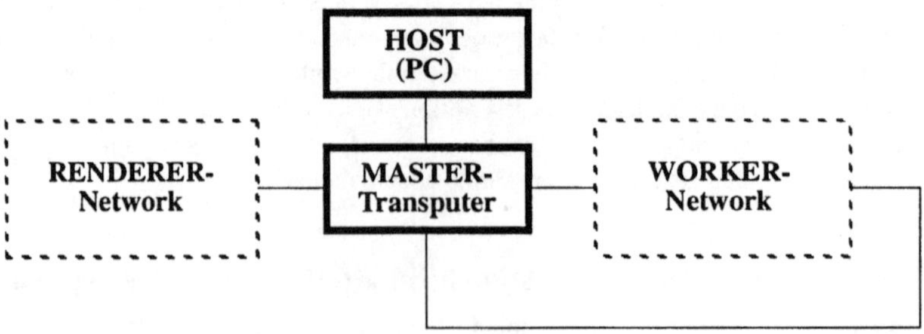

Fig. 1: System structure

4 Calculating the Radiosity Solution

The radiosity solution is calculated by a set of worker transputers supervised by a controller process running on the master. Each worker computes the form-factors of a shooting patch independently by a hemi-cube and distributes the energy of the shooting patch to the receiving patches. However, this concept can be used for a more precise analytical form-factor computation [Baum et al. 1989] and for the faster single-plane, modified single-plane and modified hemi-cube algorithms [Recker et al 1989] without any change (under implementation).

Since the choice of the best strategy, e.g.hemicube, analytical computation, single-plane etc., depends on the shooting patch, the computation times of shooting patches differ significantly. Therefore the workers are not synchronized (in contrast to the full-matrix radiosity algorithms described in [Price and Truman 1989] and [Purgathofer and Zeiller 1990]).

Since it is not possible to store complex scenes at each transputer, the scene description is split up into parts of equal size, so that each worker holds one part. The maximum number of patches that can be stored in the local memory of one worker depends on the size of memory reserved for hemi-cube computation: at a maximum hemi-cube resolution of 256x256x128, each worker can hold up to 2000 patches, for a resolution of 64x64x32 up to 4500 patches.

All read or write operations of a worker on data stored at another transputer have to be performed by message exchange. To minimize the routing effort, a minimum path length network has been adopted for the workers (used for full-matrix radiosity in [Chalmers and Paddon 1989]). Fig. 2 shows the configuration of 8 workers.

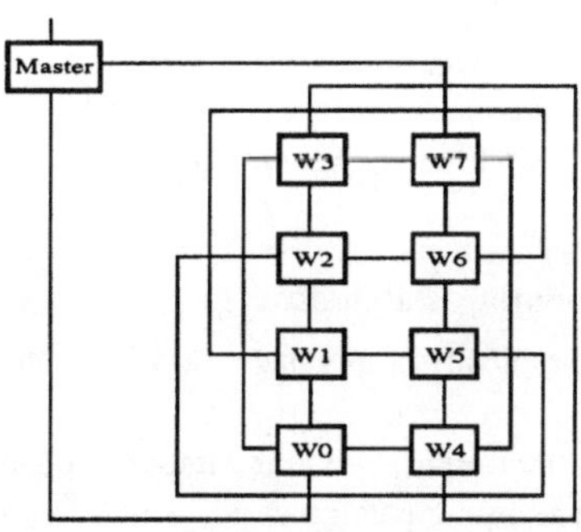

Fig. 2: Minimum path-length network for 8 workers

The maximum path-length between any pair of workers is two links for this topology. An even more important feature is the average path-length, which determines the communication overhead for routing data packets through the network. The network shown in fig. 2 has an average path-length of 1.46. It can be seen very easily that this topology is optimal for 8 workers under the constraint to connect two links to the master: each worker can communicate with four adjacent workers (W0 and W7: three adjacent workers) directly; messages to other workers must be routed over a single intermediate transputer.

The workers' algorithm can be described in three basic steps:

1. shooting patch selection
2. form-factor computation
3. distribution of the shooting patch energy

4.1 Shooting Patch Selection

A worker can only determine its locally best candidate for the next shooting patch, i. e. the patch i with the local maximum of $\Delta B_i A_i$, where ΔB_i is the unshot radiosity and A_i is the patch area. However, the controller process running on the master maintains a table of the best candidates of all workers. If a worker needs a shooting patch, it sends the locally best patch to the controller to update its table. Based on this table the controller tells the requesting worker which transputer holds the globally best patch. The description of this patch is then fetched from this serving worker to the requesting worker directly. The serving worker marks the shooting patch as "active" so that it cannot be selected again while it is processed, determines a new local maximum of $\Delta B_i A_i$, and updates the controller's table.

4.2 Form-Factor Computation

A worker needs the vertices and normal vectors of all patches to generate the hemi-cube or to determine the form-factors by one of the alternatives mentioned above. To reduce the routing effort, all patches are arranged in blocks of e.g. size 100. While the patches of one block are mapped onto the hemi-cube, a copy of another unprocessed block is loaded from its worker into a local buffer to minimize the processor idle time.

The last block of the previous calculation or a local block can be mapped on the hemi-cube of a new shooting patch immediately. The blocks from other workers are then processed in a random order. When the transputer starts its work on one block, it already requests for the next one. If a block does not arrive before the worker has completed the previous one, a local block is processed in the meantime.

4.3 Distribution of the Shooting Patch Energy

To preserve data integrity, the radiosity of any patch may be changed only by the worker it is stored at.

When a worker has completed the form-factor computation for shooting patch i, it fetches the actual unshot radiosity of the shooting patch patch i, which might have received light from shooting patches calculated on other transputers while it was processed. This unshot radiosity is distributed by sending update messages to the workers storing the receiving patches. The messages contain energy values U_j for each receiving patch j:

$$U_j := \Delta B_i \, A_i \, F_{ij}$$

ΔB_i unshot radiosity of shooting patch i
A_i area of shooting patch i
F_{ij} form-factor

When a worker receives an update message, it multiplies the values U_j by ρ_j / A_j of receiving patch j (ρ_j and A_j are locally stored) and adds the result to the patch radiosity B_j and the unshot radiosity ΔB_j:

$$B_j := B_j + U_j \rho_j / A_j \qquad \Delta B_j := \Delta B_j + U_j \rho_j / A_j$$

Finally the controller's table of the worker's best shooting patch candidate is updated, if necessary.

5 Rendering

The image plane is divided into horizontal sectors of equal size. Each renderer generates one image sector by the z-buffer algorithm. The segmentation of the image plane is determined by a rendering manager running on the master transputer.

If a sequence of images is generated with the same camera, the z-buffers are not generated once again, only the radiosity values are updated. This is why radiosity computation and rendering tasks are executed on different transputers: if each transputer would execute both tasks, they would have to share a common z-buffer for computing hemi-cubes and for rendering because of the limitations of local memory. In any case the z-buffers would have to be computed completely for each generated image, because the previous values would be overwritten by hemi-cubes. Furthermore, during image generation the radiosity computation would be locked, and vice versa. It would not be possible to generate images continuously.

As a consequence we use the specialized rendering transputers, which generate images continuously in parallel to the radiosity computation performed by the workers. Although the radiosity of patches may change during rendering for this reason, all renderers must receive the same radiosity values, otherwise the color of patches could change at the sector borders. For this reason the rendering manager fetches the required patch data from the worker network only once during the generation of one image and broadcasts them to all renderers.

However, it is still possible that some patches are already influenced by a shooting patch when they are mapped on the z-buffer, while other patches mapped on earlier are not, but such errors are not striking to the user.

In the actual implementation the rendering transputers are configured in a pipeline. Each renderer takes a copy of the scene data received from its predecessor and forwards them to the next transputer. The rgb values of all pixels are finally run-length encoded and sent to the host via the master transputer.

6 Improvements of the Fundamental Concept

6.1 Ambient Light

The ambient term described in [Cohen et al. 1988] gives almost immediately a good a-priori estimate of the final radiosity values, based on the unshot radiosities ΔB_i of all patches:

$$\text{Ambient} := \frac{R}{\sum_{i=1}^{n} A_i} * \sum_{i=1}^{n} (\Delta B_i A_i)$$

The overall interreflection factor R and the area sum $\sum A_i$ are precalculated and stored at the controller process. Each worker determines the sum $\sum \Delta B_i A_i$ of its local patches together with the locally best shooting patch candidate and sends the result to the controller as well, where it is stored in a second list containing the local sum $\Sigma\Delta B_i A_i$ of each worker. Based on this list the controller calculates the total sum and the ambient term, which is sent back to the worker together with the shooting patch information. For display purposes the workers compute improved radiosity estimates by adding the product ambient term by diffuse reflectivity to the radiosity of all their local patches:

$$B_i^* = B_i + \rho_i * \text{Ambient}$$

(The improved radiosity values cannot be computed by the controller process, which forwards them to the renderers, because the diffuse reflectivities are stored at the workers only.)

6.2 Simultaneous Processing of Several Shooting Patches on a Single Worker

If there is enough local memory available, several shooting patches can be processed on a single worker simultaneously. The blocks of patches mapped onto each hemi-cube are not fetched for each shooting patch separately but only once for several shooting patches. Thus the communication overhead for computing several shooting patches simultaneously remains the same as for one shooting patch per worker.

6.3 Storing Patches of Distant Workers in the Local Memory

If the part of the scene description assigned to one worker does not require all the available memory, blocks of other workers can be copied and stored locally. A static selection of blocks can be made based on the network topology by copying blocks from most distant workers. An adaptive strategy is to store copies of blocks which took the longest time to fetch into local memory.

This copy of blocks can be used by the worker itself and by adjacent workers instead of the original blocks for form-factor computation only, because the radiosity value of any patch must be still maintained by a certain worker to preserve data integrity. Therefore the copies of blocks just contain the geometry of patches.

7 Results

For speed-up measurements the system was tested using 1, 2, 4, 6 and 8 workers. Form-factors were calculated using hemi-cubes with resolution 64x64x32. From the improvements described above only the ambient light calculation was used. The speed-up for four test scenes of different complexity is given in fig. 3. For 8 workers it is between 5.3 (for the most complex scene) and 7.8 (for the simple scene). More complex scenes cannot be stored at a single transputer. Therefore their speed-up cannot be compared to the single worker configuration.

The low resolution 64x64x32 was chosen for two reasons: for higher resolutions the memory requirement for the hemi-cube would reduce the size of memory available for storing the patches, and it would not be possible to use a single worker configuration for the most complex scene. On the other hand, the speed-up given in fig. 3 are worst case values, because the computational expenditure for generating the hemi-cube is small, while the communication overhead remains constant for all resolutions.

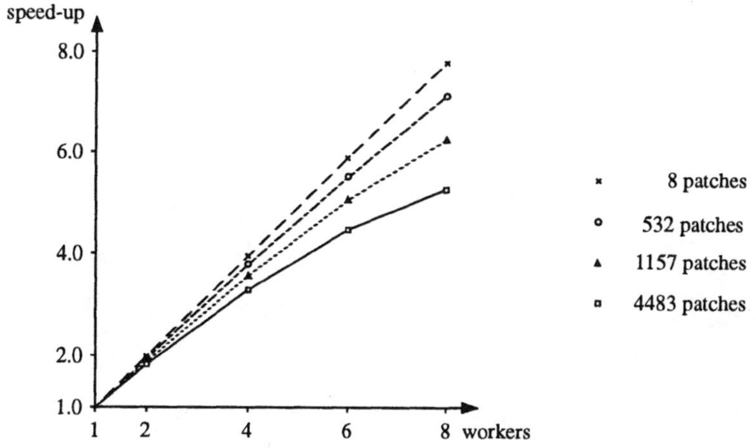

Fig. 3: Speed-up of 4 test scenes for hemi-cube resolution 64x64x32

Of course, there is a lot of aliasing problems with such a low resolution. Especially for very bright patches an increased resolution is required. A lower resolution than 64x64x32 should never be used. Fig. 4 shows the average time to compute one shooting patch using several resolutions on the 8-workers network.

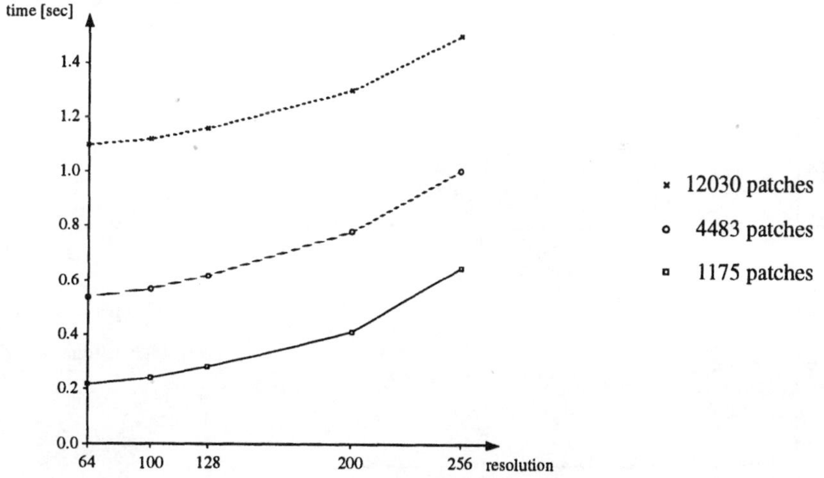

Fig. 4: Effect of hemi-cube resolution on the average computation time for one shooting patch

If the resolution is chosen in a way that there is enough memory available for several hemi-cubes, this memory should be used for simultaneous processing of shooting patches on one worker. Fig. 5 shows the speed-up of this strategy for the 8-workers network at a resolution of 128x128x64. In this case a maximum number of four hemi-cubes can be calculated.

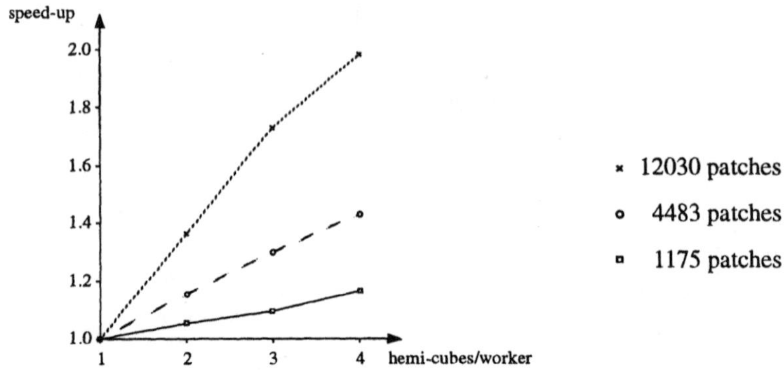

Fig. 5: Speed-up of simultaneous processing of shooting patches

This strategy works best if there are enough good candidates with almost equal unshot radiosity, which can be chosen as shooting patches. Patches with an unshot radiosity much less than other candidates should not be computed together with them simultaneously.

An important parameter for the speed of the radiosity computation is the size of the exchanged patch blocks. If a block contains few patches, many blocks must be exchanged, and the communication overhead increases. Furthermore it takes more time to fetch one block than to map the previous one on the hemi-cube, and the workers stay idle meanwhile. On the other hand, if the blocks contain many patches, a lot of memory must be used for buffering. Fig. 6 shows the impact of the blocksize on the computation time.

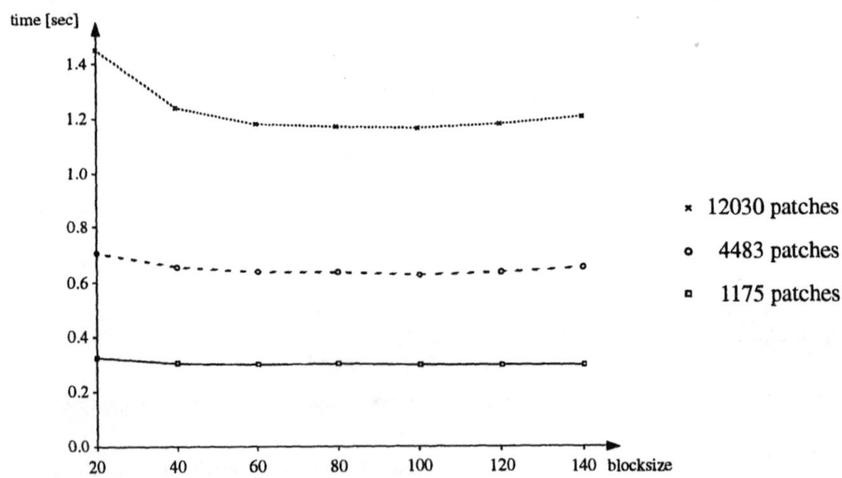

Fig. 6: Impact of the size of patch blocks on the average computation time of one shooting patch

In the current implementation the rendering transputers are configured in a pipeline. Up to 8 renderers were used so far. The speed-up of parallel rendering is given in fig. 7.

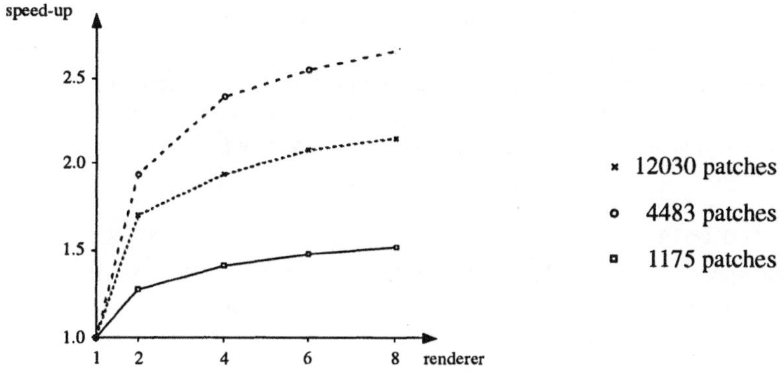

Fig. 7: Rendering speed-up

The rendering speed-up is quite poor. It takes a lot of time to forward the patch blocks down the pipeline to the last renderer and to get its results back. To reduce the length of the communication path, we will use a ternary tree configuration for the renderers instead of a pipeline.

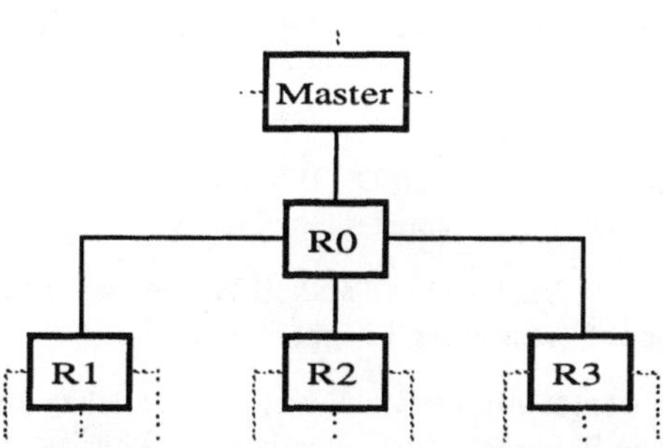

Fig. 8: Ternary tree configuration of renderers

Furthermore, the time needed by the renderers to generate their image sectors differs significantly, since some sectors contain only few patches, while others contain many. The time needed to generate the complete image is given by the slowest renderer. Thus much computation power is wasted. A better utilization of the renderers' processing power could be achieved by a segmentation of the image into sectors of non-uniform size (e.g. small sectors in the middle of the image, large sectors at the borders, because usually most patches are located in the center of the image), or by calculating line i of the image by renderer (i mod m), where m is the number of renderers.

8 Conclusion

An initial implementation of the progressive refinement algorithm is implemented on a transputer network with 8 workers and 8 renderers. All workers run asynchronously, thus almost eliminating processor idle time. Images are generated by renderer transputers continously within short intervals.

Future work includes implementation of alternatives to the hemi-cube, implementation of adaptive subdivision, improvements for rendering, and expansion of the network to 30 transputers. Interactive modifications of the scene are another important goal.

Acknowledgements

This work was sponsored by the *Jubiläumsfonds* of the Austrian National Bank. The transputer network was provided by the *Fonds zur Förderung der Wissenschaftlichen Forschung*.

References

Baum, D., Rushmeier, H., Winget, J. (1989), Improving Radiosity Solutions Through the Use of Analytically Determined Form-Factors, Computer Graphics, Vol. 23, No. 3, 1989

Baum, D., Winget, J. (1990), Real Time Radiosity Through Parallel Processing and Hardware Acceleration. Computer Graphics, Vol. 24, No. 2.

Chalmers, A., Paddon, D. (1989), Implementing a Radiosity Method Using a Parallel Adaptive System. Proceedings of the First International Conference on Applications of Transputers, Liverpool, UK.

Cohen, M., Chen, S., Wallace, J., Greenber, D. (1988), A Progressive Refinement Approach for Fast Radiosity Image Generation. Computer Graphics, Vol. 22, No. 4.

Price, M., Truman, G. (1989). Radiosity in Parallel. Proceedings of the First International Conference on Applications of Transputers, Liverpool, UK.

Purgathofer, W., Zeiller, M. (1990), Fast Radiosity by Parallelization. Proceedings of the Eurographics Workshop on Photosimulation, Realism and Physics in Computer Graphics, Rennes, France, 1990. In: Bouatouch, K., Bouville, C. (eds), Photorealism in Computer Graphics, Springer Verlag, 1992, p. 171-181.

Recker, R., George, D., Greenberg, D. (1990), Acceleration Techniques for Progressive Refinement Radiosity. Computer Graphics, Vol. 24, No. 2, 1990

Wallace, J., Elmquist, K., Haines, E. (1989), A Ray Tracing Algorithm For Progressive Radiosity. Computer Graphics, Vol. 23, No. 3, 1989

Parallel Processing of Progressive Refinement Radiosity Methods

Alan G. Chalmers, Derek J. Paddon

Abstract

Radiosity methods for accurately modelling the interaction of light within complex environments require very large amounts of computing and memory resources. Parallel processing offers one approach for solving these methods in acceptable times. To reduce the communication overheads inherent in the radiosity methods and thus improve the overall system performance of the parallel solution the processors are arranged in minimum path (AMP) configurations. AMP configurations are irregular configurations in which the distance between any two processors is minimised. This paper examines a parallel solution of the progressive refinement radiosity method on AMP configurations.

1 Introduction

Radiosity methods model the interaction of light between diffusely reflecting surfaces and can accurately predict the global illumination effects [12]. The traditional "gather" method of radiosity is a two stage process. In the first stage the hemi-cube algorithm of Cohen [7] may be used to produce a system of equations defining the interrelationships between the discrete patches in the environment. This hemi-cube algorithm can also account for occluded surfaces. In the second stage these simultaneous equations are solved for the light leaving each surface. This solution is view independent and the system of equations only has to be recalculated if the geometry of the environment is changed. A disadvantage of this gather method is that a solution is only available after all the simultaneous equations have been setup and solved.

The progressive refinement algorithm allows an approximate solution to be initially rendered. This solution then progresses continuously towards the complete radiosity solution [6]. With this "shoot" method of radiosity, the user is able to get an idea of the image and therefore manipulate the geometry without first having to wait for the full solution. In this method the incremental effect on all surfaces of all light emitted from each surface is calculated. The hemi-cube algorithm is used to calculate the form factors of the patch that is currently being used as the patch contributing to the radiosities of all other patches. Baum *et al* [1] and Wallace *et al* [15] proposed some modifications to the calculations of these form factors to improve the accuracy of this progressive refinement method.

Radiosity methods require large amounts of computing and memory resources. Despite the global dependencies that these methods exhibit, both methods may be solved on relatively large systems of parallel processors [5].

This paper examines an implementation of the progressive refinement radiosity method on a number of distributed memory processors connected in different configurations.

2 The Progressive Refinement Radiosity Method

The traditional radiosity method, or *gather* method, adapted from thermal engineering requires a set of simultaneous equations to be set up and then solved for the desired radiosities for each of the colour bands under consideration. Only after these radiosities have been obtained can the resultant image be rendered [7, 12]. For complex environments the time required to produce a solution sequentially can be significant, see for example [8]. The user is unable to see the resultant image until the entire computation is complete. If an alteration is made, the user once again has to wait until the full solution has been recomputed. The *progressive refinement* approach to radiosity put forward by Cohen *et al* in 1988 [6], attempts to compromise between the requirements of interactivity and image quality. Using the ideas of adaptive refinement of images [2, 14], the progressive refinement method of radiosity provides the user as soon as possible with an approximation of the full solution. The user is then constantly supplied with improved approximations until the full solution to the desired tolerance is achieved. The radiosity method is well suited to this adaptive refinement approach as, for diffuse surfaces, the solution of the light interactions are independent of the viewing position selected by the user and so the refinement process may proceed uninterrupted even though the user selects a different view point. An additional advantage is that the original radiosity method required order(n^2) storage for n discrete surface patches, whereas, although some recalculation may be required, the storage requirement of the progressive refinement method is order(n). This progressive refinement method is also known as the *shoot* radiosity method [5, 6].

2.1 Formulation of the method

The formulation of progressive refinement method is based on the reciprocity equation:

$$A_i F_{ij} = A_j F_{ji} \tag{1}$$

where A_i is the area of a finite surface i and F_{ij} is the form factor from surface j to surface i, that is, the fraction of radiant energy leaving surface i that arrives at surface j.

Instead of evaluating the effect that all other radiosities have on a particular patch, this method examines the effect the radiosity of that patch has on all other patches in the environment. So instead of *gathering* all the radiosities impinging on a patch, the radiosity given out by that patch is *shot* to all other patches. The difference in approach is shown in figure 1.

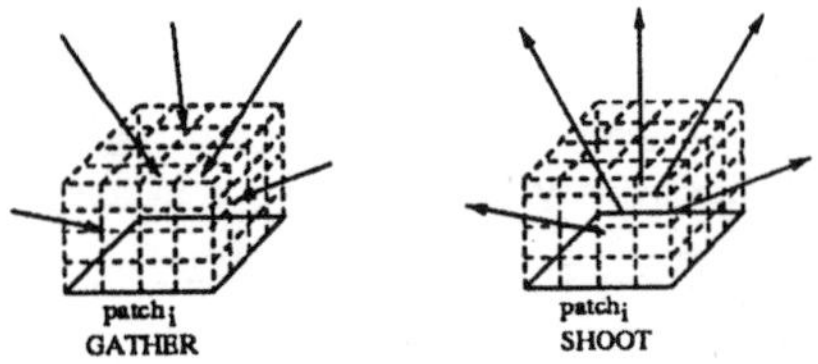

Fig 1. Different approaches adopted by the *gather* and *shoot* radiosity methods

Each step of the progressive refinement method consists of performing the hemi-cube method at a patch and then adding the contribution from that patch to the radiosities of all other patches [6]. The contribution of the radiosity from a patch i to another patch j is $\rho_j b_i F_{ij} \frac{A_i}{A_j}$, where ρ_j is the reflectivity of patch j and b_i is the radiosity of patch i. The form factor, F_{ij}, is calculated by a hemi-cube positioned at patch i.

This process is iterative with a patch possibly being considered several times. Each time the radiosity estimate of the patch will be more accurate. When a patch is revisited, the environment will already have received the contribution from the previous estimate of that patch's radiosity and so only the difference between the previous estimate and the new estimate must be "shot" to the environment. The process is shown algorithmically in figure 2 where Δb_i is the change in radiosity from the previous estimate of the radiosity of patch i to the new estimate.

The radiosities of all patches in the environment are incrementally increased with each step of this iterative process. Obviously the amount of change between the old and new estimates of a patch's radiosity will decrease as the method progress. When this change is below a certain specified tolerance for all patches in the environment then the desired full image is deemed to have been achieved.

```
(*initially b_i = Δ b_i = 0 for non light sources*)
(*initially b_i = Δ b_i = e_i for light sources*)
select patch i
WHILE Δ b_i > tolerance DO
  calculate form factors at patch i using the hemi-cube method
  FOR j := 1 TO all other patches (except i) DO (*include i if concave*)
    Δ Rad := (ρ_j Δ b_i F_ij A_i) ÷ A_j
    (*update change since last patch j ''shot'' light*)
    Δ b_j := Δ b_j + Δ Rad
    (*update total radiosity of patch j*)
    b_j : = b_j + Δ Rad
  END
  (*patch i has just ''shot'' - reset unshot radiosity to 0*)
  Δ b_i := 0
  select next patch i
END
```

Fig 2. Algorithm for the Progressive Refinement Radiosity Method

The desired solution for a patch's radiosity will be approached quickest if the largest contribution to that radiosity is added first. This largest contribution will come from those patches currently exhibiting the largest ΔbA_i values. Initially, these largest contributions will come from light sources and then subsequently from those patches that initially receive the most light from the light sources. This preferred order of selection of patches has been shown by Cohen *et al* to approach the desired solution more quickly and with less computation cost than an arbitrary patch selection schema [6].

3 A Parallel Solution

For complex environments, the calculations of the form factors for each patch in the environment can require a large amount of computing and memory resources, and take significant times to produce a full solution. Fortunately, it is possible to compute certain aspects of the radiosity methods in parallel. For example, the calculation of the form factors of a patch using the hemi-cube method, is independent of a similar calculation at another patch in the environment, and thus these computations may proceed in parallel.

Our parallel solution of the progressive refinement method commences with the initial distribution of the data describing the patches in the environment, amongst the processors according to a predetermined schema and up to each processor's memory capacity. Each processor in the system is responsible for a similar number of patches. The processing now continues with each processor determining from its set of patches the patch with the largest delta radiosity. The form factors for this patch are calculated locally from the delta form factors of the the hemi-cube method or by analytical determined form factors when necessary [1].

The projection of all patches in the environment onto the hemi-cube positioned at the centre of the receiving patch requires the data of all patches in the environment. For large problems it is unlikely that there will be sufficient storage at each processor to hold all this data, therefore it is inevitable that data will have to be passed between processors. In general, the processing of a patch will involve fetching data from all other processors. As the order in which the projections are undertaken is unimportant, the fetching of data from distant processors may be overlapped with the computation of currently available patch data. However if the time to fetch the remote data is significant then processor idle time may still occur.

When the form factor approximations by means of the delta form factors is sufficient then these precomputed values may be used directly to produce the form factor of the projected patch visible on the hemi-cube. When this approximation is insufficiently accurate, then the analytical form factor between the receiving and the projecting patch must be calculated as shown in [1]. For environments with occluded surfaces, the analytical form factors may not be computed until all patches in the environment have been projected onto the hemi-cube. As each pixel of the hemi-cube only holds the identification of the patch visible for that pixel, to calculate the element areas associated with a pixel, the data for that patch is required. Once again, memory limitations may require the fetching of the data of those patches visible on the hemi-cube.

Once the form factors for a patch have been calculated it may be advisable to store these form factors to avoid future recalculation. A packet consisting of the form factors, the delta radiosity of the patch (for each of the three colours) and the area of the patch is now "shot" to all other processors.

Having received a packet, a processor is able to update its radiosities and therefore its delta radiosities. The following packet is only released when acknowledgements are received that all other processors have received the previous packet, and then only if the largest of the delta radiosities is above a certain tolerance (the results shown will be for a tolerance of 10^{-5}). If the largest delta radiosity is less than this tolerance then the processor informs the system controller. The system controller determines when all the processing elements have achieved local "convergence" using an asynchronous convergence method [9]. The processor will continue to update its radiosities and shoot until the desired solution is achieved, or until the environment is altered by the user and the process must start from the beginning again.

4 System Configuration

An inhibiting factor to the effective solution of the progressive refinement radiosity method on systems of distributed memory MIMD processors is the global data dependencies inherent in the form factor calculations and communication overhead associated the need to shoot the resultant packet containing the form factors to all other processors in the system. The parallel solution of this method may be contemplated on a number of configurations of MIMD processors. However the communication overheads may have a major influence on the maximum number of these processors that can be interconnected in a configuration in order to achieve the maximum performance of the algorithm.

The performance of a system of distributed memory MIMD processors depends in large part on the efficiency of the message passing system that provides the interface between the co-operating processors [10]. The suitability of the underlying processor interconnection topology is an important factor affecting the performance of this message transfer system. In a fully interconnected network every processor is adjacent to every other processor. For a p processor system this requires that each processor has $(p-1)$ physical links available for interconnection. For large values of p this may not be possible, in which case processors will have to communicate with each other not directly but via intermediate processors. For a constant number of links per processor, as the number of processors in the network increases, so to does the number of intermediate processors through which a message must be routed. This increase in maximum distance between any two processors is a function of the underlying interconnection topology. For the class of problems that require every processor to communicate frequently with all other processors it is desirable that the maximum distance between any two processors be small if the system performance is not to be too severely compromised.

The topology usually recommended for dealing with a global data requirement is a ring, see for example [11]. However, in the ring configuration the maximum distance between any two processors, and therefore the maximum message path length, is large and only a simple linear arrangement of processors would give a higher total of messages on the system. Hypercubes have been proposed and widely used for algorithms that exhibit these global dependencies, for example [3]. However, for a fixed number, k, of links per processor, a maximum of 2^k processors can be included in the hypercube with a maximum distance of k between processors. Toroidal configurations are another possibility which provide, for a system of k processors a maximum distance between any two processors of order($\sqrt{k}$). These configurations are shown in figure 3.

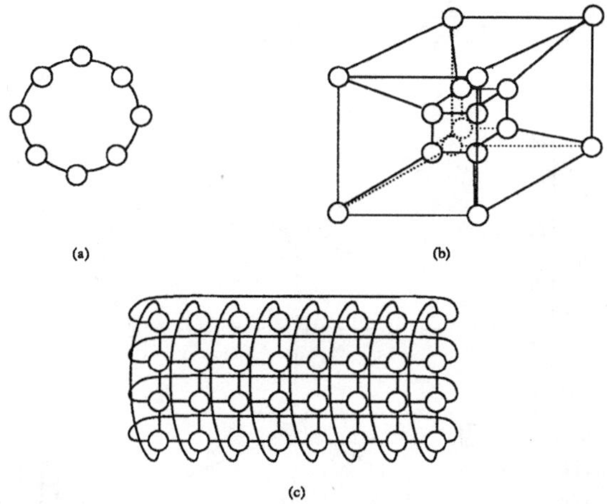

Fig 3. (a) 8 processor ring (b) 16 processor hypercube (c) 32 processor torus

We have adopted minimum path systems for dealing with problems that exhibit global data dependencies. The philosophy underlying the construction of a minimum path (AMP) configuration is to minimise the distance a message has to travel between any source processor and any other destination processor within the configuration. This principle is maintained even at the expense of the loss of symmetry in a system [4].

In order to provide a useful parallel processing platform, systems of distributed memory MIMD processors typically consist of a system controller (SC), which provides the interface between the parallel system and the I/O facilities of the host computer, and one or more processing elements which perform the desired computation.

Table 1 shows a comparison for this maximum distance between any two processors. This table has been derived for processors restricted to four links. Two processors are additionally restricted in the AMP configurations by the need to communicate with the system controller. The additional two links on the system controller are used for I/O to the host machine. Figures 4 and 5 show a 32 processor and 63 processor AMP topology respectively.

No. of processors	4	8	16	32	53	63
Ring	2	4	8	16	26	31
Hypercube	2	3	4	-	-	-
Torus	2	3	4	6	26[a]	8
AMP	1	2	3	3	4	4

[a] As 53 is prime, a 53 processor torus ≡ a 53 processor ring. A 52 processor torus will be used in the results section.

TABLE 1. Maximum distances within configurations with 4 links

The average distances between any two processors for each of the configurations is shown in table 2.

5 Results

The progressive refinement radiosity method was implemented (in occam) and run on ring, torus and AMP configurations ranging from 1 to 63 processors. A complex environment

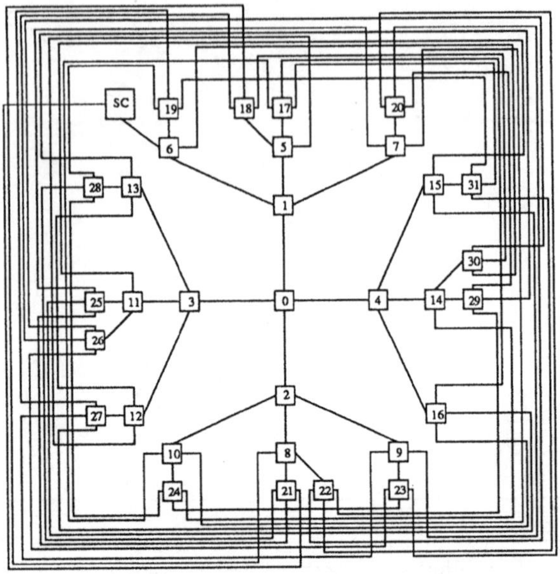

Fig 4. 32 processor AMP

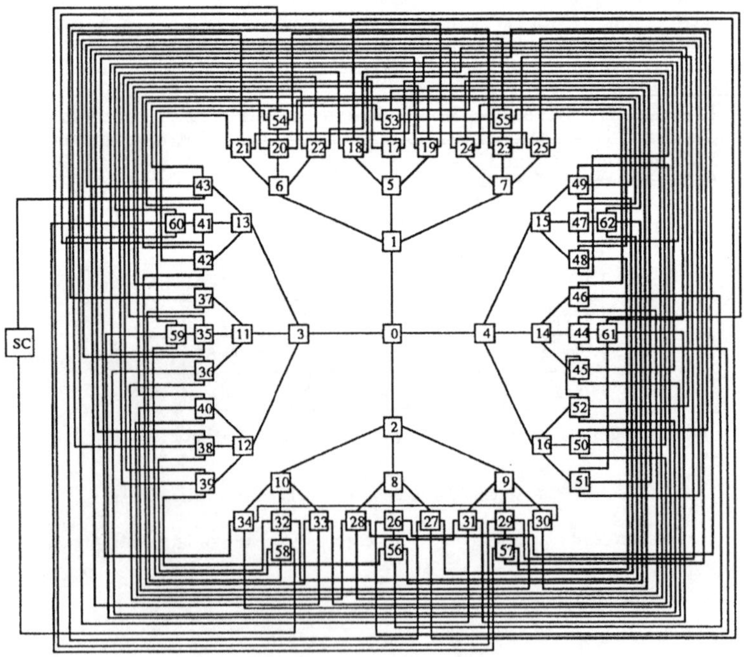

Fig 5. 63 processor AMP

No. of processors	4	8	16	32	53	63
Ring	1.0	2.0	4.0	8.0	13.25	15.75
Hypercube	1.0	1.5	2.0	-	-	-
Torus	1.0	1.5	2.02	3.0	13.25	3.94
AMP	0.75	1.28	1.74	2.31	2.77	2.92

TABLE 2. Average distances within configurations with 4 links

of 448 patches was selected as a test case as this was the largest number of patches that could be solved on a single processor without requiring further access to secondary storage. Results were also obtained for environments of up to 2268 patches. The results shown are for non analytically determined form factor calculations.

In [13] the parallelisation of the shoot method was reported on systems of Hewlett-Packard 835 workstations. The speed-up obtained from this system was low. No analysis is given, therefore, direct comparisons with this implementation are not possible. However, it is immediately evident from the speed-up graph, figure 6, that good speed-up can be achieved from shoot methods on the AMP transputer systems and that this speed-up is greater than that that may be obtained from the same implementation on either ring or torus configurations. In fact, for ring configurations of greater than 45 processors a solution of the problem to the desired tolerance of 10^{-5} was not possible due to message saturation. Figure 7 shows the corresponding time taken to converge to the full solution on each of the three configurations for this problem as well as the time taken to solve the problem using the gather method on a single processor. The time taken for the full convergence on a 42 processor AMP is 181 seconds which compares very favourably with the 2319 seconds required for the single processor.

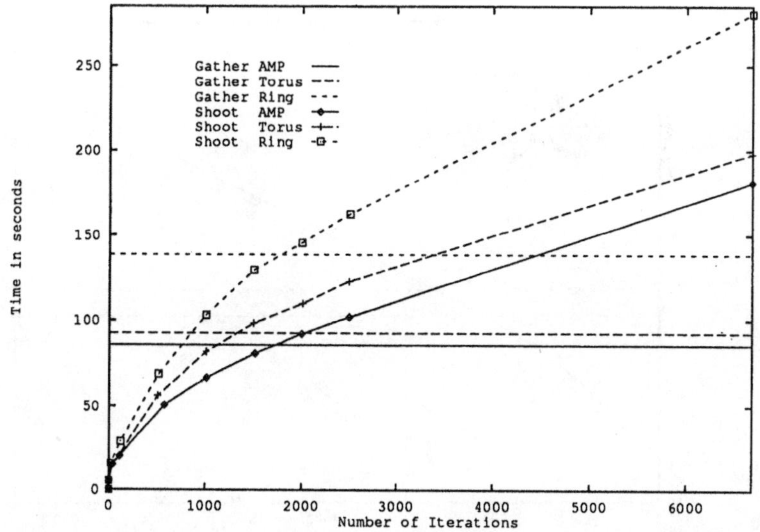

Fig 6. Speed-Up for a 448 patch problem on Rings, Tori and AMPs

Figure 8 shows the comparison in time to fully solve the 448 patch problem to the desired tolerance of 10^{-5} on a 63 processor AMP compared to the time taken to solve the same problem to the same tolerance use the gather radiosity method on 63 processor configurations arranged in an AMP, torus and ring. Significantly the time required to

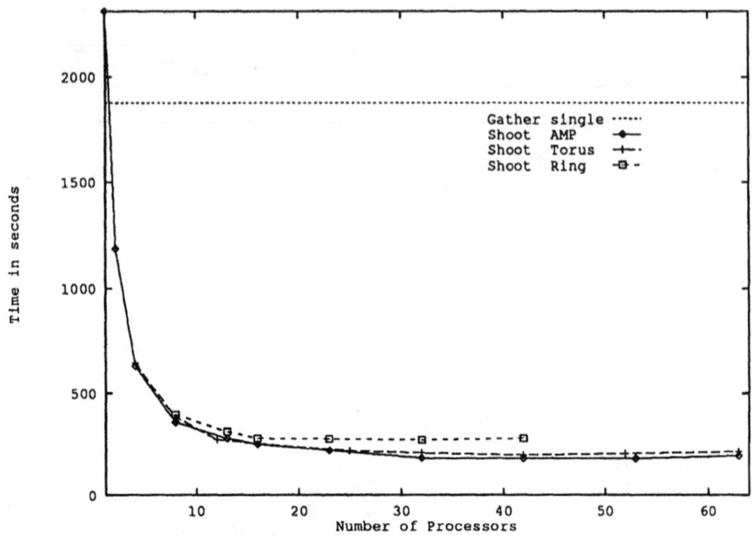

Fig 7. Time to converge for 448 patch problem

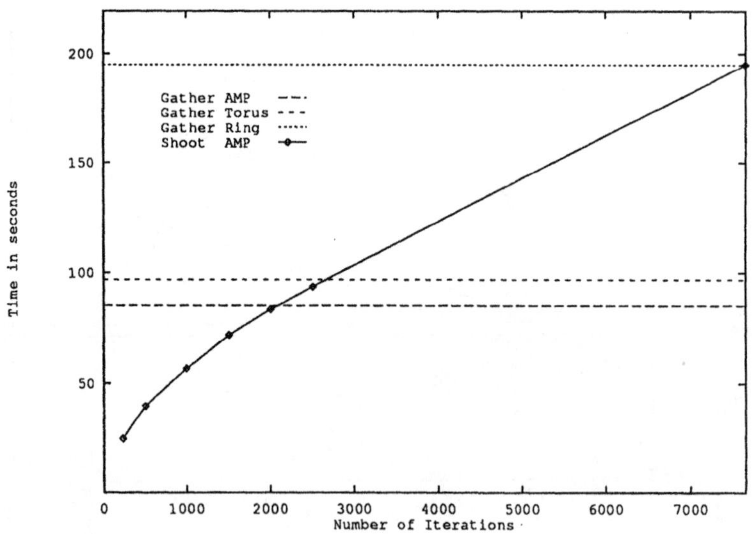

Fig 8. Time comparison for a 448 patch problem on 63 processors

perform 1000 iterations of the shoot radiosity method on a 63 processor AMP system only takes approximately 56 seconds while on a single processor 1000 iterations took 2022 seconds.

Figure 9 shows the same comparison for a environment of 2268 patches. The full solution of the shoot method required 37986 iterations and took 3322.047 seconds. For this problem approximately 5000 iterations of the shoot method are possible in the time required for the full solution using the gather method. On the other hand, 200 iterations are completed in roughly 20 seconds. Combined with Cohen *et al*'s suggestion of solving for the patches making the largest initial contribution to the environment, a good approximation to this image can thus be available to the user in seconds rather than approximately 10 minutes required before the image can be produced using the gather method. However, the full solution using the shoot method is only available long after it would have been achieved using the gather method.

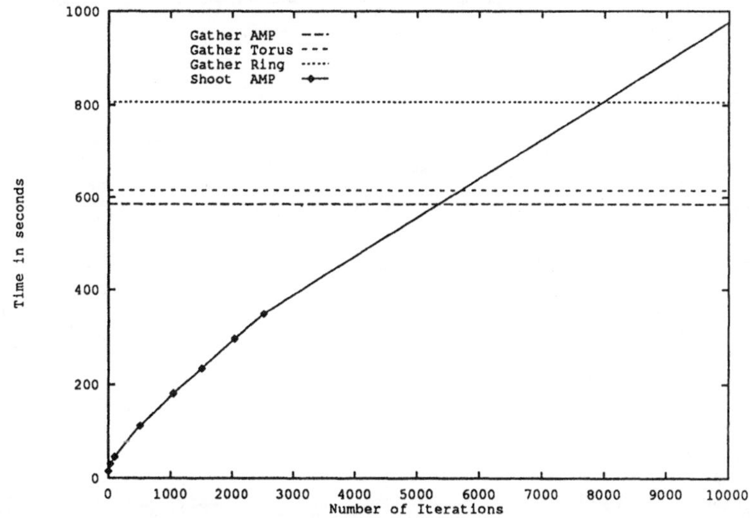

Fig 9. Time comparison for a 2268 patch problem on 63 processors

Finally figure 10 shows the time to solve the 448 patch problem on 42 processors arranged in an AMP, ring and torus, compared with the time required for the solution using the gather method for the same problem. As can be seen, the AMP configurations perform better than either the tori or rings in both the gather and shoot radiosity methods.

The introduction of the more computationally complex analytically determined form factors has implications for the computation to communication ratio of the parallel solution of this method. It requires approximately 20 seconds to solve a single iteration with the analytical computed form factors compared to the 4 seconds using only the delta form factors. On a 32 processor AMP 30 iterations have been timed at taking 38 seconds compared to the 11 required for the delta form factor solution. Less iterations will therefore be possible using this more accurate shoot radiosity method in the time that an accurate solution may be obtained using the traditional gather radiosity method.

6 Conclusions

Despite the inherent global dependencies of both the gather and shoot radiosity methods, these methods may be effectively solved on systems of distributed memory processors. The use of parallel processing will thus allow radiosity methods to be applied to the most complex environment and yet still produce results in acceptable times.

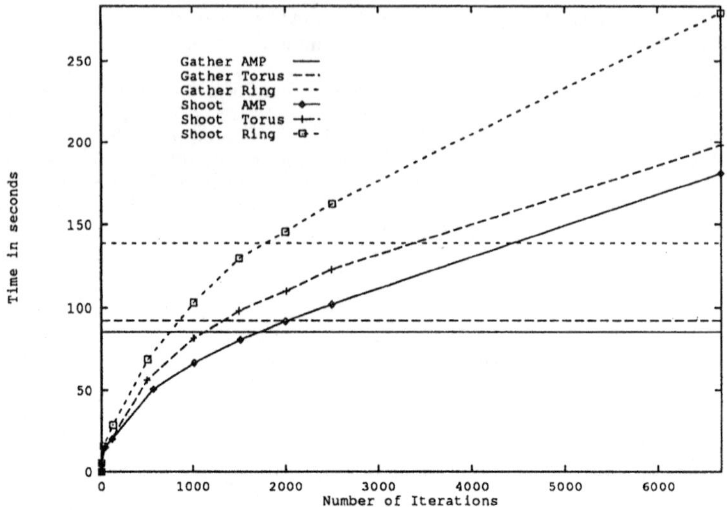

Fig 10. Time comparison for a 448 patch problem on 42 processors

The reduced interprocessor distance and consequential decrease in the message density prevelant in the system allows the AMP configurations to constantly perform better than either the torus or ring configurations. In fact in a number of instances, solutions were possible on AMPs when they were not possible using either the tori or rings due to message saturation.

In general, the solution using the analytically calculated form factors is about four times as expensive as the solution derived directly from the delta form factors. The hybrid method of Baum *et al* [1] appears to offer a good compromise between the accuracy and the cost of the solution. However, whichever method is adopted, the structure of the computation is inherently less scalable on distributed memory parallel processors than the gather method [5]. The progressive refinement method suffers from message congestion on even medium size configurations. These undesirable message densities are further exacerbated for analytically calculated form factors by the need to refetch patch data after the hemi-cube projections have been completed. The numerical characteristics of the progressive refinement method are similar to those of the gather method when a Jacobi iterative scheme is adopted for solving the system of radiosity equations. The iterative scheme in the parallel solution of the gather method can of course be accelerated by using direct solvers such as the conjugant gradient method.

The results reported here indicate that the progressive refinement method is of considerable value if interim results are needed for modelling the requirements of the environment. However, when the full solution is required, it is suggested that the interim solution so far obtained should now be used as the initial solution in the gather method, and therefore the gather method is used to complete the calculation. The form factors so far calculated with the progressive refinement method can, of course, be used within the gather method.

The gather method has two factors in its favour:

1. with the appropriate choice of parallel matrix solver the convergence rate of the method is higher than can be obtained by the progressive refinement method; and

2. the gather method has been shown to be more scalable to large numbers of processors particularly if the matrix solver is chosen to minimise communication overheads [5].

Future work will examine ways in which the time required by the analytical solution can be improved upon, specifically ways in which the possible need to refetch data can be avoided, and will also study communication efficient parallel matrix solvers.

7 Acknowledgements

We would like to thank Dr Roger Miles and the Bristol Transputer Centre for the use of their equipment.

References

[1] D. R. Baum, H. E. Rushmeier, and J. M. Winget. Improving radiosity solutions through the use of analytically determined form factors. *Computer Graphics*, 23(3):325–334, July 1989.

[2] L. Bergman, H. Fuchs, E. Grant, and S. Spach. Image rendering by adaptive refinement. *ACM Computer Graphics*, 20(4):29–38, Aug. 1986.

[3] C. J. Catherasoo. Separated flow simulations using the vortex method on a hypercube. In *AIAA 8th Computational Fluid Dynamics Conference*, Honolulu, 1987.

[4] A. G. Chalmers and D. J. Paddon. Communication efficient MIMD configurations. In *4th SIAM Conference on Parallel Processing for Scientific Computing*, Chicago, 1989.

[5] A. G. Chalmers and D. J. Paddon. Parallel radiosity methods. In D. L. Fielding, editor, *4th North American Transputer Users Group*, pages 183–193, Ithaca, NY, Oct. 1990. IOS Press.

[6] M. F. Cohen, S. E. Chen, J. R. Wallace, and D. P. Greenberg. A progressive refinement approach to fast radiosity image generation. *ACM Computer Graphics*, 22(4):75–84, Aug. 1988.

[7] M. F. Cohen and D. P. Greenberg. The hemi-cube: A radiosity solution for complex environments. *ACM Computer Graphics*, 19(3):31–40, July 1985.

[8] M. F. Cohen, D. P. Greenberg, D. S. Immel, and P. J. Brock. An efficient radiosity approach for realistic image synthesis. *IEEE Computer Graphics and Applications*, 6(3):26–35, Mar. 1986.

[9] I. S. Duff. The influence of vector and parallel processors on numerical analysis. In A. Iserles and M. J. D. Powell, editors, *The State of the Art in Numerical Analysis*, Oxford, 1987. Clarendon Press.

[10] T. Feng. A survey of interconnection networks. *IEEE Computer*, pages 12–27, Dec. 1981.

[11] G. Fox et al. *Solving problems on concurrent processors - Volume 1*, volume 1. Prentice Hall International, 1988.

[12] C. M. Goral, K. E. Torrance, D. P. Greenberg, and B. Battaile. Modelling the interaction of light between diffuse surfaces. *ACM Computer Graphics*, 18(3):213–222, July 1984.

[13] R. J. Recker, D. W.George, and D. P. Greenberg. Acceleration techniques for progressive refinement radiosity. *ACM Computer Graphics*, 24(2):59–66, Mar. 1990.

[14] K. R. Sloan and S. L. Tanimoto. Progressive refinement of raster images. *IEEE Transactions on Computers*, 28(11):871–874, Nov. 1979.

[15] J. R. Wallace, K. A. Elmquist, and E. A. Haines. A ray tracing algorithm for progressive refinement. *ACM Computer Graphics*, 23(3):315–324, July 1989.

Two Parallel Approaches for a Progressive Radiosity

Pascal Guitton, Jean Roman and Christophe Schlick

1 Introduction

In recent years, great progress in creating realistic synthetic images has led to increased interest in the interactive production of such images. For example, they are used in architecture or in town planning to simulate a new building and to estimate the consequences of these creations. CAD is also a customer for a realistic representation of the internal aspects of certain elements (such as engines) to detect any conceptual anomalies. The most important restriction for more intensive use of these techniques has been the very large image computation time and the memory cost.

Methods based on Ray Tracing (RT), which was first presented in 1980 ([WHIT80]), have led to a significant increase in realism. Unfortunately, RT deals, in a satisfying way, only with specular properties of scene elements (like mirrors); moreover, any modification of the observer's position necessitates a whole new image computation.

In 1984, the Radiosity concept was first presented ([GORA84]); it is based on the energy conservation principle and gives solutions to these two problems. But it requires lengthly execution time to build and to solve an equational system, and it deals only with diffuse light properties. This first drawback was reduced by the apparition of the Progressive Radiosity (PR) in 1988 ([COHE88]). Progressive radiosity treats, by a successive way, each patch beginning by the most energetic ones in order to obtain, as soon as possible, a preliminary image. Then, the treatment of the other patches provides refinements of the first image, up to the final image.

The mixed use of RT and PR, called ray tracing based progressive radiosity (RT based PR), allows us to deal with both specular and diffuse light properties. There are two ways to realize a RT based PR: first, stochastic methods performing a Monte Carlo solution to solve the rendering equation ([MALL89], [SHIR90]) and second, deterministic methods performing an analytical solution ([WALL89], [LESA90]).

These extensions satisfy our requirement expression: realism of computed images and interactivity of their production. Unfortunately, this interactivity remains potential with most sequential computers; we think that parallel architectures give interesting responses for a better exploitation of this potential. Moreover, these methods present two major drawbacks: lengthy computation time and the manipulation of large data collections. Consequently, we are studying parallel solutions which provide time savings and data repartition, and thus allow the treatment of more complex scenes.

After Ray-Tracing domain ([BAD90], [PRIO89]), parrallelism has met Radiosity; so, several parallel solutions to Radiosity have already been proposed, which can be grouped in three different families, in respect with the chosen architecture:

- first, networks of specialized graphic workstations (DEC 3100, HP 835/TSRX) linked by a local area network (Ethernet); the global program is decomposed into master-slaves modules

which communicate using shared memory and the whole scene description is duplicated on every workstation. The main drawback of these solutions is the bottleneck which results from the fact that the data are sent by only one module (master); so the efficacy decreases when the slave number increases ([PUEC90], [RECK90]).
- Second, workstations based on a multi-processor architecture (Silicon Graphics Iris 4D280) with shared memory. The decomposition is also based on a master-slaves principle and the scheduling is performed by the operating system ([BAUM90]).
- Finally, some well-suited architectures for parallelism (transputer networks) for which each processor has a local memory and communicates by message exchanges ([CHAL90], [PURG90]).

Our method belongs to the third family and will be implemented on a transputer based T.Node machine. Unlike the other methods which are based on the calculation of form-factors by the hemicube technique, our algorithm uses ray-tracing both for the computation of the rendering equation, and for the visualization of the scene according to the position of the observer.

In this paper, we detail the basic ideas of this method currently under implementation at the LaBRI. In the second paragraph, we describe the main characteristics (action and data organization) of our method from a general point of view. Then, in the third (resp. fourth) paragraph, we present the parallelization of the stochastic (resp. deterministic) method. Finally, we propose, in the fifth paragraph, some implementation characteristics like termination test or load balancing.

2 Global Description

2.1 Data Structures

The Radiosity method is based on an energy balancing of the visible spectrum for the considered scene. To perform the computation, the elements of the scene are decomposed into a set of elementary surfaces named patch(es) and the energy balance between these patches is expressed.

Then, the interactions between the patches are determined by computing the form-factors (F_{ij} is the ratio of energy sent by the patch P_i which is received by the patch P_j), and finally, the linear system obtained from all the balanced equations written for each patch is solved.

The drawback of a model based on patches lies in the fact that these patches are used for sampling both the geometrical and the optical characteristics. Now, an ideal sampling for these two aspects may be very different; for example, a wall in a room needs only one patch to model the geometry, but it needs several dozens of patchs in order to take into account all the subtle shading variations.

Hence, the use of a unique sampling for these two aspects consists either in a compromise between the ideal optical and geometrical samplings (inducing a loss of precision in the computations), or in the choice of the sampling with the maximal rate (inducing an increasing for the memory and computation costs).

On the other hand, the choice of two fully independent samplings does not work better. As a matter of fact, during the computation we often need to find the optical sample corresponding to a given point of a geometrical sample, and conversely. For example, the computation of the contribution to a surface of the energy carried by a ray requires the following informations: the geometrical facet intersected by the ray, the characteristics of the intersection point (position, normal vector, ...), and lastly the optical facet containing this point in order to compute the energetic contribution of the ray. All these steps can be very expensive when the two samplings are completely independent.

Therefore, we define a new model based on tiles. Tile modeling has the following characteristics: the geometrical and optical samplings are independent, but they overlap one another in some particular points of the scene.

A tile is defined as a part, not necessarily planar, of an element of the scene. If this element is defined using facets, the tile is an agglomerate of adjacent facets; in the case where the element has a parametric definition f(u, v), the tile is the restriction of f to a domain $[U_0, U_1] \times [U_0, U_1]$. On this tile, we put two samplings (geometrical and optical) more or less fine according to the characteristics of the element (cf. figure 1).

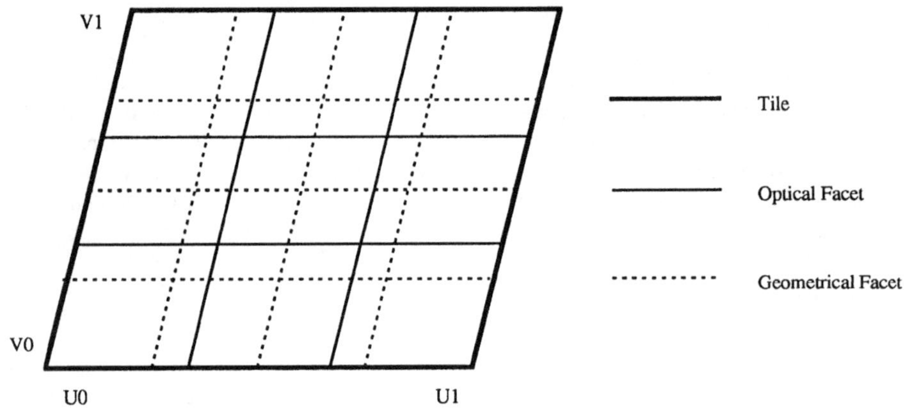

Fig. 1: Tile description

As the two samplings have in common at least the four vertices of the tile, we can easily compute, by a simple linear relation, the two geometrical and optical samples containing a given point M(u,v) (cf. Appendix). So this model allows the use of independent samplings having each its own rate, but allows also an efficient transition from one to the other. Hence it fullfills the desired conditions described above.

In the next sections, the word tile will be used differently according to the context (geometry or optic). When we consider the intersection between a ray and a tile, it means the intersection with one of the geometrical facets of this tile, and when we add a contribution to a tile, it means the addition of this energetic contribution to a facet of this tile. Lastly, the energy of a tile will represent the summation of the energies of the whole set of samples contained in this tile.

2.2 Action Organization

The method is based on the emission of rays from a tile T, the detection of intersected tiles T', and then, the computation of light interaction between T and T'. This work is done for each tile T of the scene, beginning with the most energetic.

More precisely, during the first step, the method distributes the data set (geometrical and optical tile descriptions) on the process set. In order to obtain this distribution, the scene is decomposed into voxels to optimize the intersection computations which are subsequently performed ([FUJI86]). Then, the voxel set is cut out in slices and each slice is associated with a process (cf. figure 2); all the data contained in the slice are stored on it.

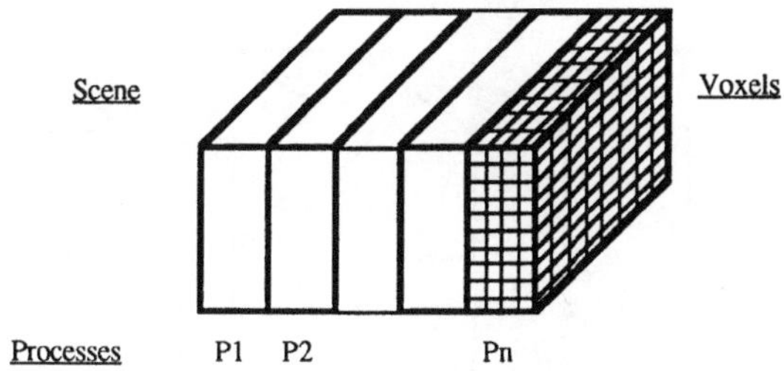

Fig. 2: Scene decomposition

If a tile T is extended on several voxel slices, and consequently on several processes, its geometrical properties are duplicated on these processes. In such a case, one process is chosen to centralize the optical properties of T; it is called the proprietor process. Therefore, the intersection computation can be performed locally, whereas the light contribution computation needs message exchanges.

Then, we propose two methods to achieve ray casting: a stochastic and a deterministic one. In the stochastic method, rays are cast in randomly chosen directions, whereas in the deterministic one, rays are cast towards each interesting patch.

3 Parallelization of a Stochastic Method

3.1 Description of Principles

After the data distribution between the processes, we loop over the tiles of the scene, selecting the most illuminating one, T, at each iteration. Then, significant interactions between T and the other tiles are computed by casting a set of rays, which are randomly selected, so as to obtain a sampling of the visible directions from the T tile. For each selected tile T, it is necessary:

- to decompose the visible direction hemisphere into N samples;
- to cast a ray R from T tile towards each sample;
- to search for the potential intersection between R and a tile T';
- if an intersection is detected, to compute the contribution of T to T' and to update the energy of T';
- to iterate these steps by choosing another tile with maximal energy;
- to stop when this maximal energy is less than a fixed threshold value.

As the processes locally know only their scene slice, to trace a ray R from a tile T implies the emission and the receipt of messages containing the properties of R. These messages run from process to process, and are analysed with respect to the data stored locally by the process.

To cast a ray R from an origin tile T, involves, for the process P_i which possesses the T description, to search for a potential intersection between R and the tiles locally stored. If this search fails, P_i sends towards the next process a message asking for an intersection search with its own tiles. This message runs in the network, as long as R does not reach any tile. As soon as a process Pj detects an intersection between R and a tile T', it computes the light interaction between T and T', and then it updates the energy of T' (cf. figure 3).

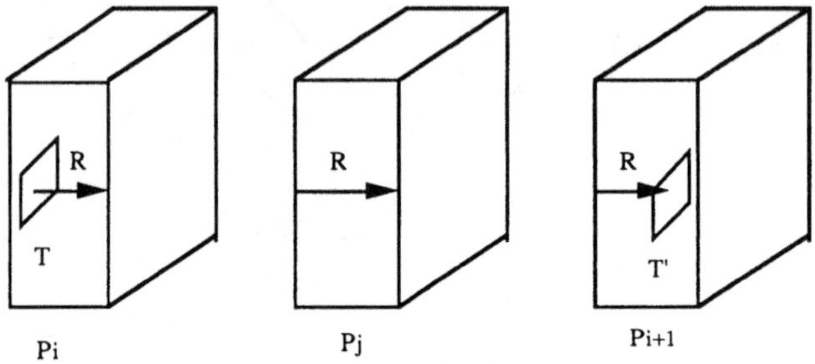

Fig. 3: Ray transfer

The data base distribution leads to a problem which does not exist in a classical sequential solution: the intersection point and the vertex which stores the energy may be on two different processes P_i and P_j. This problem is solved by the emission, from P_i towards P_j, of a message which carries the computed contribution.

The communication required by the method determines the network topology. Due to the decomposition into adjacent slices, a process P_i communicates with only two processes P_{i+1} and P_{i-1} which are associated with the two neighbouring slices. Therefore, each process needs to be linked with only two other processes. A bidirectional bus perfectly meets the requirements of this kind of communication (cf. figure 4). This topology will be modified by the implementation of an algorithmic termination test which needs a particular structure.

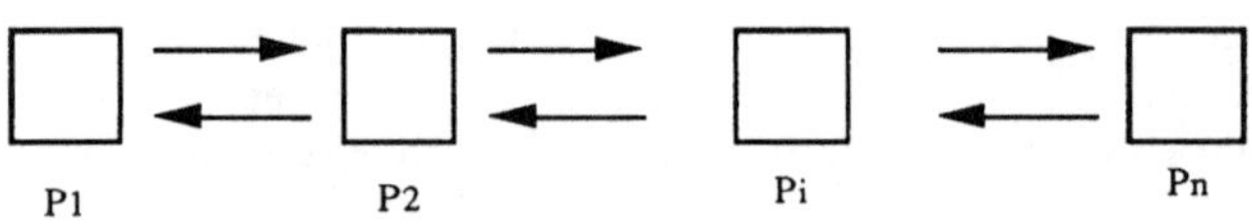

Fig. 4: Bidirectional bus

The concurrent execution, by each process, of these functions carries out the method in parallel. So, at a given time, n tiles are simultaneously treated to compute their interaction with the whole scene.

3.2 Description of Computation

To optimize the treatments, energy contributions which are below a fixed value, called energy threshold Σ, are omitted. So, it is useless to compute such an interaction; the knowledge of the energy E of a tile T allows us to compute N (= E / Σ) which is the number of "useful" rays to cast from T. Moreover, this value allows us, during the choice of a tile T, to stop the global treatment as soon as the energy of the selected tile T is inferior to Σ.

The computations are performed according to these different steps:

- the choice of the most energetic tile: it is the tile which emits/reflects the largest energy quantity (product of the average radiosity B of tile T with the surface S of T);
- the visible direction sampling: as soon as the total energy E (= B * S) leaving T is calculated, N the number of useful ray cast from T is computed. Then, the hemisphere of the visible directions from T is decomposed into N samples so that, for each sample, the projected solid angle is constant ([LESA90]). Afterwards, a ray carrying an initial energy Σ is cast in a direction which is randomly chosen in each sample (stratified sampling);
- the ray contribution calculation: when a ray R reaches a facet, the nearest vertex V is located and the energy of R is added to the current value of the counter associated with V. This energy may be inferior to Σ, if it has suffered some attenuations with respect to the distance, the respective orientations of the sending and receiving tiles ([LESA90]).

4 Parallelization of a Deterministic Method

The second method we present leaves this stochastic approach for an analytical evaluation of the rendering equation. It computes every interaction which is greater than the energy threshold Σ. In this way, this method can be seen as a direct parallelization of the sequential method described in [LESA90].

Let us consider, on P_i, the most energetic tile T with energy greater than Σ; the search for the tiles T' on which T has a contribution greater than Σ is done in two steps:

- the contribution C of T to T' is calculated using the geometrical and optical properties of T and T',
- if C is greater than Σ and if T' is visible from T, then C is added to the energy of T'.

During the first step, the description of an illuminating tile T is transfered through the communication graph to consider every scene tile. After having received from the process P_i the characteristics of the tile T, the process P_j computes the contributions of T on local tiles T'. In fact, as in the previous method, the proprietor process concept is used to solve the problem of tiles which extend over several scene slices; so, P_j calculates only contributions to local tiles T' of which it is proprietor.

The communication graph is a monodirectional ring linking all the processes; when T has run round the ring and has come back to P_i, all its contributions greater than Σ have been treated. This method is repeated while there exist some tiles T with energy greater than Σ.

The second step is performed using a ray tracing from T' towards T. This result determines whether T' is visible from T. This result is carried back from P_i to P_j by a message which contains the amount of energy really transfered between T and T': the contribution C if there is no occlusion, and 0 if there is one. When this message comes back to P_j, the contribution stored in it is added to the energy of T'.

For the tiles T' belonging to P_j, these operations are performed locally without any interaction with the exterior. For all other tiles T', it is necessary to use:

- a bidirectional bus linking the n processes to transfer the rays;
- a bidirectional ring linking the n processes to transmit the visibility flags.

These two structures are bidirectional in order to minimize the length of the inter-process traversal. This optimization is performed using the process numbering deduced from the slice decomposition. As for the first method, there exists a communication logical sub-network for each kind of message; a process contains three concurrent tasks. Each of them deals with the computation and the transmission of one message kind. Without the process structure which performs the termination, the communication graph looks like the following figure:

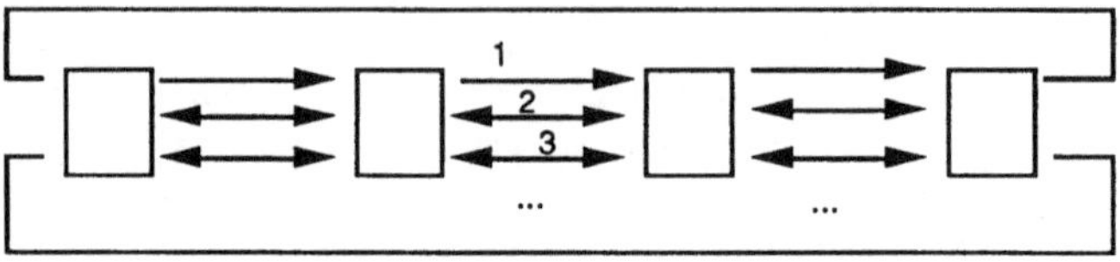

1 monodirectional ring for the most energetic tiles
2 bidirectional bus for rays
3 bidirectional ring for visibility flag

Fig. 5: Network topology

5 Propositions for a Parallel Implementation

5.1 Process Structure

Before we describe the process structure, let us recall the major characteristics of the formal algorithm expression model we use ([BAL89], [COUN90]): the specification is based on a set of processes which does not share any data and which communicates by exchanging messages on communication ports. A process is made of a set of entities called tasks, which are concurrently executed on the processor associated with the process. The tasks can share the local data of the process; in fact, a process is the mother task which activates all other ones. A task is interruptible by the clock or by communication: emission (resp. receipt) request with a full (resp. empty) channel. In the two methods, there exists a logical communication subnetwork for each kind of message; a process is based on several concurrent tasks, each of them dealing with the computation and the communication for a kind of message.

5.1.1 Stochastic Method

The global working of the *StochasticCalculation* process is based on the selection of an origin tile T, then on the calculation of the interactions of T to the scene. To achieve its goals, each process *StochasticCalculation* (P_j) contains two entities: the *RayInit* function which casts a ray, computes the local intersections and then sends the ray towards the next process, and the *RayTreatment* task which deals with a ray sent by the previous process.

After the selection of the origin tile T, *RayInit* begins to compute N the number of useful rays to cast from T. In order to obtain a uniform distribution, the visible direction hemisphere is divided into N samples. Then, in each sample E_i, *RayInit* chooses randomly a direction D_i. The selection of a point P belonging to T enables us to define a ray R_i starting at P towards the sample E_i in the direction D_i (cf. figure 6).

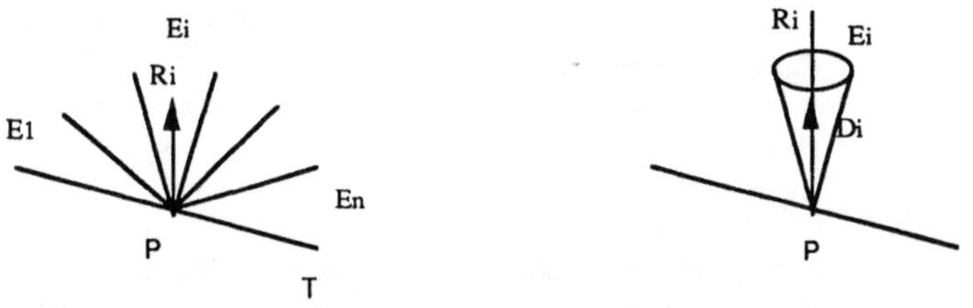

Fig. 6: Ray start

Before the emission of a ray R_i, *RayInit* searches for a potential intersection between R_i and one of the tiles locally stored in P_i. This search is based on a incremental progression of R_i through voxels. As in the classical sequential method, it is necessary to continue this search in the neighbouring voxels if it fails in the current one. The distribution of the voxel description constitutes the main difference. *RayInit* sends towards one of the two neighbouring processes (P_{i-1}, P_{i+1}) a message containing information useful for the continuation (Point P, Direction D_i,...). The choice between P_{i-1} and P_{i+1} is solved by the incremental progression which indicates the next voxels to examine, and so the process which contains their descriptions. This kind of progression enables us to detect when a ray "leaves" the scene and so to stop the message traversal.

The *RayTreatment* task of P_{i+1} (resp. P_{i-1}) performs the receipt of the message coming from P_i, searches for an intersection, and then sends the message towards P_{i+2} (resp. P_{i-2}) if it fails. If an intersection with a tile T' is detected, *RayTreatment* computes the interaction between T and T', and then updates the energy of T'.

5.1.2 Deterministic Method

As in the stochastic method, the global working of *DeterministicCalculation* process starts by selecting an origin tile T. Each process P_i contains four modules: the *TileEmission* function which sends tiles on the monodirectional ring, the *TileTreatment* task which deals with tiles sent on the first ring and which sends rays on the bus, the *RayTreatment* task which deals with rays sent on the bus and which sends energies on the bidirectional ring, and the *EnergyTreatment* task which receives energies and updates the radiosity of tiles.

The *TileEmission* function begins to compute the potential contribution C between T and the local tiles T'. If C is greater than the minimum threshold Σ, a visibility test between T and T' (by ray tracing) is performed; if there is no occlusion, C is added to the energy of T'. Then, the characteristics of T are sent to the neighbouring process on the monodirectional ring.

The *TileTreatment* task of P_{i+1} (resp. P_{i-1}) performs the reception of the message coming from P_i. Then, it computes the potential contribution C between T and every tile T' which belongs to P_{i+1} (resp. P_{i-1}) ; when C is greater than Σ, a ray is cast towards T in order to detect an eventual occlusion. That ray casting leads to a message sending from P_{i+1} (resp. P_{i-1}) to the process owing T on the bidirectional bus.

The *RayTreatment* task receives a ray message on the bus, searches for an intersection between the ray and the local tiles. If an intersection is found, the process creates a null energy message that is sent back to the owner of T, on the bidirectional ring. If there is no intersection, the ray is sent to the neighbouring process on the bus. Finally, when the ray comes on tile T without having been occluded, an energy message is created (containing the contribution of T on T') and sent to the owner of T on the bidirectional ring.

The *EnergyTreatment* task of P_i simply receives an energy message, tests if the message is intended to P_i and adds the energy to the tile which is referenced in the message.

5.2 Termination Test

In contrast to classical sequential machines, the termination problem needs a specific solution for parallel architectures. It is not sufficient to detect the inactivity of all the network processes: for example, a message may be waiting in the network and not yet be accepted by the receiver; then its receipt and treatment may reactivate a sleeping process.

We propose a termination test based on a token method; we use a token whose logical structure describes, at any moment, the set of active processes in our communication graph. This purely algorithmic solution provides a termination test which does not depend on the characteristics of the machine and is valid for the two methods presented above.

5.3 The Data and Load Balancing

As seen in paragraph 2.2, the data description of the scene is decomposed into slices which have a constant section and a variable thickness and which are associated with processes. This decomposition results from a BSP algorithm which uses cost functions. In order to distribute the computation and communication loads on the processor network, as fairly as possible, these functions try to approximate, a priori, the treatment cost.

More precisely, the cost of each voxel V is evaluated with respect to the number of tiles having V as proprietor, to the energy of the source tiles and the brightness of the reflective tiles. Indeed, we think that the energy conservation in the system induces, on average, a constant treatment on the processes, and so induces a conservation of the load repartition. Then, these costs are summed for each slice of one voxel thickness. Finally, a binary recursive decomposition is performed to optimize the cost balance between two resulting areas. The axis performing the better balance is then chosen.

6 Perspectives

In this paper, we present two parallel approaches of a progressive radiosity for MIMD computers.

Concerning the graphic point of view, the main characteristics of our work are a ray-tracing method for the computation of the form-factors (in that sense, we have a more general technique than the hemicube one; moreover, ray-tracing allows a natural integration of general reflectance functions [LESA90]), and a data structure based on the tile model which optimizes both the cost and the precision of the computations.

Now, concerning the parallel point of view, we propose a loosely coupled implementation well suited for medium grain computers with a statical data distribution for load balancing, and a purely algorithmic termination test.

The two described methods are currently under programmation in the DESIR project context at LaBRI. We use a T.Node machine which contains 32 transputers at 20 MHz with 2Mb.

Acknowledgements

Work granted by Conseil Régional d'Aquitaine.

References

[BADO90] D. Badouel, Schémas d'exécution pour les machines parallèles à mémoire distribuée. Une étude de cas: le lancer de rayon, Université de Rennes 1, Thèse d'Université, Octobre 1990

[BAUM90] D.R. Baum & J.M. Winget, Real time radiosity through parallel processing and hardware acceleration, Computer Graphics, vol 24, no 2, pp 41-50, 1990

[CHAL90] A.G. Chalmers & D.J. Paddon, Parallel radiosity methods, Transputer Research and Applications, no. 4, pp. 183-193, 1990

[COHE88] M.F. Cohen, S. Chen, J. Wallace & D.P. Greenberg, A progressive refinement approach for fast radiosity image generation, Computer Graphics, vol. 22, no. 4, pp. 75-84, August 1988

[COUN90] M.C. Counilh & J. Roman, Expression for massively parallel algorithms - description and illustration example, Parallel Computing, vol. 16, pp. 239-251, 1990, North-Holland

[FUJI86] A. Fujimoto, T. Tanaka & K. Iwata, ARTS: Accelerated ray-tracing system, IEEE Computer graphics and applications, vol. 6, no. 4, pp. 16-26, 1986

[GORA84] C.M. Goral, K. Torrance, D.P. Greenberg & B. Battaile, Modeling the interaction of light between diffuse surfaces, Computer Graphics, vol. 18, no. 3, pp.213-222, July 1984

[KAJI86] J.T. Kajiya, The rendering equation, Computer Graphics, vol. 20, no. 4, pp. 143-150, August 1986

[LESA90] B. Le Saec & C. Schlick, A progressive ray-tracing based radiosity with general reflectance functions, Eurographics Workshop on Photosimulation, Realism and Physics in Computer Graphics, pp. 103-116, Rennes, June 1990

[MALL89] T.J. Malley, A shading method for computer generated images, University of Utah, Master of Sc. Thesis, 1989

[PRIO89] T. Priol, Lancer de rayons sur des architectures parallèles: étude et mise en oeuvre, Université de Rennes 1, Thèse d'université, Juin 1989

[PUEC90] C. Puech, F. Sillion & C. Vedel, Improving interaction with radiosity-based lighting simulation programs, Computer Graphics, vol. 24, no.2, pp. 51-57, March 1990

[PURG90] W. Purgathofer & M. Zeiller, Parallel radiosity methods, Eurographics Workshop on Photosimulation, Realism and Physics in Computer Graphics, pp. 173-183, Rennes, June 1990

[RECK90] R.J. Recker, D.W. George & D.P. Greenberg, Acceleration techniques for progressive radiosity, Computer Graphics, vol. 24, no. 2, pp. 51-60, 1990

[SHIR90] P.S. Shirley, Physically based lighting calculations for computer graphics, University of Illinois, PHD Thesis, 1990

[SILL89] F. Sillion & C. Puech, A general two-pass method integrating specular and diffuse reflection, Computer Graphics, vol. 23, no. 3, pp. 335-344, July 1989

[WALL87] J.R. Wallace, M.F. Cohen & D.P. Greenberg, A two-pass solution to the rendering equation: a synthesis of ray-tracing and radiosity method, Computer Graphics, vol. 21, no. 4, pp. 311-320, July 1987

[WALL89] J.R. Wallace, K.A. Elmquist & E.A. Haines, A ray-tracing algorithm for progressive radiosity, Computer Graphics, vol. 23, no. 3, pp. 315-324, July 1989

[WHIT80] T. Whitted, An improved illumination model for shaded display, Communications of the ACM, vol. 23, no. 6, pp. 343-349, June 1980

Appendix

A tile is the part of a given surface for which the parametric coordinates are limited by a domain $[U_0, U_1] \times [V_0, V_1]$:

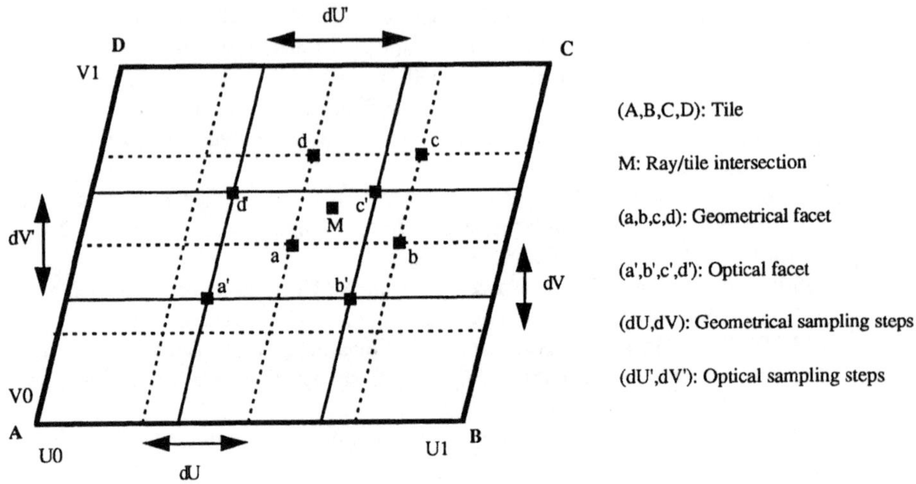

Fig. 7: Tile computation

When a ray hits a tile (A, B, C, D), the intersection function returns the intersected point M, the intersected geometrical facet (a, b, c, d) and the coefficients (u, v) defined by : $M = (1\text{-}u)\,(1\text{-}v)\,a + u\,(1\text{-}v)\,b + u\,v\,c + (1\text{-}u)\,v\,d$

Thanks to the regularity of the geometrical sampling, the parametric coordinates (U_a, V_a) of point **a** can be easily found (in our figure, $U_a = U_0 + 2\,dU$ and $V_a = V_0 + 2\,dV$). That gives the parametric coordinates (U,V) of point M :

$U = U_a + u\,dU$ and $V = V_a + v\,dV$

In the same manner, the regularity of the optical sampling enables to find the optical facet (a', b', c', d') containing M as well as the parametric coordinates (U_a', V_a') of point **a'** (in our figure, $U_a' = U_0 + dU'$ and $V_a' = V_0 + dV'$). Finally the coefficients (u', v') defined by $M = (1\text{-}u')\,(1\text{-}v')\,a' + u'\,(1\text{-}v')\,b' + u'\,v'\,c' + (1\text{-}u')\,v'd'$ can be found by :

$u' = (U - Ua') / dU'$ and $v' = (V - Va') / dV'$

These coefficients (u', v') will serve as weights to spread the energy brought by the incoming ray, over the four counters (a', b', c', d'). So, an efficient implementation will be able to compute the contribution of an incoming ray on a tile using only six multiplications.

An Extended Radiosity Using Parallel Ray-Traced Specular Transfers

Jean-Pierre Jessel, Mathias Paulin, René Caubet

Abstract

The realism in image synthesis needs complex illumination models. In this paper, our goal is to describe a parallel extended radiosity method with general reflectance functions. This approach will allow us to produce realistic images.

At first, we analyse existing extended radiosity methods to explain the energy transfer principles and how to compute them. Then we study theoretical frameworks on radiance and luminance transfers in a close environment to deduce a progressive extended radiosity method with parallel ray-traced specular transfers. Then we describe our implementation of this method in the VOXAR machine, parallel architecture dedicated to the ray-tracing algorithm.

Key words: Extended radiosity, Parallelism, Specularity, Ray-tracing.

1 Introduction

In the image synthesis world, the quest for realism has led to several illumination computation methods. The first techniques produced comprehensive images with hidden surfaces removal and elementary shading effects. Then new algorithms have appeared, more effective but also more expensive according to computation time, such as the ray-tracing, introduced by Whitted [20], or the radiosity method presented by Goral et al. [3], then by Cohen and Greenberg [2]. This increasing need for computation power introduced parallel hardware developments in order to decrease computation costs. In this paper, we describe a radiosity method allowing general reflectance functions, and its implementation on the VOXAR machine [1, 10]. VOXAR is a parallel hardware developed to decrease the ray-tracing cost which belongs to the "object-oriented" machine set according to the Priol's classification [12].

Wallace et al. showed that to add a ray-tracing to a basic radiosity is not sufficient to keep the advantages of both the methods [18]. They introduced a two-pass method: the first pass is equivalent to a basic radiosity where the specularity is integrated in the equations. The second pass is a ray-tracing using the first pass results. Sillion extended the Wallace's method using a matricial formalism [16]. The rendering equation is solved using “extended form-factors”, taking into account the specular reflects. In the second pass, the global specular operator is applied to the diffuse component to compute the reflection and refraction effects from the observer's location. This is done by a simplified ray-tracing without shadow rays, as sources have been treated in the first pass.

Shirley presented a three-pass method in [17]. First, feeler rays sample the scene, to find where to send dense energy carrying rays. Rays hitting a diffuse object after at least one bounce on a specular one are tabulated in the illumination map of the diffuse object. The second pass is a radiosity in which the direct lighting is omitted. The direct lighting is treated by a distributed ray-tracing where the illumination maps and the radiosity values are used instead of the ambient term. This method is dependent on the space sampling rate for the feeler rays, and it is possible to miss little objects.

The Rushmeier's method is an extension of the radiosity accounting for specularly reflecting surfaces in a otherwise diffuse scene, and specular or ideal diffuse transmitters [15]. Rushmeier gives an energetic presentation of the ray-tracing to fit the energy transfers in the radiosity method and to allow its extension. New form-factors are introduced, and a virtual scene is built behind each mirror. This approach is very attractive, but the use of ray-tracing to compute form-factors would avoid the creation of virtual worlds.

Our approach for a parallelization of the radiosity is different from others methods such as Purgathofer [13], where a pipeline of processors is used to solve the classical radiosity equations set and to render the scene. Recker and Greenberg use a workstations network to implement a progressive radiosity [14, 3], where the server distributes the work to slaves at each step. This is closer to our method but we have not a centralized management and almost we use an "object oriented" parallelism into a space partition.

2 Design of an Extended Radiosity Method

The study of previous extended radiosity methods allowed us to present a radiosity including all the transfers composed from the four light transfers described by Wallace [8], in which we used a specularity sphere [7]. Here we present a more accurate method that needs less data storage. This method allows to palliate classical progressive radiosity defaults (specular transfers not included), because it uses a ray-tracing to determine energy transfers.

2.1 The General Radiance Formula

In a space closed by N surfaces such as in figure 1, the surface radiance is given by following equation (1):

$$I_{o,n}(x_n,\theta_{o,n}) = I_{e,n}(x_n,\theta_{o,n}) + \frac{\rho_{d,n}}{\pi}\left\{\sum_{m=1}^{N}\int_{A_m} V_r(x_n,x_m) I_{o,m}(x_m,\theta_{o,m})\cos\theta_{if}\, d\omega_{if,n}\right\}$$
$$+ \rho_{s,n} I_{o,n}(x_n,\theta_{o,n}) + \frac{\tau_{d,n}}{\pi}\left\{\sum_{m=1}^{N}\int_{A_m} V_t(x_n,x_m) I_{o,m}(x_m,\theta_{o,m})\cos\theta_{ib}\, d\omega_{ib,n}\right\}$$
$$+ \tau_{s,n} I_{o,n}(x_n,\theta_{o,n})$$

with:

X_n, X_m: point in patch **n, m.**

θ_{if}, θ_{ib}: front, backward, incident angle.

$\theta_{o,n}$: propagation direction.

V_r, $V_t(X_n,X_m)$: visibility function of X_m by X_n by reflection, transmission:
1 if Xm is visible from Xn, else 0.

$I_{e,n}$, $I_{o,n}(X_n,\theta_{o,n})$: emitted, outgoing, intensity at X_n in the direction $\theta_{o,n}$.

$\rho_{d,n}$, $\rho_{s,n}$: diffuse, specular, reflectance of patch n.

$\tau_{d,n}$, $\tau_{s,n}$: diffuse, specular, transmittance of patch n.

$d\omega$: solid angle substained by patch A_m.

N: number of patches.

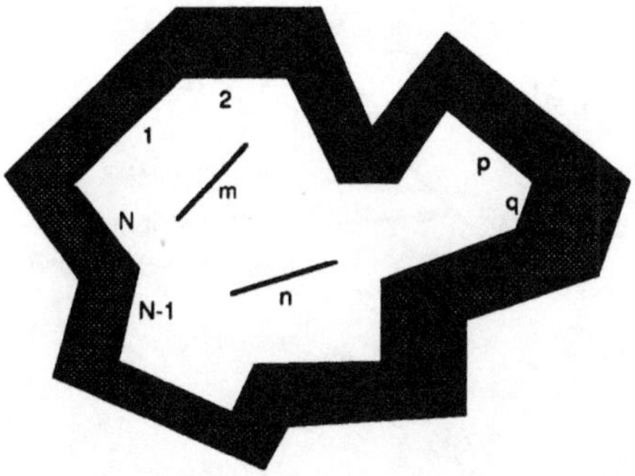

Fig. 1: A closed space

This equation is available only when reflectance and transmittance are constant over the patch n. Then, as shown by Rushmeier in [15], there are two distinct equations describing the diffuse component I_{od} and the specular component $I_{os}(X,\theta)$ of the radiance is:

$$I_{os,n}(X_n,\theta_{o,n}) = \tau_{s,n} I_{o,n}(X_n,\theta_{o,n}) + \rho_{s,n} I_{o,n}(X_n,\theta_{o,n}) \qquad (2).$$

The diffuse component can be expressed independently of the outgoing angle $\theta_{o,n}$ by introducing a geometric factor describing the part of intensity leaving a patch to another. This factor is the form factor between the two patches and can be extended to include specular reflection or transmission available on the environment. Sillion described an extended form factor computation method by ray-tracing [16] when Rushmeier used mirror form factors to express the visible patches through a patch [15].

2.2 The Progressive Extended Radiosity Method

To distribute progressively the energy into the scene, we must define the elementary quantities of energy distributed at each step of processing. The following equation (3) expresses the contribution of the patch m to the intensity received by the patch n according as the sum of this equation over the environment is equal to the diffuse component of the equation (1):

$$\begin{aligned} I_{od,n} = {} & \rho_{d,n}\left\{ I_{od,m} F_{nm} + \tau_{s,m} \sum_{p=1}^{N} I_{od,p} T_{f,nmp} + \rho_{s,m} \sum_{p=1}^{N} I_{od,p} F_{f,nmp} \right\} + \\ & \tau_{d,n}\left\{ I_{od,m} T_{nm} + \tau_{s,m} \sum_{p=1}^{N} I_{od,p} T_{b,nmp} + \rho_{s,m} \sum_{p=1}^{N} I_{od,p} F_{b,nmp} \right\} \end{aligned} \qquad (3)$$

with:

$I_{od,n}$: outgoing **d**iffuse **I**ntensity of patch **n**.
F_{nm}: **F**acial form factor between patches **n** and **m**.
T_{nm}: **T**ransparent form factor between patches **n** and **m**.
$F_{f,nmp}$: **f**ront **F**acial mirror form factor between patches **n** and **p** through **m**.
$F_{b,nmp}$: **b**ackward **F**acial mirror form factor between **n** and **p** through **m**.
$T_{b,nmp}$: **b**ackward **T**ransmit mirror form factor between **n** and **p** through **m**.
$T_{f,nmp}$: **f**ront **T**ransmit mirror form factor between **n** and **p** through **m**.

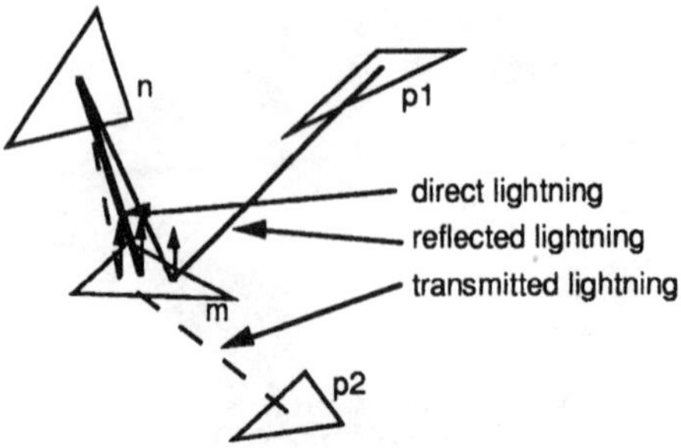

Fig. 2: The lighting modes

We can extract from this equation three types of energy transfer from patch m to patch n: direct transfer from m to n, reflection on m of the environment's emission and transmission by m of the environment's emission. These three kinds of participation are shown in figure 2. As the three transfer types are independent, and correspond to separate and independent terms in equation (3), we can compute them separately.

The method is based upon a progressive radiosity. Each step is made of two phases: the gathering of the energy directly emitted from primary sources (or secondary ones in latter steps), and its distribution over the scene by reflections and transmissions. Thus, the energy of every patch directly or indirectly visible from the sources is updated at each step.

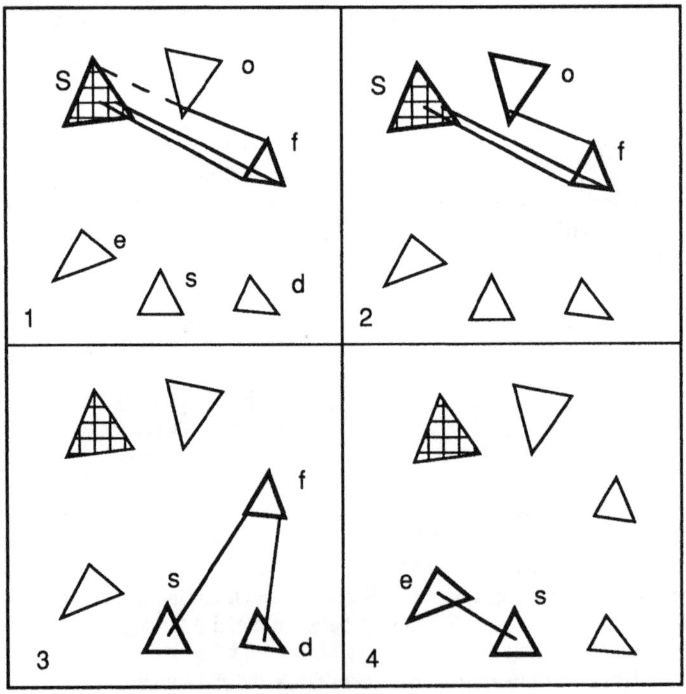

Fig. 3: The energy gathering and distribution

As in a classical progressive radiosity, at each step, the current source is elected among potential sources. The different phases of the algorithm for one step are shown in figure 3. Only few rays are traced to get the picture more comprehensive. In this simple example, the face 'f' is the current receiver, 'S' the current source; 'o', 'd', 's', and 'e' are faces in the environment.

The energy sent by source 'S' to 'f' is collected by 'f' with rays sent to sampling points laying on the source surface, in figure 3.1. The rays that reach the source return the corresponding energy, else a negative answer is received from the occluding face 'o', figure 3.2. If we suppose 'f' to be specular, the distribution uses rays traced from directly visible face 'f' in the specular directions according to source sampling rays, figure 3.3. These rays bounce into the environment and compute simultaneously needed form factors and energies to send. This is done for every specular patch, such as face 's', until the ray path reaches a perfectly diffuse patch such as face 'd', or the carried energy is insignificant as in 'e'. If 'f' were a diffuse reflector, the distribution would be stopped as for 'd'. If 'f' were transparent, additional rays (refracted) would be sent through the face. Only one ray is sent for an incoming sampling ray, as we consider that patches are perfectly specular (or perfectly diffuse). Reflected and transmitted rays may be sent simultaneously, as their carrying energies are independent after leaving the surface.

For an incoming energy, we distribute immediately the specular component computed by (2) and we store the diffuse component computed by (3), which will be emitted if the patch becomes a secondary source, see figure 3. When needed, elementary form factors are computed by the Wallace's formula for differential surface d_S to differential surface d_f form-factors: $dF_{dS\text{-}df} = \cos\theta_S \cos\theta_f \, d_f / \pi r^2$, where d_S and d_f are elementary surfaces [19]. They are computed while ray-tracing, step by step, as needed parameters are known. In figure 3.1, when the ray is sent, 'f' knows $\cos\theta_f$, and when it comes back after reaching the source, the ray carries the incoming energy e_i, $\cos\theta_S$, and r. Energy to be stored and energy to be sent are computed, and a ray carrying $\rho_s \, e_i \cos\theta_S \cos\theta_f \, d_f / \pi r^2$, the re-emitted energy. Once on 's', $\cos\theta_s$ is computed and energies are computed, if needed, a new ray is sent, carrying the corresponding energy and $\cos\theta_s$.

3 The Implementation

Even if this method is available for any multi-processor hardware, we first describe some of the VOXAR's features that lead to some particularities for the application.

3.1 VOXAR's Features Influencing the Implementation

VOXAR, **VOX**el **AR**chitecture, is a parallel machine, made of two set of processors, the cells network organized in hypertorus and the host processor. The VOXAR's purpose is to compute images by a ray-tracing method which implements an incremental integer logic algorithm into a regular space partition [8, 9]. VOXAR is built on a Transputers network [10]. As the Transputer is designed for multi-tasking, our applications are cut in set of tasks where each task is involved for a treatment. A task starts a local treatment when it receives a message and may send messages to other tasks. This is built on a communication shell allowing two any tasks (on two any processors) to dialogue in a transparent way. The control and the management are assumed by a set of specialized tasks on the host processor.

For the scene representation, the space is cut into voxels that are space unity for the incremental ray-tracing. Voxels are spatial active elements in which treatments to apply are detected. With regards to space-processors linkage, voxels are joined into metavoxels (cubes of contiguous voxels) to avoid too much processors. VOXAR defines a cyclical three-dimensional multi-processor architecture in which each processor manages a set of metavoxel, as shown in figure 4:

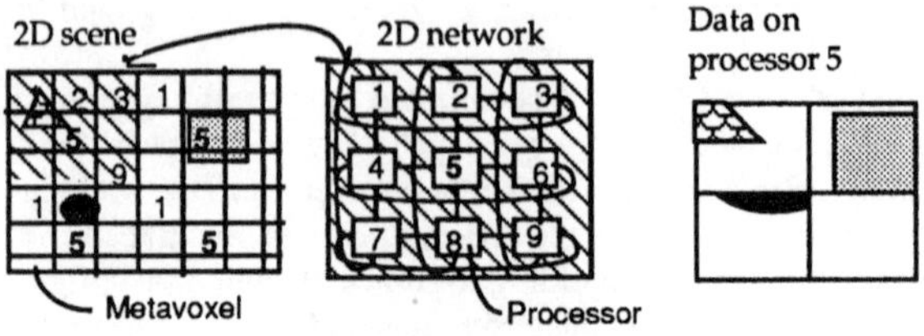

Fig. 4: A 2D data distribution

We introduced an unified objects modeling, the Atom mesh model, which is a discretization of the object's surface with local storage capabilities (texture, light energy) [5, 6]. Objects handled by VOXAR are sets of meshes loaded in the host processor then distributed in the network. This distribution is based on the space partition among the processors, the cyclical links allowing space continuity in all the directions. A mesh is attributed to the processor managing the part of space in which it is located. As a mesh may cross several metavoxels, it is known in all of concerned processor, but to ensure the treatment unicity, the following location rule is applied: a mesh is stored in the processor owning the location of the vertex with the smallest coordinates.

3.2 The Implementation as a Set of Tasks

Our implementation of this extended radiosity method on the VOXAR machine is realized by two passes: the first one computes the distribution of the energy into the scene, and the second one is a fitted ray-tracing giving the view dependant specularity. Each step of the first pass works in parallel on every processor, with a centralized synchronization task on the host processor. Each particular treatment is assigned to a particular task. Figure 5 shows the tasks map for the first pass.

The tasks are arranged in the following sets:

- The Control set, made of the Control task itself, on the host processor, and on each cell, the Termination task and the Regulation task. Tasks of this set manage the application.
- The Election set, made of the Election task and the Update task. These tasks are on each cell and manage the election phase.
- The Illumination set, made of, on each cell, the Send_direct task, the Recv_direct task, the Distribution task an the Send_spe task. They compute the illumination phase.
- The Rendering set, made of, on each cell, the Send_prim task, the Distribution task and the Send_spe task. These tasks compute the final image.
- The Rays management set, made of, on each cell, the Collision task, the Following task and the Scene_exit task. They manage the rays' life.
- Service set made of the Atom_server task.

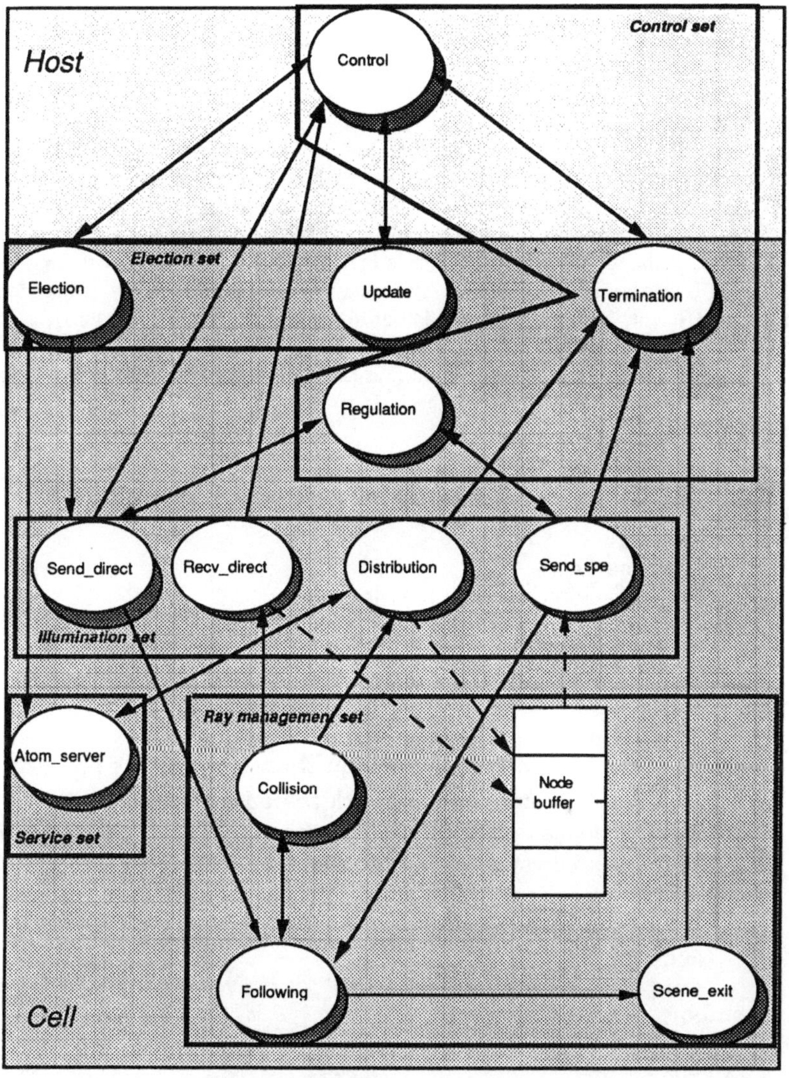

Fig.5: The tasks map for the first pass

The first pass general algorithm is:

begin
election of more energetic sources;
    while (sources to be treated)
        sources sampling; sources energy gathering;
        specular transfers; patches energies update;
        election of more energetic sources;
    endwhile;
end.

The Election task scans every patch in the cell then, for each, computes the energy to send by adding atom's (vertex) energy (using the Atom_server task for atoms belonging to other cells). The patch with the more energy to send is elected and broadcasted to other cells, else a message "no source" is broadcasted. The Control task is also signaled to know if patches are elected to detect the termination.

When a cell knows all the current sources, the acquisition phase begins: the Send_direct task computes sampling points on the sources and send acquisition rays to each sampling point to gather energy. An acquisition ray is a message sent to the Following task where it is tracked trough the scene until a collision is detected [8]. A message is sent to the Collision task to detect if the aimed sampling point is reached. Then, a message with an energetic information is sent to the Recv_direct task on the cell which emit the ray. If the incoming energy is sufficient, (null if the source is not hit) and if the patch is specular, a node is created in the node buffer for a ray to be sent in the specular direction. It is the same for transmitted rays. When all the acquisition rays are sent, and when they are come back, a signal is sent to the control task on the host to start the detection termination of the distribution step.

The acquisition phase algorithm is:

parfor each processor
for each vertex i
for each source s
for each point k sampling s
send a ray from i to k;
if k is hit **then**

$$E_k = \frac{\cos\theta_{1k} \cos\theta_{2k}}{\pi r^2 + \frac{A_s}{n}} * \frac{B_s A_s}{n}$$

if the ray comes from the front **then** $\Delta rad = \Delta rad + \rho d,i * E_k$;
if the ray comes from the back **then** $\Delta rad = \Delta rad + \tau d,i * E_k$;
if i belongs to a specular patch **then**
send a reflected ray carrying ($\rho s,i * E_k$);
if i belongs to a transparent patch **then**
send a transmitted ray carrying ($\tau s,i * E_k$);
endfor;
$\Delta Ri = \Delta Ri + \Delta rad$;
endfor;
$\Delta Ri = \Delta Ri + \Delta Si$;
$\Delta Si = 0$;
Energy(i) = Energy(i) + ΔRi ;
if i is a source **then** $Bi = \Delta Ri$ **else** $Bi = Bi + \Delta Ri$;
endfor;
endparfor.

where:

E_k: Energy collected at point k.
Δrad: Energy collected on source s.
ΔRi: Energy received in one iteration by vertex i.
ΔSi: Energy received in a specular way by vertex i.

Energy(i): Total energy at vertex i.
Bi: Energy no yet sent by vertex i.
As: Area of source s.
n: number of sampling points.
cosθ1k: cosine of incident angle of ray on the source at point k.
cosθ2k: cosine of outgoing angle of ray from i to point k.

Then, in each cell, the Send_spe task takes nodes into the node buffer to create rays that will bounce into the scene until they die: the distribution phase begins. The Send_spe task acts in the same way for distribution and rendering: consuming a node to create a ray, then sending it to the Following task to be tracked. The node may be either for a reflected or for a transmitted ray. When a collision occurs, the Collision task sends a message to the Distribution task which dispatches energies and creates as many rays as necessary (zero for diffuse surfaces or insufficient energy or maximal bounce reached, one for specular surfaces, two for transparent surfaces). Each time a ray is created or killed, a signal is sent to the Termination task. When a ray escapes from the scene (when it hits the surrounding diffuse surface) the ray is killed, as for other diffuse surfaces.

The "Distribution" task algorithm is:

begin
    if collision detected on patch f **then**
        distribute the energy E given by the ray to vertices of f
    if i belongs to a specular patch **then**
        send a reflected ray carrying (rs,i * E)
    if i belongs to a transparent patch **then**
        send a transmitted ray carrying (ts,i * E)
end.

3.3 Control and Termination Detection

The control task is the master of the application. It orders to start the first election, validates the end of each step with Update task starting. It detects the end of first pass then manages the rendering pass as well as in the ray-tracing application [11]. When each of the acquisition tasks, Send_direct and Recv_direct, are finished, they send a signal to the Control task which now starts the Termination task. At this time, the active tasks are: Distribution, Send_spe, element of ray-management set and Atom_server (when needed). Time to time (after a delay estimated according to the number of rays still alive and the average life), the Control task gather ray's births and deaths recorded by Termination task. When the difference is zero, there is no more rays alive or to be born, so the step is finished.

4 Conclusion

Ray-tracing and multi-processing techniques used in the implementation have been well-tried in previous applications. The method described in this paper will allow to compute realistic images, because it accounts for the different kinds of patch participations to the scene enlightenment. Indeed, a patch gathers directly the energy from sources, and then re-emits either immediately, for the specular component, either latter, for the diffuse component, if the patch becomes a secondary source. This may be sufficient as the specular component accuracy is more important for the image quality because the diffuse component does not vary very much inside a patch, even though a few rays can product an important specular effect. In order to be able to manage non perfect specular objects, we should send a lot of rays around the specular direction, according to material features. In this first implementation, this is a sufficient enlightenment approximation because a spray of incoming rays on direct enlightened patches produces an outgoing spray.

Several improvements are studied. A lot of sources may be elected in each processor in the same step: the time needed for multiple elections is saved, but intermediate pictures are unusable because the ordering given by the sequential treatment is disturbed. It is possible to introduce a pipe-line for illumination and rendering treatments: the rendering is applied on a frozen copy of the scene, while the illumination become more accurate. It is also possible to compute in the same time several pictures of the same scene from different point of view, after the complete illumination: when this other pipe-line is primed, a picture may be rendered at each cycle, and so, viewer animations may be less expensive.

Acknowledgments

We want to express our thanks to the reviewers for their helpful comments and remarks which aided us to improve this paper, and to Mr Kadi Bouatouch, for his precious advices about the extended abstract.

References

[1] Caubet R, Duthen Y, Gaildrat-Inguimbert V: A tridimentional architecture for fast realistic image synthesis. Proccedings of Computer Graphics International, May 1988.

[2] Cohen M, Greenberg D: The hemi cube, a radiosity solution for complex environments. Computer Graphics, Vol 19, n°3, July 1985.

[3] Goral C, Torrence K, Greenberg D: Modelling the interaction of light beetween diffuses surfaces. Computer Graphics, Vol 18, n°3, July 1984.

[4] Greenberg D: Global illumination: the radiosity approach. Course supports PIXIM89.

[5] Jessel JP, Cipres P, Pitot P, Caubet R, Duthen Y: Une modelisation unifiee intégrant des informations locales: la discrétisation en atomes. Actes de PIXIM 89, Paris, September 1989.

[6] Jessel JP, Cipres P, Pitot P, Caubet R, Duthen Y: Déformations d'objets modélisés par réseaux d'atomes. Journées Grosplan d'Informatique Graphique, Strasbourg, Décembre 1989.

[7] Jessel JP, Caubet R, Duthen Y: A progressive radiosity including specularity: a parallel approach. Conference Proceedings, Computer Graphics, London, November 1990.

[8] Pitot P, Caubet R, Duthen Y, Gaildrat V: Le suivi analytique de rayons: un algorithme incrémental rapide pour la machine VOXAR. Actes de MICAD89, Paris, February 1989.

[9] Pitot P, Caubet R, Duthen Y: A parallel architecture for the ray-tracing.
Conference Proceedings, Computer Graphics, London, October 1989.

[10] Pitot P, Moisan B, Caubet R, Duthen Y: A Transputer based implementation of the VOXAR project. Euromicro'90, Amsterdam, 1990.

[11] Pitot P: Conception et réalisation d'une machine parallèle dédiée à la synthèse d'images réalistes: la machine VOXAR. PhD Thesis, Université Paul Sabatier, Toulouse, January 1991.

[12] Priol T: Lancer de rayons sur des architectures parallèles: étude et mise en oeuvre. PhD Thesis, Université de Rennes 1, June 1989.

[13] Purgathofer W, Zeiller M: Fast radiosity by parallelization. First European Workshop on Photosimulation, Rennes, June 1990.

[14] Recker R, George DW, Greenberg DP: Acceleration techniques for progressive radiosity. Computer Graphics Symposium on Interactive Graphics, 1990, 24-2.

[15] Rushmeier HE, Torrance KE: Extending the radiosity method to include specularly reflecting and translucent materials. ACM Transactions on graphics, Vol 9, n°1, January 1990.
[16] Sillion F: Simulation de l'éclairage pour la synthèse d'images: réalisme et interactivité. PhD Thesis, Université de Paris Sud, 1989.
[17] Shirley P: A ray traced method for illumination calculation in diffuse-specular scene. Proceedings of Graphics Interface 90.
[18] Wallace JP, Cohen M, Greenberg D: A two pass solution to the rendering equation: a synthesis of ray-tracing and radiosity methods. Computer Graphics, Vol 21, n°4, July 1987.
[19] Wallace JP, Elmquist KA, Haines E: A ray-tracing algorithm for progressive radiosity. Computer Graphics, Vol 23, n°3, July 1989.
[20] Whitted T: An improved illumination model for shaded display. Communication of ACM, Vol 23, June 1980.

Subdivision of 3D Space Based on the Graph Partitioning for Parallel Ray Tracing

Veysi İşler, Cevdet Aykanat, Bülent Özgüç

Abstract

An approach for parallel ray tracing is to subdivide the 3D space into rectangular volumes and assign the object descriptions with their related computations in each volume to a different processor. The subdivision process is critical in reducing the interprocessor communication overhead, and maintaining the load balance among processors of a multicomputer. In this paper, after a brief overview of parallel ray tracing, a heuristic is proposed to subdivide the 3D space by converting the problem into a graph partitioning problem. The proposed algorithm tries to minimize the communication cost while maintaining a load balance among processors.

1 Introduction

Realistic images are used widely in many computer graphics applications such as computer-aided design (CAD), animation and visualization, simulation, education, robotics, architecture, advertising, medicine, etc. Ray tracing is the most powerful algorithm to produce realistic images by finding the interaction of light sources and the objects in an environment. Although ray tracing is a simple algorithm, it requires excessive floating point operations. The amount of computation mainly depends on two parameters: the total number of pixels in the generated image and the number of objects in the scene. In the naive algorithm, the number of objects has a great effect on the total computation time, since each ray is tested with all objects in the scene to find the first intersection point. Several methods have been developed to reduce the computation time by testing only the objects on the ray's path for intersection. Spatial subdivision and the use of a hierarchy of bounding volumes are two of the well known techniques that aim to generate ray traced images independent of the complexity of the objects in the scene.

Even if we could reduce the total computations for a ray to a constant time, it is still necessary to process all of the pixels independently. Additionally, we may wish to shoot more than one ray for each pixel to increase the accuracy of the image as well as adding diffuse effects. This means that the algorithm has another bottleneck due to the number of rays traveling in the scene. Thus, parallelism is essential in speeding up the ray tracing algorithm. Furthermore, ray tracing is easily amenable to parallelization on a multiprocessor, since each primary ray is traced independently.

In recent years, research on ray tracing has been mostly concentrated on speeding up the algorithm on multiprocessors. There are mainly two approaches to parallelize ray tracing. One of them is an image space subdivision in which the computations related to different rays are distributed to the processors. The other approach studies the parallel ray tracing on a distributed-memory message-passing multiprocessor (multicomputer). In a multicomputer, there is no global memory and synchronization, and coordination between processors are achieved through message exchange. For an efficient parallelization on a multicomputer, the object space data as well as computations should be distributed among processors of the multicomputer. Since the model database (scene description with the auxiliary data structure) may not fit into the local memory of each processor, object space data need to be distributed.

In this paper, our concern is to speed up the ray tracing on a multicomputer for complex scenes that require large amounts of memory. Thus, both computations and the entire database are distributed fairly among the processors. Section 2 contains a short overview of parallel ray tracing algorithms. In Section 3, we propose a subdivision scheme which maintains the load balance among processors and minimizes the interprocessor communication cost. The scheme transforms the subdivision of 3D space problem to a graph partitioning problem with some imposed constraints as discussed in Section 3.1. A Kernighan-Lin-like [11] algorithm is presented as a solution in Section 3.2.

2 Parallel Ray Tracing

A large number of parallel systems have been proposed to exploit the inherent parallelism in the algorithm. Most of these are special-purpose systems that require the construction of custom hardware using VLSI. The recent developments in the VLSI technology have made it feasible to design and implement special-purpose hardware for the ray tracing algorithm [5, 10, 16]. In spite of the gain obtained in this way, these special purpose architectures have several disadvantages. First, there are still studies to improve the algorithm itself. Researchers should thus work on general purpose machines in order not to be restricted by the hardware. Second, special purpose hardware is expensive and often restricts the applications that require other computer graphics algorithms.

The other approach that exploits speedup through the inherent parallelism in ray-tracing investigates the algorithm on a general purpose parallel architecture independent of the hardware configuration [4, 8, 12, 13, 14]. The effective parallelization of the ray tracing algorithm on a multicomputer requires the partitioning and mapping of the ray tracing computations and the object space data. This partitioning and mapping should be performed in a manner that results in low interprocessor communication overhead and low processor idle time. Processor idle time can be minimized by achieving a fair load balance among the processors of the multicomputer. Two basic schemes exists for parallelization. In the first scheme, only ray tracing computations are partitioned among the processors. In the other scheme, both ray tracing computations and object space data are partitioned among the processors.

In the first scheme, the overall pixel domain of the image space to be generated is decomposed into subdomains. Then, each pixel subdomain is assigned to and computed by a different processor of the multicomputer. However, each processor should keep a copy of the entire information about the objects in the scene in order to trace the rays associated with the pixels assigned to itself. Hence, an identical copy of the data structure representing the overall object space is duplicated in the local memory of each processor. This scheme requires no interprocessor communication since the computations associated with each pixel is independent of each other.

The image space subdivision achieves almost a linear speedup. No communication is needed between processors. The only overhead is the communication between the scheduler and the processors of the multicomputer. On the other hand, each processor should have access to the whole scene description, since ray-object intersection tests may be carried out with any object in the scene. This is a big disadvantage. Furthermore, sometimes a large amount of storage is needed to hold the object definitions and other related information. Therefore, processors cannot store the entire information about the objects in the scene.

The subdivision of the object space necessitates interprocessor communication, because each processor owns only a portion of the database. During the execution, a processor may need some portion of the database that exists in the local memory of another processor. In this case, either the needed portion is sent to the requesting processor [2, 10, 9] or the ray with the other relevant information is passed to the processor that has the needed part of the database [3, 4, 5, 13].

In object-based type of algorithms, the load imbalance is the major problem to deal with, since some processors may contain objects that are more likely to be intersected than others. Additionally, it is not easy to achieve a linear speedup as in image space subdivision where object space data is duplicated in the local memory of each processor. The communication overhead between processors might drastically degrade the performance.

3 Subdivision of 3D Space

A new scheme is proposed in order to reduce the negative effects of object-space subdivision techniques mentioned above. The proposed scheme consists of three phases. In the first phase, the problem is converted to a graph partitioning problem. The graph obtained in this phase is a directed graph with weighted nodes and edges. In the second phase, this graph is partitioned into P clusters where P is the number of processors in the multicomputer. The objective in this partitioning is to minimize the weighted sum of inter-cluster edges, subject to the constraint that the cluster weights are balanced to within a specified tolerance. The actual assignment of the clusters to the processors of the multicomputer is done in the third phase. The objective in the assignment of these P clusters to P processors is to minimize the distances of interprocessor communications.

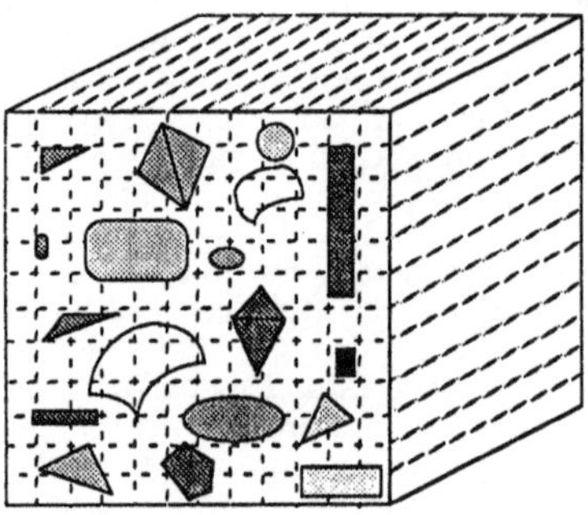

Fig 1. Regular subdivision of 3D scene.

3.1 Graph Generation

We start by subdividing the 3D space into disjoint rectangular volumes by planes perpendicular to x-y plane as shown in Figure 1. Then, a primary ray is shot into the 3D space for each region (called coarse grid) in a very similar way Bouatouch, et. al. proposed [1]. The primary rays are traced until they reach a certain level. During this sub-sampling phase, an account of rays entering and leaving each subvolume is collected. Additionally, the number of rays processed in each subvolume is stored. The subvolumes and adjacency relations correspond to the nodes and edges of the graph, respectively. Figure 2 illustrates the resulting graph for a 12 × 12 coarse grid. Nodes of the graph represent the rectangular subvolume in the 3D coarse grid. Each node u of the graph is associated with a weight $\omega(u)$ where $\omega(u) = \alpha\omega_{CL}(u) + \beta\omega_{MR}(u)$. Here, computational load $\omega_{CL}(u)$ is proportional to the number of rays processed in subvolume u during the subsampling phase. The complexity of the intersection tests for each ray processed in subvolume u is also taken into account during the computation of $\omega_{CL}(u)$. The second term $\omega_{MR}(u)$ represents the storage requirement for the objects resident in subvolume u. The α, β parameters are introduced to achieve a balance between the computational and memory load in assigning weights to the nodes of the graph. The node weights will be used to determine the load (computational + memory) of each partition. The directed edge uv from node u to node v represents the adjacency relation between the associated subvolumes. Each directed edge uv is associated with a weight $d(uv)$ where $d(uv)$ is proportional to the number of rays entered into subvolume v from subvolume u during subsampling phase. The edge weight $d(uv)$ represents the additional number and volume of communications to be performed, due to the rays originating from subvolume u and entering subvolume v, by processor P_i to processor P_j if subvolumes u and v are mapped to different processors P_i and P_j, respectively. Hence, edge weights will be used to determine the interprocessor communication load.

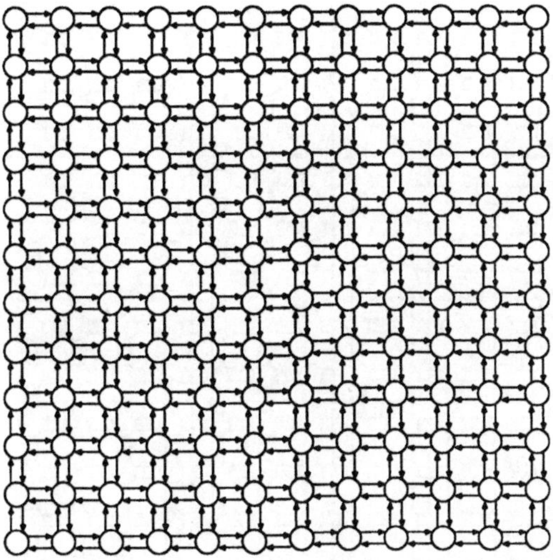

Fig 2. Resulted graph from the first step.

3.2 Graph Partitioning

It is known that partitioning a graph into a number of clusters with predetermined sizes and with minimum number of inter-cluster edges is an NP-complete problem [7]. The algorithm proposed by Kernighan and Lin (KL) is a popular non-trivial heuristic to solve the graph partitioning problem [11]. This heuristic provides hill-climbing ability to strictly local search methods by finding a favorable sequence of swaps of nodes between partitions rather than a single favorable swap of groups of nodes. Fiduccia and Mattheyses [6] improved the KL algorithm by introducing efficient data structures and node moves across clusters instead of swaps between clusters.

The partitioning algorithm proposed in this paper resembles the one proposed by Fiduccia and Mattheyses. However, the application problem that is of interest here imposes a major constraint on the spatial shape of the clusters. The constraint imposed is the convexity requirement for the image subspace associated with each cluster. The reason for this constraint is the problem that arises due to the traversal of the rays in 3D space in order to utilize spatial coherence. For spatial subdivision into non-convex subvolumes, movement to the next subvolume on the ray's path is complex and time consuming; besides the same ray might exit and reenter a subvolume thus causing extra tests and interprocessor communication for intersections and other operations. The convexity constraint for clusters can easily be maintained by restricting a move to a move of a row or a column of nodes at a time instead of a single node move. The simplest way to achieve this is to introduce horizontal and vertical virtual lines to partition the graph into clusters. Assume that the nodes of the graph are to be mapped onto a multicomputer with $P = P_x \times P_y$ processors. The algorithm partitions the graph into P clusters by $P_x - 1$ vertical and $P_y - 1$ horizontal lines. Figure 3 illustrates the partitioning of a 12×12 coarse grid for a $P = 4 \times 4 = 16$ processor multicomputer. In this scheme, a move is considered to be the move of a whole horizontal and a vertical line effectively resulting in the move of groups of nodes across clusters. The resulting clusters will be in rectangular shape thus satisfying the convexity requirement and making the use of simple traversal algorithms feasible. This scheme also ensures the 2D mesh adjacency relation among clusters.

An alternative approach is to move either vertical or horizontal boundary line segments of the clusters instead of whole lines dividing the mesh graph into two mesh subgraphs.

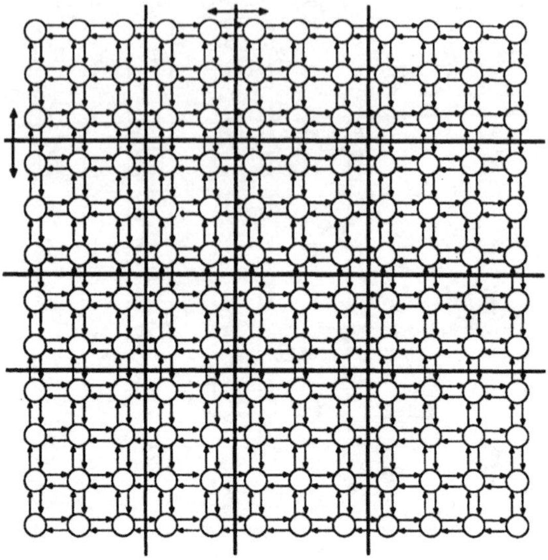

Fig 3. Line moves.

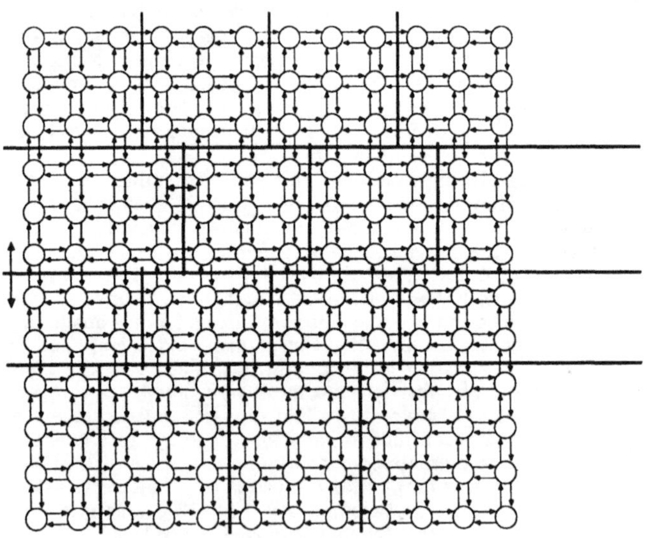

Fig 4. Line segment moves.

That is, either whole vertical lines and horizontal line segments or whole horizontal lines and vertical line segments are allowed to move. Vertical line segment moves are illustrated in Figure 4 for partitioning a 12×12 coarse grid for a $P = 4 \times 4 = 16$ processor multicomputer. This approach increases the search space by increasing the total number of possible moves. In this approach, the convexity of the clusters is ensured by restricting the moves to either horizontal or vertical line segment moves. The only drawback in this approach is that the mesh adjacency relation among the resulting clusters is not ensured. The problem of mapping of the clusters to the processors of the hypercube in this case is not straightforward and necessitates heuristics discussed in [15].

Assume that the nodes of the graph are to be mapped onto a multicomputer with P processors, where $P = P_x \times P_y$. The algorithm initially partitions the graph into P clusters by $(P_x - 1) \times P_y$ vertical line segments and $P_y - 1$ horizontal lines. The vertical and horizontal lines are aligned with the horizontal and vertical coarse grid lines such that each cluster is assigned equal number of coarse grid pixel areas. Each line is associated with two moves. Two moves associated with a horizontal line (vertical line segment) correspond to the action of sliding the horizontal line (vertical line segment) up or down (right or left) by one position on the 2D coarse grid. A line move may increase the load of some clusters by enlarging them spatially and decrease the load of some other clusters by contracting them spatially. The steps of the algorithm are given below.

1. **a)** Compute the weight $W_{C_{ij}} = \sum_{u \epsilon C_{ij}} \omega(u)$ for each cluster C_{ij}.
   b) Find all feasible moves associated with horizontal lines and vertical line segments. A move is feasible if the weight of any cluster affected by the move does not deviate from the average weight within a specified tolerance.
   c) Compute the gain of each feasible move. The gain of a move corresponds to the decrease in the sum of the inter-cluster edge weights if the move is realized.
2. Unlock and unmark each move.
3. Choose the move M_i with maximum gain g_i among the feasible unlocked moves.
   a) Mark move M_i, lock the reverse move of M_i. Reverse move is the other move associated with the same line segment as M_i but in reverse direction.
   b) Recompute the gain of move M_i.
   c) If move M_i is a horizontal line move then:
      i) Update the gains of the moves associated with its $2(P_x - 1)$ adjacent vertical line segments.
      ii) Unlock these adjacent vertical line segment moves (if they were previously locked).
   d) If move M_i is a vertical line segment move then unlock the moves associated with its two adjacent horizontal lines (if they were previously locked).
   e) Update the weights of the clusters that are affected by the move M_i. Then, update the feasibility status of the moves adjacent to these clusters.
4. Step 3 is repeated until no move can be selected, resulting in the marked moves $M_1, ..., M_n$ with gains $g_1, g_2, ..., g_n$.
5. Choose k $(1 \leq k \leq n)$ which maximizes $G = \sum_{i=1}^{k} g_i$. If $G_{max} > 0$, realize the sequence of moves $M_1, M_2, ..., M_k$ and then go to step 1. Otherwise exit.

In the first step of the algorithm, the initializations related to the cluster weights, the feasibility status and the gains of moves are carried out. There are $P_x \times P_y$ clusters which are divided by $P_y - 1$ horizontal lines and $(P_x - 1) \times P_y$ vertical lines. Since each line is associated with two moves, the number of possible moves is $2 \times ((P_x - 1) \times P_y + P_y - 1) = 2 \times (P - 1)$. These initial computed values are updated incrementally during the inner loop at step 3.

The second step of the algorithm is used to clear all flags before the execution of the inner loop. A locked move is not taken into consideration until it is unlocked during the iteration of the inner loop. On the other hand, a marked move is still considered during the iteration of the inner loop.

The third step picks a move with the highest gain from feasible and unlocked moves. This move is marked at step 3.(a) for a possible realization at step 5. On the other hand, the reverse move associated with the same line as the marked move is locked to prevent oscillations. The move M_i of a horizontal line or a vertical line segment does not prevent the further move of the same line or line segment in the same direction. This new candidate move is automatically relabeled as M_i again, and its gain is recalculated as indicated at step 3.(b).

A horizontal line move alters the nature of the adjacent vertical line segment moves. A vertical line segment and a horizontal line are considered to be adjacent if they form a corner of a cluster. Therefore, there are two successive rows of vertical line segments adjacent to an individual horizontal line. If a horizontal line move is realized, then each adjacent vertical line segment move corresponds to the move of a different group of nodes. That is, either one node is to be added or deleted from the original group of nodes. For this reason, the moves associated with these adjacent vertical line segments should be unlocked if they were locked and their gains should be updated incrementally as indicated at step 3.(c).

Similarly, a vertical line segment move alters the nature of the moves associated with the two successive horizontal lines which are adjacent to that vertical line segment. Hence, these adjacent horizontal lines should be unlocked if they were previously locked as indicated at step 3.(d). However, the gains of the moves associated with these two adjacent horizontal lines are not affected at all.

A vertical line segment move alters the weights of only two adjacent clusters which share this vertical line segment as a right or left boundary. A horizontal line move modifies the weights of the two successive row of clusters sharing this horizontal line as an up or down boundary. Hence, the weights of these clusters should be updated incrementally as indicated at step 3.(e). The change in the weight of a cluster may affect the current feasibility status of the moves associated with the lines or line segments which are adjacent to that cluster. A line or line segment is considered to be adjacent to a cluster if it constitutes a boundary for that cluster. Hence, the feasibility status of the adjacent line or line segment moves should be updated accordingly as indicated at step 3.(e).

Step 3 is iterated until no feasible and unlocked move remains as indicated at step 4. Then, at step 5, the sequence of moves which maximize the overall gain is selected. If this maximum gain $G_{max} > 0$, it means that the sequence of marked line or line segment moves $M_1, M_2, ..., M_k$ result in a further reduction in the overall interprocessor communication overhead. In this case, this sequence of moves is realized and then the outer loop (steps 1-5) is iterated once more. Note that, the selected moves are not really realized at step 3. In fact, the effects of these selected moves are simulated as if these moves are realized. If $G_{max} \leq 0$ then the algorithm is terminated since no further reduction in the overall interprocessor communication can be obtained.

Note that, at step 3, a selected move M_i with maximum gain may have a negative gain. Such a move is still considered for realization since it may result in moves with positive gains in the following cycles of the inner loop. This selection scheme enhances the KL algorithm with hill-climbing ability which is not available in other strictly local search methods.

Figure 5 illustrates a subgraph (two middle successive cluster rows) of a sample graph to be partitioned. Assume that, move M_j (left move of vertical line segment L_x) has a maximum gain and hence, selected at the kth iteration of step 3. Note that move M_j corresponds to the move of group of nodes $u_{5,4}, u_{6,4}, u_{7,4}$ from cluster $C_{y2,x1}$ to $C_{y2,x2}$. Hence, the gain of M_j is $g_j = \sum_{i=5}^{7} D(u_{i,4}, u_{i,5}) - \sum_{i=5}^{7} D(u_{i,3}, u_{i,4})$. Here, $D(u,v)$ is the total weight of the directed edges between node u and v. That is, $D(u,v) = d(u,v) + d(v,u)$. At step 3.(a), move M_j is marked as M_k with gain $g_k \leftarrow g_j$. The reverse move M_{j+1} (right move of vertical line L'_x) is also locked at step 3.(a). The gain of move M_j is recalculated as $g_j = \sum_{i=5}^{7} D(u_{i,3}, u_{i,4}) - \sum_{i=5}^{7} D(u_{i,2}, u_{i,3})$ L_{y-1} are adjacent to L_x. Hence, these lines are unlocked (if they were previously locked) at step 3.(d). The weights of the two clusters $C_{y2,x1}$ and $C_{y2,x2}$ are updated at step 3.(e) as follows: $W_{y2,x1} = W_{y2,x1} - \sum_{i=5}^{7} \omega(u_{i,4})$ and $W_{y2,x2} = W_{y2,x2} + \sum_{i=5}^{7} \omega(u_{i,4})$. The feasibility status of the moves associated with hori-

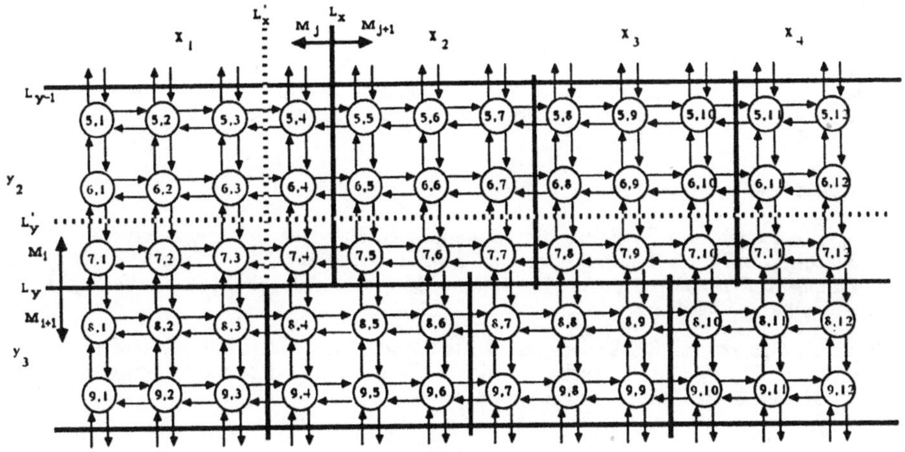

Fig 5. Sample graph to be partitioned.

zontal lines L_{y-1} and L_y and all vertical line segments lying between these two successive horizontal lines are updated, accordingly.

Assume that, move M_i (up move of horizontal line L_y) is selected in a later iteration of step 3. The dotted horizontal line L_y' denotes the new position of L_y if move M_i is realized. Move M_i as marked and the reverse move M_{j+1} is locked similarly at step 3.(a). The gain of move M_i is recalculated at step 3.(b) as $g_i = \sum_{j=1}^{12} D(u_{8,j}, u_{7,j}) - \sum_{j=1}^{12} D(u_{7,j}, u_{6,j})$. All vertical line segments seen in this subgraph are adjacent to the horizontal line L_y. Hence, the gains of the moves associated with all the vertical line segments are updated at step 3.(c). For example, $g_j = g_j - D(u_{7,3}, u_{7,4}) + D(u_{7,2}, u_{7,3})$. Similarly, these moves are unlocked if they were previously locked. For instance, previously locked move M_{j+1} is unlocked at this step. All clusters shown in the subgraph (Fig. 5) are affected by move M_i. Thus, the weights of these clusters are updated accordingly at step.3(e). All moves shown in the subgraph are adjacent to the updated clusters. Hence, the feasibility status of all possible moves shown in the subgraph are also updated at step 3.(e).

4 Conclusion and Future Work

The proposed heuristic algorithm explicitly tries to minimize the overall sum of inter-cluster edge weights while maintaining the load balance within a specified tolerance. The minimization of the sum of inter-cluster edge weights corresponds to the minimization of the interprocessor communication overhead since different clusters will be mapped to different processors.

Since the implementation of the proposed heuristic partitioning algorithm is still in progress, no performance results can now be presented. It seems to converge to a solution as in the original graph partitioning algorithm given by Fiduccia and Mattheyses. However, we are not sure of the ratio between the preprocessing phase time and the observed image generation time. In the near future, the efficiency and speed-up curves for the parallel ray tracing algorithm mapped to the hypercube using this proposed heuristic partitioning algorithm will be obtained.

In order to enlarge the search space for line moves in the given graph, the number of horizontal line moves might be increased as vertical moves. That is, both horizontal and vertical line segments can be allowed to move. However, in this case efficient graph theoretical methods should be developed to restrict the line segment moves in order to maintain the convexity of the resulting clusters.

Acknowledgements

This project is supported by the following grants and funds: Bilkent University Research Funds, The Scientific and Technical Research Council of Turkey (TÜBİTAK) Research Grant MAG917-EEEAG5, Intel Corporation Grant SSD100791-2.

References

[1] K. Bouatouch and T. Priol. Parallel space tracing: An experience on an ipsc hypercube. In N. M. Thalmann and D. Thalmann, editors, *New Trends in Computer Graphics, Proceedings of CG International'88*, pages 170–188. Springer-Verlag, Berlin Heidelberg, 1988.

[2] M. B. Carter and K. A. Teague. Distributed object database ray tracing on the intel ipsc/2 hypercube. Technical report, Dept. of Electrical and Computer Engr., Oklahama State University,USA, 1990.

[3] E. Caspary and I. D. Scherson. A self-balanced parallel processing for computer vision and display. In *International Conference*. University of Leeds, UK, 1988.

[4] J. G. Cleary, B. M. Wyvill, G. M. Birtwistle, and R. Vatti. Multiprocessor ray tracing. *Computer Graphics Forum*, 5:3–12, 1986.

[5] M. Dippé and J. Swensen. An adaptive subdivision algorithm and parallel architecture for realistic image synthesis. *ACM Computer Graphics*, 8(3):149–158, July 1984.

[6] C. M. Fiduccia and R. M. Mattheyses. A linear-time heuristic for improving network partitions. In *Proc. 19th Design Automation Conf.*, pages 175–181, 1982.

[7] M. R. Garey and D. S. Johnson. *Computers and Intractability: A Guide to the Theory of NP-Completeness*. W. H. Freeman and Company, New York,, 1979.

[8] S. A. Green and D. J. Paddon. Exploiting coherence for multiprocessor ray tracing. *IEEE CG&A*, pages 12–26, November 1989.

[9] S. A. Green and D. J. Paddon. A highly flexible multiprocessor solution for ray tracing. *The Visual Computer*, (6):62–73, 1990.

[10] S. A. Green, D. J. Paddon, and E. Lewis. A parallel algorithm and tree-based computer architecture for ray-traced computer graphics. In *International Conference*, University of Leeds, UK, 1988.

[11] B. W. Kernighan and S. Lin. An efficient heuristic procedure for partitioning graphs. *The Bell System Technical Journal*, pages 291–307, February 1970.

[12] H. Kobayashi and T. Nakamura. Parallel processing of an object space for image synthesis using ray tracing. *The Visual Computer*, (3):13–22, 1987.

[13] K. Nemoto and T. Omachi. An adaptive subdivision by sliding boundary surfaces. In *Proceedings: Graphics and Vision Interface '86*, pages 43–48. Canadian Information Society, Toronto, 1986.

[14] D. J. Plunkett and M. J. Balley. The vectorization of ray-tracing algorithm for improved execution speed. *IEEE CG&A*, pages 52–60, August 1985.

[15] T. Priol and K. Bouatouch. Static load balancing for a parallel ray tracing. *The Visual Computer*, (5):109–119, 1989.

[16] R. Pulleyblank and J. Kapenga. The feasibility of a vlsi chip for ray tracing bicubic patches. *IEEE CG&A*, pages 33–44, March 1987.

Divide and Conquer Radiosity

Robert van Liere

Abstract

This paper presents a coarse-grain parallel algorithm for solving the radiosity method. It describes a technique that partitions a large scene into a number of independent subscenes. The well known progressive refinement solution process will be applied to each subscene. After a fixed number of iterations each subscene will transfer accumulated energy to its neighbor. Although we have limited ourselves to only diffuse scenes, the algorithm can easily be extended to specular scenes.

1 Introduction

During the past few years the field of image synthesis has been dominated by two trends: visual realism and interactivity. On one hand, considerable research is being done on producing increasingly realistic global illumination models. On the other hand, mainly due to the high expectations of future workstations, a trend towards interactive algorithms has been set. Within this context, novel techniques are being developed that will allow global illumination models to be solved at interactive speeds.

The radiosity method, [5], has recently become a very popular solution to a limited, but in many cases adequate, global illumination model. Originally, the radiosity method was restricted to environments consisting of only ideal diffuse reflectors and emitters. Later extensions, [6] and [8], have relaxed these restrictions somewhat so that specular and translucent reflectors and emitters can now also be taken into account.

Implementations of the initial radiosity method resulted in the so-called hemi-cube methods, [4], and its superior progressive reformulation, [3]. Subsequent extensions resulted in various two pass approaches which use hemi-cube method to determine the diffuse to diffuse surface reflections and ray-tracing techniques to take the specular surfaces of the environment into account, [9] and [8].

Recently, some researchers have also focussed their attention to parallel implementations of the radiosity method. Reker et. al., [7], use multiple workstations to solve a number of steps of the progressive refinement method in parallel. This method applies the server/client approach in implementing a coarse-grain parallel solution. It relies on future network technology to solve the high communication band-width between clients and server. Baum et. al., [2], is an example of a finer grain parallel solution. This approach is unique in that it allows the end user to steer the radiosity calculations during the solution process. Finally, a completely alternative approach has been taken in a recent publication by Xu et. al., [10], in which a mathematical formulation is given of a method that subdivides a scene into multiple smaller scenes.

In this paper we present a parallel algorithm for solving the radiosity equations. We describe a technique to partition the complete set of radiosity equations into a number of independent smaller sets. Although we have limited ourselves to only diffuse environments, the algorithm can easily be extended to specular environments.

This paper is organized as follows: we first review the governing radiosity equations which are relevant for our environment. Next, we discuss the algorithm. Finally, we discuss convergence and error analysis of the method.

2 The Radiosity Equations

In [8], Rushmeier and Torrance extend the radiosity method to include specularly reflecting and translucent surfaces. Applying their results to surfaces that are *only* diffuse or ideal specular transmitting, we arrive at the following intensity equations:

- for ideal specular transmitting surfaces:

$$I_{o,n}(\theta_{o,n}, \phi_{o,n}) = I_{o,n(t)}(\theta_{o,n(t)}, \phi_{o,n(t)})$$

  Intuitively, this equation says that the outgoing intensity of surface n is equal to the outgoing intensity of the surface, denoted as $n(t)$, that intersects a ray specularly transmitted through surface n.

  Note that the intensities of ideal specular transmitting surfaces are direction dependent. This is indicated by the two angles, $\theta_{o,n}$ and $\phi_{o,n}$, denoting the polar and azimuthal angles.

  The equation is illustrated in figure 1, which also provides insight in the used notation.

- for diffuse surfaces:

$$I_{o,m} = I_{e,m} + \rho_m \sum_{n=1}^{N} \{I_{o,n}F_{mn} + \tau_n \sum_{q=1}^{N} I_{o,q}T_{f,nmq}\}$$

  Intuitively, this equation states that the outgoing intensity of diffuse surface m is equal to its emission plus a term due to other diffuse surfaces plus a term due to the other ideal specular transmitting surfaces.

  The term due to other diffuse surfaces is the same as in the original radiosity equations for diffuse surfaces, [3]. It includes the form-factor, denoted as F_{mn}, that specifies the fraction of the energy leaving one surface which lands on another.

  The term due to the ideal specular transmitting surfaces is equal to the sum of the outgoing intensities times the so-called forward window form factor, denoted as $T_{f,nmq}$. Window form factors denote the fraction of energy leaving the transmitting surface, n, from its *back* side which impings on surface m. Window form factors are direction dependent.

  The terms, ρ_n and τ_n, denotes the reflectance resp. transmittance function of surface n. Since we consider only diffuse surfaces this functions are independent of direction.

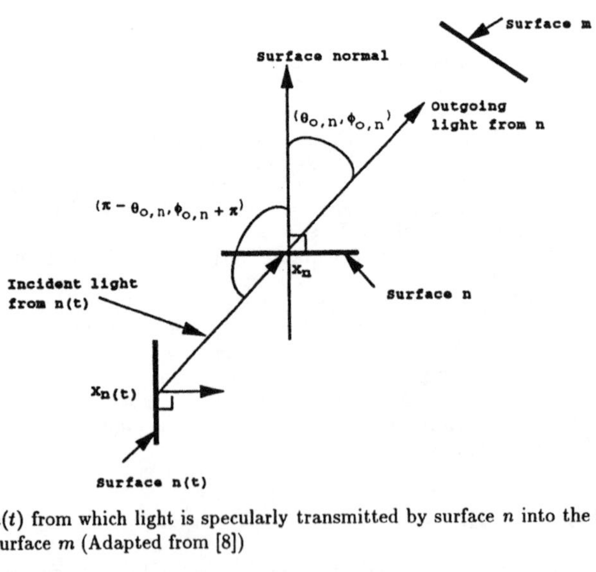

Fig 1. Surface $n(t)$ from which light is specularly transmitted by surface n into the direction $\phi_{o,n}$ and impinging onto surface m (Adapted from [8])

3 The Algorithm

3.1 An Intuitive View

Before going into the details, we first give a step-wise intuitive idea of how the algorithm works:

1. The algorithm starts by partitioning the scene to be rendered into a number of disjunct subscenes. Intuitively, this is done by placing a number of ideal translucent surfaces into to the scene. These translucent surfaces act as boundaries for individual subscenes. Each boundary has two sides, the front and back sides.
2. Next, apply in each subscene a number of iterations of the progressive refinement solution process. All energy that impings a boundary surface is stored in the front side of the boundary surface and will not be shot back into this subscene. Both the impinging energy and impinging angle are stored in the accumulation buffer. Note that since the subscenes are independent of each other that this step can be done in parallel for each subscene.
3. Transfer the accumulated boundary energy from one subscene to the boundary in the adjacent subscene. This is done by transferring the impinging energy from one accumulation buffer (stored in the front side) to the corresponding angle in the adjacent buffer (which will be stored in the back side of the boundary). The energy stored in the back side of a boundary will subsequently be shot into the subscene.
4. Display all subscenes without displaying the boundary surfaces. Strictly speaking, this step can be done in parallel with the energy transferring step, or after each iteration of the progressive refinement solution process in each subscene. We choose to display the scenes sequentially simply for clarity reasons.
5. Go back to step two and repeat the process.

3.2 A Closer Look

The more detailed discussion of the algorithm can best be described through the following fragments of pseudo-code.

```
subDivideScene ()

while (notConverged)
{
    foreach subScene
       doIterations (n, subScene)

    foreach boundary
       transferEnergy (boundary)

    displayScene ();
}
```

Pseudo-code fragment 1. - The main loop

This fragment of code contains the main loop which forks off a number of "workers" each executing the function *doIterations*. Note that during execution of *doIterations* there is no communication between workers. Once the workers have completed their tasks, the energy that has accumulated on each boundary is passed on to the adjacent subscene, *transferEnergy*.

```
doIterations (n, subScene)
{
    while (n--)
    {
        shootPatch (selectPatch (subScene))
    }
}
```

Pseudo-code fragment 2. - The worker

doIterations is a straightforward extension to the original sort and shooting algorithm. It selects a patch for shooting (the one with the greatest energy), and shoots its energy into the scene. The difference is that directions will have to be taken into account for boundary patches.

```
transferEnergy (boundary)
{
    foreach patch
    {
        transferEnergy (patch-> twin1, patch-> twin2)
        transferEnergy (patch-> twin2, patch-> twin1)
    }
}
```

Pseudo-code fragment 3. - The "synchronizer"

A boundary is divided in a number of patches. Each patch has two parts called *twins* . Twins are defined as: [1]

```
struct twin
{
    Radiosity incoming [SOLIDANGLES];
    Radiosity outgoing  [SOLIDANGLES];
}
```

in which *SOLIDANGLES* is a constant that determines the number of solid angles of the hemisphere. Note that the amount of memory used in a *twin* is proportional to the discretization accuracy of the hemisphere.

The following figure depicts a two dimensional view of the basic idea. The (progressively accumulated) incoming radiosity of *twin1* will be transferred to *twin2* which, in turn, will become a "shooting" candidate in its subscene.

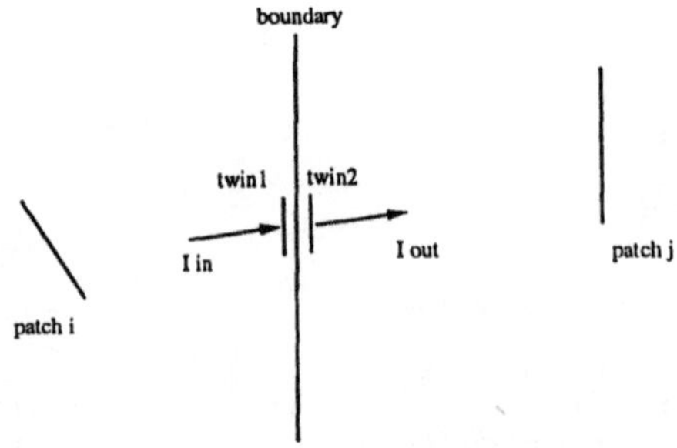

Fig 2. Twins on a boundary

Immel et. al., [6], have published an implementation technique that determines the radiosity in a non-diffuse environment. Aside from the convergence speed, Immel's method

[1] There is substantial room for memory optimization here.

suffers from the vast amounts of storage needed to store directional form-factors and directional radiosity. Our method also requires a vast amount of memory. Two comments must, however, be made:

1. memory needed for the boundaries can be distributed over a number processors/ workstations instead of only one central workstation.
2. additional memory is needed only for boundaries. This allows the additional memory needed to be a function of the available hardware resources and is independent of the complexity of the scene itself.

4 Convergence Analysis

In [3] , Cohen et al. introduce a metric for comparing convergence rates of various radiosity solution methods. The given metric provides a quantitative measure of the overall radiosity inaccuracy after each iteration in the progressive refinement solution process. The square root of the area weighted mean of the square of the the individual errors (RMS error) was used as a indication of the error made at each iteration; i.e.

$$RMS\ error = \sqrt{\frac{\sum_{i \in Env} ((I_i^* - I_i)^2 A_i)}{\sum_{i \in Env} A_i}}$$

where I_i^* is the converged intensity and I_i is the intermediate intensity of a patch i in an environment Env.

The following graph shows a hypothetical graph of the normalized RMS error set out against the number of iterations during the solution process. Function S denotes the convergence curve of the progressive refinement solution process in one environment as a function of the number of iterations; i.e. the sequential case. Function P denotes the convergence curve of the progressive refinement solution process in multiple subenvironments as a function of the number of iterations; i.e. the parallel case. Both functions will converge to the same result.

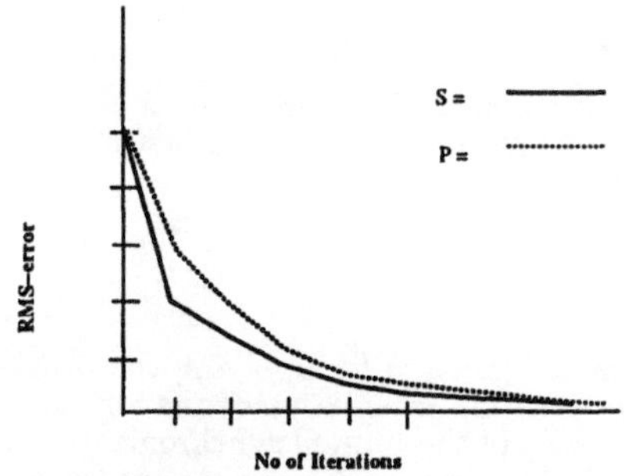

Fig 3. Plot of normalized RMS error for the sequential (thick) and parallel (dotted) cases

In general, the following statements can be made about the behavior of S and P:

1. For small n : $S(n) < P(n)$; i.e. after a small number of iterations the mean error made by the single environment method is smaller than in the multiple subenvironment case. This is due to a) extra boundary patches that are introduced in the subenvironment method, and b) a partial ordering is used in the case of multiple subenvironments to determine the shooting patch. Recall that patches in each subenvironment are sorted in increasing order. Partial orderings do not guarantee that a patch with the maximum accumulated energy is chosen as the shooting patch. Note that n is the sum of the iterations in all subenvironments.

2. For large n : $S(n) = P(n)$; i.e. both cases converge to the same mean error. This is because both methods converge to the same result.
3. For small n and $k > 1$: $S(n) > P(kn)$ with k as the number subenvironments; i.e. after n iterations in a single environment the mean error is greater than after n iterations in each subenvironment in the multiple environment case. This is because k patches are shot simultaneously. Hence, k times the number of iterations are made.

5 Error Analysis

The assumption made by the divide and conquer method is that introducing additional boundaries does not influence the results of light energy calculations in any way. In practice, however, additional errors are made during the form-factor calculations. The quantity of the errors depend in a complicated way on the scene being rendered. In practice, worst case errors are not very useful and "average case" errors are not well defined. We will, therefore, restrict ourselves to a qualitative analysis of errors that may occur.

Baum et al. [1] have shown that three types of errors can occur when the hemi-cube method is used to calculate form-factors:

1. errors due to the violation of the *proximity* assumption; i.e. the distance between patches i and j is small compared to the diameter of patch i. The proximity assumption is violated when, for example, i and j are adjacent patches that share a common edge. Since introduction of boundaries decreases the average distance between patches, the proximity assumption will be violated more often.
2. errors due to the violation of the *visibility* assumption; i.e. the visibility of a projected patch j does not alter across the shooting patch i. The visibility assumption is violated if another patch partially occludes the projected patch. Due to the additional boundaries, the visibility assumption will tend to be violated less frequently since there are a reduced number of patches per subenvironment.
3. errors due to the violation of the *aliasing* assumption; i.e. that patch j projects exactly onto whole pixels of patch i's hemi-cube. Introduction of boundary patches cause additional violations of the aliasing assumption. These errors can be minimized by, for example, using a larger hemi-cube.

Note that the divide and conquer algorithm does not mandate that form-factors be calculated with the hemi-cube method. Other form-factor calculation methods should be equally applicable.

6 Conclusion

We have presented a parallel algorithm for applying the progressive refinement method to the radiosity equations. The implementation of the algorithm is well suited for multiprocessor architectures because of the limited synchronization between processes.

We have also discussed convergence / error analysis of the algorithm. The convergence curve is far better than the sequential case. Additional errors are due to artifacts caused by the hemi-cube method for calculating form-factors.

The algorithm has been implemented in C++ on a network of Personal Iris workstations using SGI's graphics library for the low level rendering.

Future work will concentrate on: determining heuristics for optimal subscene division, dynamic subscene reconfiguration, minimizing memory consumption of boundaries.

Acknowledgements

Many thanks to Henk Schouten who contributed substantially to initial ideas and implementation issues. Thanks go also to Paul ten Hagen who provided me with the time and patience to work on these ideas.

References

[1] D.R. Baum, H.E. Rushmeier, and J.M. Winget. Improved radiosity solutions through the use of analytically determined form-factors. *Computer Graphics (SIGGRAPH '89 Proceedings)*, 23(3):325–334, 1989.

[2] D.R. Baum and J.M. Winget. Improving interaction with radiosity-based lighting simulation programs. *Computer Graphics (Interactive 3D Graphics)*, 24(4):51–57, 1990.

[3] M.F. Cohen, S.E. Chen, J.R. Wallace, and D.P. Greenberg. A progressive refinement approach to fast radiosity image generation. *Computer Graphics (SIGGRAPH '88 Proceedings)*, 22(4):75–84, 1988.

[4] M.F. Cohen and D.P. Greenberg. The hemi-cube : A radiosity solution for complex environments. *Computer Graphics (SIGGRAPH '85 Proceedings)*, 19(3):31–40, 1985.

[5] C.M. Goral, K.E. Torrance, D.P. Greenberg, and B. Bataille. Modelling the interaction of light between diffuse surfaces. *Computer Graphics (SIGGRAPH '84 Proceedings)*, 18(3):213–222, 1984.

[6] D.S. Immel, M.F. Cohen, and D.P. Greenberg. A radiosity method for non-diffuse environments. *Computer Graphics (SIGGRAPH '86 Proceedings)*, 20(4):133–142, 1986.

[7] R.J. Recker, D.W. George, and D.P. Greenberg. Acceleration techniques for progressive refinement radiosity. *Computer Graphics (Interactive 3D Graphics)*, 24(4):59–66, 1990.

[8] H.E. Rushmeier and K.E. Torrance. Extending the radiosity method to include specularly reflected and transluctent materials. *A.C.M. Transactions on Graphics*, 9(1):1–27, 1990.

[9] J.R. Wallace, M.F. Cohen, and D.P. Greenberg. A two-pass solution to the rendering equation: A synthesis of ray tracing and radiosity techniques. *Computer Graphics (SIGGRAPH '87 Proceedings)*, 21(4):311–328, 1987.

[10] H. Xu, Q-S. Peng, and Y-D. Liang. Accelerated radiosity method for complex environments. In W. Hansman, F.R.A. Hopgood, and W. Strasser, editors, *EUROGRAPHICS '89*, pages 51–61, Hamburg, 1989. North-Holland.

On the Division of Environments by Virtual Walls for Radiosity Computation

Bruno Arnaldi, Xavier Pueyo and Josep Vilaplana

Abstract

Based on previous works oriented to reduce the complexity of big environments, the paper presents a feasibility study of a dividing strategy. The original environment is divided into local environments introducing virtual walls. Then, the local environments are treated separately and local results are transfered to the neighbouring local environments.

1 Introduction

The work we present is based on the strategies first proposed by Xu et al. [7] . Also Neumann et al. [3] pointed to a solution based on the same idea: Given that the radiosity problem is $O(n^2)$, the efficiency of the algorithm could be improved by means of subdividing the environment into *local environments* (*l.e.*), using *virtual walls* (figure 1). In this way some aspects may be solved locally in each *l.e.*. This kind of solution is also useful for animation and parallel applications.

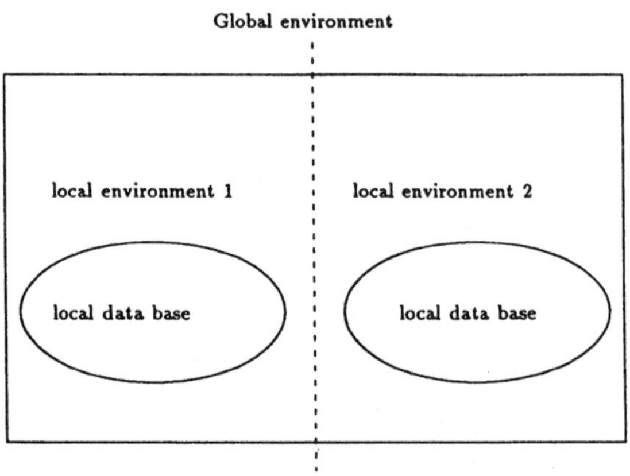

Fig 1. Local environments

In [7] the authors propose to consider virtual walls as being transparent, whereas in [3] the authors propose a method (for special cases) where virtual walls are modelled by black surfaces. To solve the equations, Neumann et al. propose to compute an approximated solution for each subsystem considering the others as light emitters. When all the subsystems have been treated, the process starts again with the first system using the previous approximation of neighbouring subsystems. Nevertheless, Xu et al. use the division technique to express form factors between a patch i of a *l.e.* and patch j of another as a combination of form factors of i and j to a virtual patch k ($F_{ij} = f(F_{ik}, F_{kj})$). Finally they solve the whole global equation.

We present below a feasibility study of the use of virtual walls for diffuse environments. Our goals are:

- To find a model of virtual wall that is not too expensive.

- To obtain a useful radiosity solution.
- To analyse the possibilities offered by this technique for parallel solutions.

2 Algorithm Overview

Let's consider the 2D environment in figure 1. The form factors are locally computed using a classical technique like the hemicube strategy, so dividing the computational cost by four if the data are well balanced in the two *l.e.*. But, each *l.e.* must take into account the influence of its neighbour. This influence is modelled by the virtual wall transfer of light and increases the computational cost of the *l.e.* strategy.

After computing the form factors of each *l.e.* and the model of the virtual wall, we obtain two local systems of linear equations to be solved accounting on the light transfer through the virtual wall in order to obtain a global energetical equilibrium. We show in figure 2 a scheme of the process.

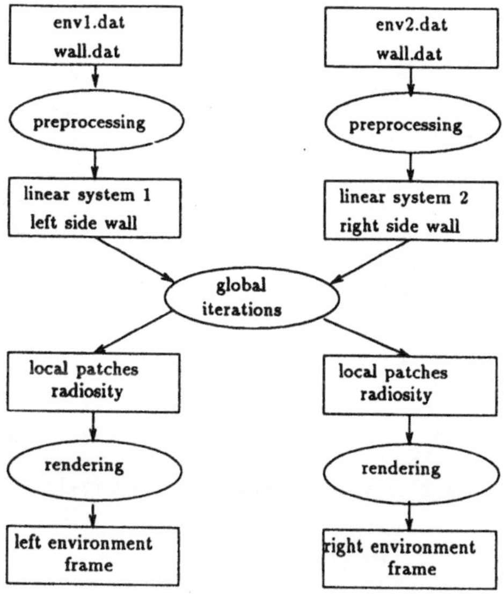

Fig 2. Process of the local environments

3 Virtual Walls

Virtual walls are composed of a set of patches (*virtual patches*) which have two different behaviors: as receiver and as emitter. Let's j be a patch of the *l.e.*1, i a patch of the *l.e.*2 and v a virtual patch.

The behavior of patch v as receiver is the same as for real patches [2]. So, F_{vj} will be computed as usual(respec. for F_{vi}) and two different B_v are computed at each side of the virtual patch v. Now, each B_v must be transferred to the neighbour *l.e.*. Let us consider the radiosity B_v received by the patch in the *l.e.*1. This radiosity must now be transfered to the *l.e.*2 in a directional way. As shown in figure 3, the contribution of patch i to B_v must be transfered in such a way that only affects a portion of *l.e.*1. So, we will transfer B^1_{vi} (contribution of patch i to B_v) for every patch i of the *l.e.*2 to the *l.e.*1. Thus the behavior of patch v as emitter will be the same as a set of emitters located in v and we need to compute a set of directional form factors f^i_{jv} for every patch i in *l.e.*1.

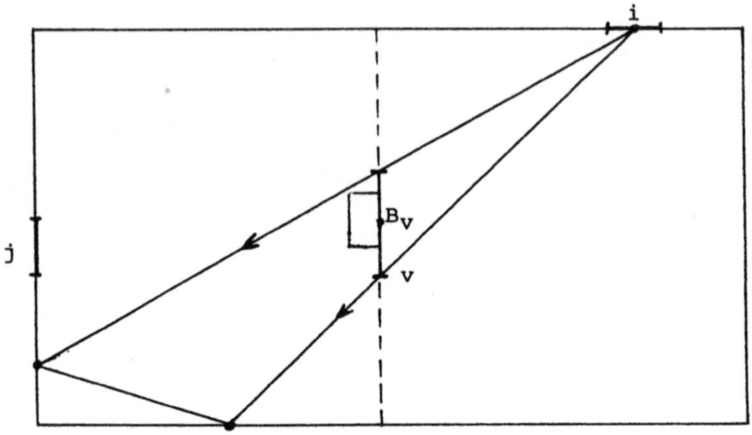

Fig 3. Directional transfer of light

4 Radiosity Equations

From section 3 we may derive a linear equation system, for each *l.e.*, with $n_1 + n_v n_2$ unknowns, where n_1 and n_2 are the number of patches of *l.e.*1 and *l.e.*2 and n_v is the number of virtual patches. These unknowns are the n_1 radiosities of the real patches and the $n_v n_2$ transfered by the n_v virtual patches in the n_2 directions defined by the patches of *l.e.*2. Thus for the environment of figure 3, the algorithm could be,

```
while not global convergence {
   local_resol(environment_1,.... );
   /* exchange from one environment to the other */
   energy_exchange(virtual_wall, environment_1, environment_2);
   local_resol(environment_2,.... );
   energy_exchange(virtual_wall, environment_2, environment_1);
}
```

This resolution of the local environments would be too expensive but its analysis shows a simpler solution. Let's first solve,

$$[\mathbf{F}|\mathbf{f}]\begin{bmatrix}\mathbf{B}\\ \mathbf{B}_d\end{bmatrix} = [\mathbf{E}]$$

where

$$\mathbf{F} = \begin{bmatrix} 1 & -\rho_1 F_{12} & \cdots & \cdots & -\rho_1 F_{1n_1} \\ -\rho_2 F_{21} & 1 & \cdots & \cdots & -\rho_2 F_{2n_1} \\ \cdots & \cdots & \cdots & \cdots & \cdots \\ \cdots & \cdots & \cdots & \cdots & \cdots \\ \cdots & \cdots & \cdots & \cdots & \cdots \\ -\rho_{n_1} F_{n_1 1} & -\rho_{n_1} F_{n_1 2} & \cdots & -\rho_{n_1} F_{n_1 n_1 - 1} & 1 \end{bmatrix}$$

$$\mathbf{f} = \begin{bmatrix} -\rho_1 f_{11}^1 & \cdots & -\rho_1 f_{11}^{n_2} & \cdots & -\rho_1 f_{1n_v}^1 & \cdots & -\rho_1 f_{1n_v}^{n_2} \\ \cdots & \cdots & \cdots & \cdots & \cdots & \cdots & \cdots \\ \cdots & \cdots & \cdots & \cdots & \cdots & \cdots & \cdots \\ \cdots & \cdots & \cdots & \cdots & \cdots & \cdots & \cdots \\ -\rho_{n_1} f_{n_1 1}^1 & \cdots & -\rho_{n_1} f_{n_1 1}^{n_2} & \cdots & -\rho_{n_1} f_{n_1 n_v}^1 & \cdots & -\rho_{n_1} f_{n_1 n_v}^{n_2} \end{bmatrix}$$

$$\mathbf{B} = \begin{bmatrix} B_1^1 \\ B_2^1 \\ \vdots \\ B_{n_1}^1 \end{bmatrix} \qquad \mathbf{B}_d = \begin{bmatrix} B_{11}^2 \\ \vdots \\ B_{1n_2}^2 \\ \vdots \\ B_{n_v 1}^2 \\ \vdots \\ B_{n_v n_2}^2 \end{bmatrix} \qquad \mathbf{E} = \begin{bmatrix} E_1 \\ E_2 \\ \vdots \\ E_{n_1} \end{bmatrix}$$

for the *l.e.*1 in order to obtain the result of the interreflections between the real patches of this *l.e.*.

B_{vi}^2 ($i = 1, 2..., n_2$, $v = 1, 2, ..., n_v$) are the radiosities transfered from the *l.e.*2, so constants at each transfer (emitter's behavior). These radiosities affected by the corresponding directional form factors will then be added to the emissivity of the real patches of *l.e.*1 before solving the local system with n unknowns B_j^1. After computing these radiosities, we compute the B_{vj}^1 to be transfered from *l.e.*1 to *l.e.*2 being,

$$B_{vj}^1 = F_{vj} B_j^1 \qquad j = 1, 2, \ldots, n_1, \quad v = 1, 2, \ldots, n_v$$

5 Evaluation and Experimentation

In this section we present a simplified evaluation of the computational cost of the local environments algorithm as well as the results of a first implementation. This evaluation is based on the environment of figure 3. Each *l.e.* is composed by 5 surfaces (real walls) of the same size and the virtual wall have also the same size. Both, real walls and virtual wall are divided into the same number of patches.

5.1 Form Factor Calculation

As well known, the form factor computation is assumed to be $C_g = (n_1 + n_2)^2.h$ (for our test environment) where h is the cost of the hemicube process.

Let's call v_w the cost due to the process of one side of the virtual wall. Then the cost of processing the two *l.e.* is,

$$C_{le} = n_1^2 h + n_2^2 h + 2v_w$$

and, for *l.e.*1,

$$v_w = n_v h(n_1 + n_2) + n_v n_2 4c_i$$

where $4c_i$ is the cost of computing the area of influence of a patch of *l.e.*2 in the *l.e.*1, being c_i the cost of the computation of the four vertices of the area of influence.

From the specification of our test environment,

$$n = n_1$$

$$n_v = n/5$$

On the other hand we call n_d to n_2 given that it represents the number of directional transfers for each virtual patch. Then,

$$C_g = 4n^2h$$

and

$$C_{le} = \frac{12}{5}n^2h + n_dn(\frac{2}{5}h + \frac{8}{5}c_i)$$

In order to get $C_{le} < C_g$,

$$\frac{12}{5}n^2h + n_dn(\frac{2}{5}h + \frac{8}{5}c_i) < 4n^2h$$

so,

$$n_d < \frac{4nh}{h + 4c_i}$$

Assuming,

$$c_i \simeq n_sh = \frac{n}{r_d}h$$

where, n_s is the number of surfaces of a *l.e.* and r_d is the environment discretization rate ($r_d = n/n_s$),

$$n_d < \frac{nr_d}{\frac{1}{4}r_d + n} \simeq \frac{nr_d}{n} = r_d$$

Thus, in order to obtain a reasonable cost for the *l.e.* strategy in front of the global environment solution, each virtual patch must transfer the light in $n_d < r_d$ directions. This may be done by grouping patches on macropatches in the *l.e.*2 in order to define the n_d directions.

From an error analysis of the *l.e.* solution, we could obtain a n_d^e which minimizes the error produced by the virtual wall. Then, from n_d and n_d^e we could derive the right values of the parameters (size, n_v, n_d) defining an efficient virtual wall.

5.2 Equation Solution

In radiosity algorithms this step is considered of $O(n^2)$ so the cost of solving the complete system of our test environment is considered,

$$C_g^e = (2n)^2$$

and for the two *l.e.*,

$$C_{le}^e = 2n^2 + 2n^2n_v$$

Thus, $C_{le}^e > C_g^e$; but a big amount of f_{jv}^i is zero. If we call r_s the sparse ratio of the submatrix f containing the directional form factors, then,

$$C_{le}^e = 2n^2 + 2n^2n_vr_s$$

where,

$$r_s = \frac{nn_vn_p}{n^2n_v} = \frac{n_p}{n}$$

being n_p the average number of patches affected by a given direction of n_d.
Thus, in order to obtain $C_{le}^e < C_g^e$,

$$2n^2 + 2n^2n_vr_s < 4n^2$$

concluding that the resolution of the *l.e.* systems is less time consuming than the resolution of the system corresponding to the global environment if,

$$r_s < \frac{1}{n_v}$$

or

$$\frac{n_p n_v}{n} < 1$$

5.3 Experimentation

For our test we have used a radiosity package offered by S.N.Pattanaik to obtain a solution of the global environment presented in figure 5 (on color plates). The room contains 300 patches with one light source located on the upper left corner of the front plane.

In order to test the *l.e.* strategy we have divided the room by means of a virtual wall and used the above package modified to obtain the scheme of figure 2. The result is shown in figure 6 (on color plates) where the right environment has 143 patches, the left environment has 157 patches (4 light patches) and the virtual wall is divided into 36 virtual patches. In this figure, the colors of the right environment have been scaled to compensate for the error introduced by the technique.

6 Parallelization

One of the aims of our work is to find an algorithm allowing a mapping scheme for parallel distributed systems. Classic radiosity algorithms may decompose into independent tasks (for example, the task of hemi-cube computation for a form-factor matrix's row evaluation) requiring all the environment data-base for their computation. With this decomposition there is not an obvious way to distribute the data-base amongst the local memories of the network. In [5] all the database is replicated in the local memory of each processor and in [4] the database is accessed by the processors as messages flowing in a ring network.

The parallel scheme of the local radiosity computation can be performed in a natural way. Each processor takes a local data base and the virtual wall structures lying in the concerning area (one side of the wall). The local form factors, the virtual wall structures and local equation solution can be computed in parallel for each environment.

The energy exchange may be seen as a synchronization point between processors. In this step, they transfer directional energy to their neighbours. The figure 4 shows an example with two environments :

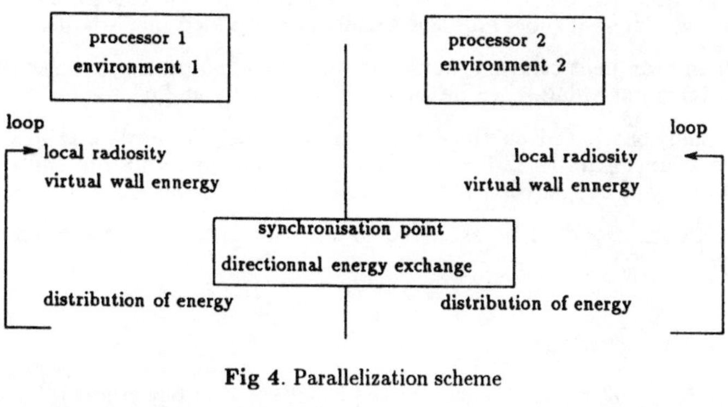

Fig 4. Parallelization scheme

Mapping nearest local environments to nearest local memories favours a local message flow amongst nearest-neighbour processors.

Before a parallel implementation, some points must be particularly studied :

Load balancing amongst processors depends on the data base distribution related to the virtual walls location strategy. Size of local database is the significant factor in the cost of a local radiosity computation. Similar sizes of local environments and virtual walls in all the local memories could guarantee a good load balancing. The division of the environment would be made according to the size of its spatial subdivisions.

The synchronization point involves a message passing between processors. In current technology, communication and computation in one processor can be done in parallel. If the communication time between processors is greater than the computation time to process it, idle times will be present affecting the performance. The amount of information or the time needed to transfer (directional energy) depends on the number of discrete directions and the number of patches in the virtual wall at the same time as these numbers influence the precision of the final solution.

7 Conclusions

From our study we conclude:

From a first analysis, the sequential solution of the *l.e.* strategy seems to be more time consuming than the classic solution.

The selection of an adequate number of transfer directions (n_d) at the virtual patches reduces the cost of the *l.e.* algorithm. On the other hand, the efficiency of the method may be also improved exploiting the coherence between virtual patches (*surface coherence* [6]).

The characteristics of the *l.e.* radiosity equation make it easy to get on with.

The relation of the cost parameter with the error introduced by the virtual wall must be studied in order to determine whether virtual walls may or not be useful. In this sense, we think that virtual walls will be suitable when located at doors, windows, ...

The location of the virtual walls is essential for both load balanced distribution of the environment and reduction of the visibility problem.

The transfer directions on a virtual patch will not be uniformly distributed; but a selection will be made based on the emissivity of real patches. This is automatically done if more emissive surfaces are divided into smaller patches.

The *l.e.* environment strategy would be improved computing, for each *l.e.*, an ambient term like value [1] to be added to its neighbour *l.e.*.

Parallel solutions based on the *l.e.* techniques must be developed in order to compare against parallel radiosity solutions without virtual walls and to study the suitability of cooperation between both parallel techniques.

The *l.e.* algorithm could be extended to the use of the progressive refinement strategy.

8 Acknowledgements

We would like to thank S.N.Pattanaik, of the NCST of Bombay (India), for his radiosity implementation which is a good experimentation tool due particularly to the intermediate data storage capabilities.

References

[1] M. F. Cohen, S. E. Chen, J. R. Wallace, and D. P. Greenberg. A progressive refinement approach to fast radiosity image generation. *ACM Computer Graphics*, 22(4):75–84, Aug. 1988.

[2] M. F. Cohen and D. P. Greenberg. The hemi-cube: A radiosity solution for complex environments. *ACM Computer Graphics*, 19(3):31–40, July 1985.

[3] L. Neumann and A. Neumann. Photosimulation: interreflection with arbitrary reflectance models and illumination. *Computer Graphics Forum*, 8(1):21–34, 1989.

[4] W. Purgathofer and M. Zeiller. Fast radiosity by parallelization. In K. Bouatouch and C. Bouville, editors, *Photorealism in Computer Graphics*, pages 171–181. Springer-Verlag, 1992.

[5] R. J. Recker, D. W.George, and D. P. Greenberg. Acceleration techniques for progressive refinement radiosity. *ACM Computer Graphics*, 24(2):59–66, Mar. 1990.

[6] J. Vilaplana and X. Pueyo. Coherence for clipping and view transformations in radiosity algorithms. In K. Bouatouch and C. Bouville, editors, *Photorealism in Computer Graphics*, pages 137–149. Springer-Verlag, 1992.

[7] H. Xu, Q. Peng, and Y. Liang. Accelerated radiosity method for complex environments. *Proc. of Eurographics'89*, pages 51–61, sep 1989.

Texture Shaping: A Method for Modeling Arbitrarily Shaped Volume Objects in Texture Space

Georgios Sakas, Bertram Kernke

Abstract

Texturing methods developed in the past few years are well suited for 'solid objects', which possess a well-defined and sharp surface, but they are inadequate for texturing 'foggy' or 'fuzzy' volume objects, which do not possess such surfaces. Here we propose an approach for quick generating density distributions defining volume objects of arbitrary shape. Our approach is a modified spectral synthesis method operating on the spectral representations of a texture and a main body and generates discrete voxel fields of the desired body shape. For that purpose, a spectral representation of the main body and the texture are calculated first, e.g., by means of a Fourier transform. After this, the amplitude and/or phase spectra of the body and texture are first filtered adequately and then 'merged' together. The resulting spectrum is back-transformed to the (Euclidean) voxel space by means of inverse Fourier transformation. With this approach it is possible to generate continuous transitions of an amorphous and random texture to a desired shape at low computational cost.

1 Introduction and Motivation

Texturing is the most efficient and low-cost technique for adding details and enhancing the optical complexity of computer-generated objects. It is also an inexpensive method for simulating the details of an object without modeling them explicitly. Usually, the complete texturing process can be seen as consisting of two steps: texture mapping and texture sampling ([5], [8], [4]). During texture mapping texture coordinates are calculated from the object coordinates at a particular location of interest (e.g., the location of the ray-object intersection during ray tracing). The sampling step involves evaluating the texture values at the location of the calculated texture coordinates. By means of this method, the definition of the object geometry and the generation of the texture data are totally uncoupled; these two attributes of a body, i.e., geometry and texture, remain 'alien' during the modeling phase and are linked together 'a posteriori' through the texture mapping step, which is evaluated during the rendering process. Nevertheless, this method works satisfactorily for texturing solid object surfaces using 2-D textures (color patterns, photographs, bump maps, etc.).

A slight variation of the above-mentioned method is solid texturing ([17], [16], [15], [10], [19]). In this case the texture is specified as a spatial 3-D pattern (e.g., a voxel field or function in the x, y, and z coordinates) defining a 'unity block' from which the desired body is 'sculptured'. This method is well-suited for texturing bodies whose surface pattern is conditioned by an internal spatial structure, such as in wood and marble. As in the previous case, solid texturing is also suited for texturing bodies with a well-defined (solid) surface.

In recent years a new category of bodies has become popular: so-called volume objects. These 'cloudy' objects (e.g., smoke, clouds, steam, etc.) are best described as spatial density variations, and therefore do not provide anything like a well-defined surface. Usually, volume objects are defined by a solid shape, which defines the space occupied by the object, and a 3-D texture attached to the shape, which defines the spatial distribution of the volume object within the solid. Such 3-D textures are generated as a deterministic or stochastic 3-D patterns (e.g., Saturn rings function in [1], stratified fog density in [13], fractals in [22], 'turbulence function' in [17], etc.), and rendered by illumination models that simulate the absorption and scattering of light within participating media ([1], [9],

[18], [3], [20]). Although the *all-round spatial occupation* of the object is defined over a solid body, also called bounding volume, the *volume object itself* does not have a well defined surface. This absence of a sharp or solid surface makes the use of the usual modeling and texturing methods mentioned above difficult or even impossible. Therefore, the traditional object definition method (solid object & 3-D texture) does not always allow for an efficient control of the shape of the generated density distribution. As an example, suppose one wants to generate a cloud having roughly the shape of an ellipsoid by means of 3-D texturing. Using the methods described in [22], one can generate a fractal of the appropriate resolution which closely approximates the density distribution of a real cloud within the unity cube. Further, by means of solid texturing an ellipsoid can be filled with this cloud texture. As can be seen in fig.1, the defined cloud has very sharp boundaries due to the underlying solid object used (in this case a unity cube or sphere). This makes its appearance rather unrealistic. This paper will introduce a method which permits the fast generation of volume objects of any shape and thus avoids the shortcomings mentioned above.

2 Texture Shaping

Analyzing the reason for the unrealistic appearance of the ellipsoid cloud mentioned in the previous section, one can see that neither the geometric shape nor the used texture are responsible for the poor result. As Gardner showed in [6], ellipsoids are well-suited for defining the rough shape of a cloud. Further, Mandelbrot, Voss, and Lovejoy ([12], [22], [11]) showed that fractals closely approximate the vapor density distributions occurring in real clouds. Therefore, the reason for the poor result lies in the inadequate method of combining them by means of traditional solid texturing. Speaking about clouds as typical representatives of volume objects, such objects show a roughly defined abstract all-round shape (such as an ellipsoid in our case) with a rather wide, diffuse (as opposed to sharp), and usually random or turbulent boundary zone, which 'dies out' at some distance from the main body (cluster). The width, or thickness, of the zone around the main object in which this transition takes place varies at different locations and shows a random character itself. The definition of this zone is more difficult when a thin and non-symmetric, or even irregular, shape is involved, like the one shown in fig.5. Due to this irregular surface, the definition of a 'bleeding-out' function, like those suggested in [15], is usually not possible. Therefore, scaling down the texture density within a transition area around an object with a well-defined surface by means of a bleeding-out function will produce good results only in limited cases, where regular, or symmetrical, objects are involved (see fig.2). In addition, the continuous and smooth 'metamorphosis' of an amorphous cloud to a well-defined shape is not straight-forward.

The basic idea of texture shaping involves *forcing the process, which generates the density distribution of the volume object, to produce directly data having the desired all-round shape.* In other words, the target shape is concerned prior to the texture generation (a priori). This is in contrast to the usual modeling and texturing methods which generate object geometry and texture data independently of each other and link them together (a posteriori) during the rendering process.

The method introduced here is used for generating natural 'foggy' volume objects, like smoke or clouds, as discrete voxel representations. We use a spectral synthesis method which operates on the texture and body representations in the frequency domain. Euclidean space and frequency domain representations are equivalent to each other, and are coupled together by means of the Fourier transformation [2]. Since we deal with voxel data, we only use the discrete Fourier transformation DFT. For computational efficiency we restrict ourselves to quadratic power-of-two fields and invoke the Fast-Fourier-Transformation FFT.

The shape of the texture data generated by spectral synthesis depends on the form of the amplitude and phase spectrum in the frequency domain. Speaking in terms of Fourier analysis, the rough geometry of a body is mainly given by the low frequencies around the spectrum origin, whereby the boundary structure and sharpness are defined by the high frequencies [7]. One can isolate these frequencies from a given spectrum by using a low-pass or a high-pass filter.

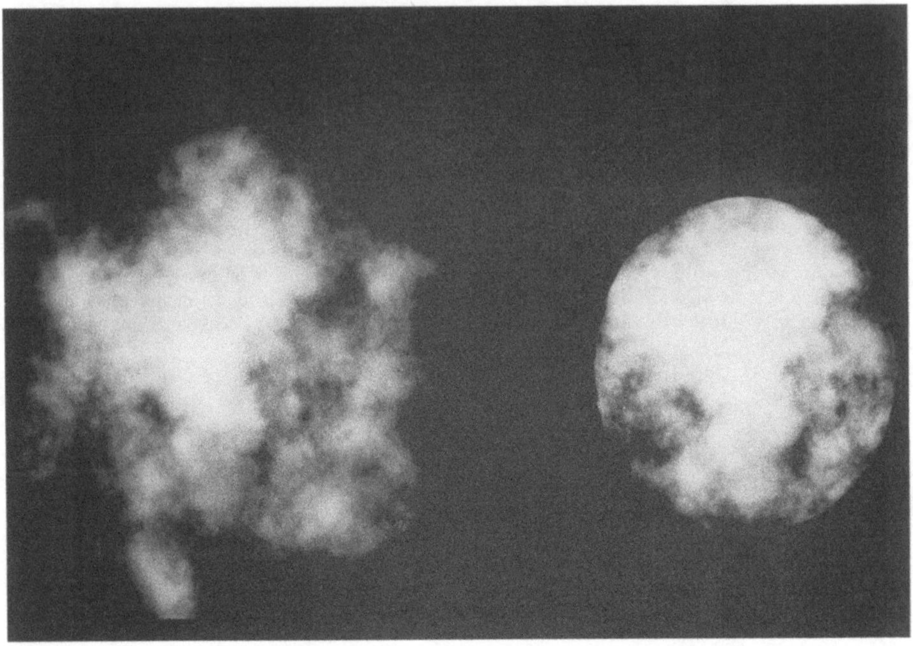

Fig. 1. Solid texturing of a unity cube (left) and a sphere (right).

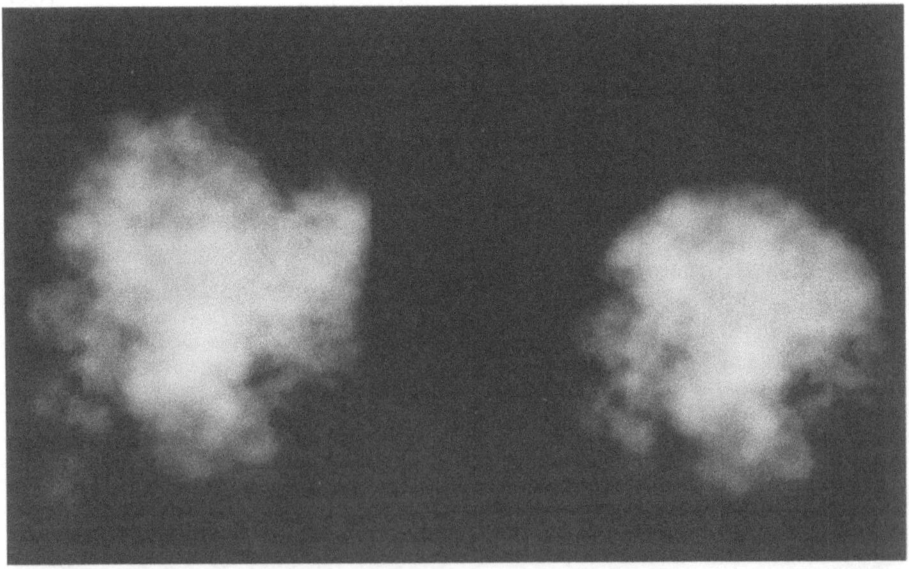

Fig. 2. Scaling the texture from fig. 1 along an axis (left) and around a point (right).

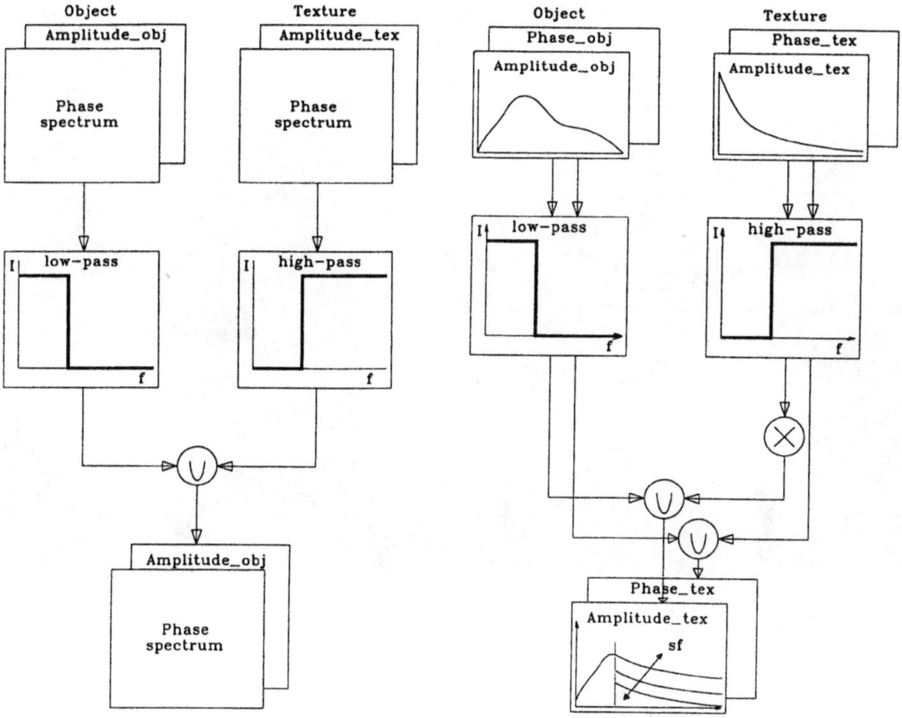

Fig 3. Phase substitution. Phase and amplitude substitution.

With this in mind, the approach consists in 'mixing' the spectra of the solid body and the density texture. For that purpose we first generate the discrete spectra for the main body and for the texture at the appropriate resolution. In our case, for the cloud texture we use the well-known $1/f^{\beta}$ amplitude spectrum and a $[0, 2\pi)$ randomly distributed phase spectrum [22]. The main characteristic of this spectrum is that the majority of the energy is concentrated around the origin. We tested only this type of spectrum, but other texture spectra may be used as well. In order to gain the spectral representation of the main body, we first generate a voxel approximation of it and then transform it by using the FFT. For the sake of simplicity, we use the same voxel resolution for the texture and body spectra. We examined two possibilities for mixing the two spectra: mixing only the phases, and mixing amplitudes and phases. Both cases are discussed below.

3 Phase Spectrum Manipulation

In this case we operate only on the phase spectra of the two bodies; the amplitude spectrum of the texture remains thereby unchanged. First we isolate the phases of the low frequencies of the main body by means of a low-pass (see fig.3 left). In the same way, we filter the cloud by using a high-pass. Both filters have the same cut-off frequency. Next, we add both phase spectra together building their union. The result is then back-transformed to voxel space.

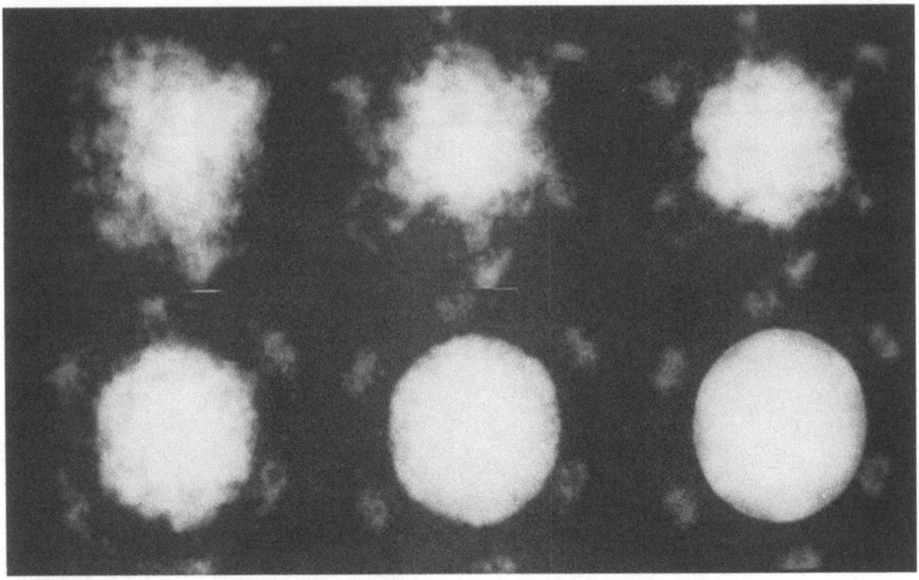

Fig 4. Varying the cut-off frequency for the phase substitution method. The cut-off is from upper-left to lower-right 1, 2, 3, 5, 10, and 32; the texture resolution is 64^3.

The idea behind these operations is that one wishes to prevent the foggy characteristics of the texture, but wants to shape the cloud in the form of the main body. Therefore, the $1/f^\beta$ amplitude spectrum is not changed, and the frequencies around the origin are only shifted to the locations defined by the phase spectrum of the main body. Since these locations are gained by a low-pass filter from the main body spectrum, and the corresponding cloud frequencies include the majority of the spectrum energy, the body shape is approximated.

The accuracy of the approximation can be influenced by means of the cut-off frequency. The higher the cut-off, the more accurate the approximation; i.e., the boundaries of the texture become sharper. This effect can be seen in fig.4; the cut-off frequency is increased from 0 to 100%. Please note that even when 100% is used, the generated texture is not identical to the original body because the amplitude spectrum $1/f^b eta$ is, in general, not identical to the amplitude spectrum of the main object. This can be seen, in particular, in the errors at the corners of the original cloud cube (the last motive in fig.4). In addition, the transition of the cut-off frequency from 0 to 100% can be done gradually from frame to frame as indicated by the snapshots.

4 Amplitude and Phase Spectrum Manipulation

In the second case the amplitude spectrum is manipulated in the same way as the phase spectrum. The operations are identical with those in the preceding case, but include both types of spectra (see fig.3 right). Due to the consideration of both the amplitudes and the phases, the shape of the main object can be approximated more accurately (see fig.5). Again, the cut-off frequency influences the approximation accuracy. In comparison to the previous method, the errors at the cube corners are greatly reduced; due to the amplitude substitution the shape of the object is approximated more precisely. Again, by means of a continuous change of the cut-off frequency, a smooth and continuous transition between the 'key-frames' is possible.

In addition to the cut-off frequency we now have a second parameter to influence the texture appearance. As can be seen in fig.3 right, the amplitudes of the filtered spectra

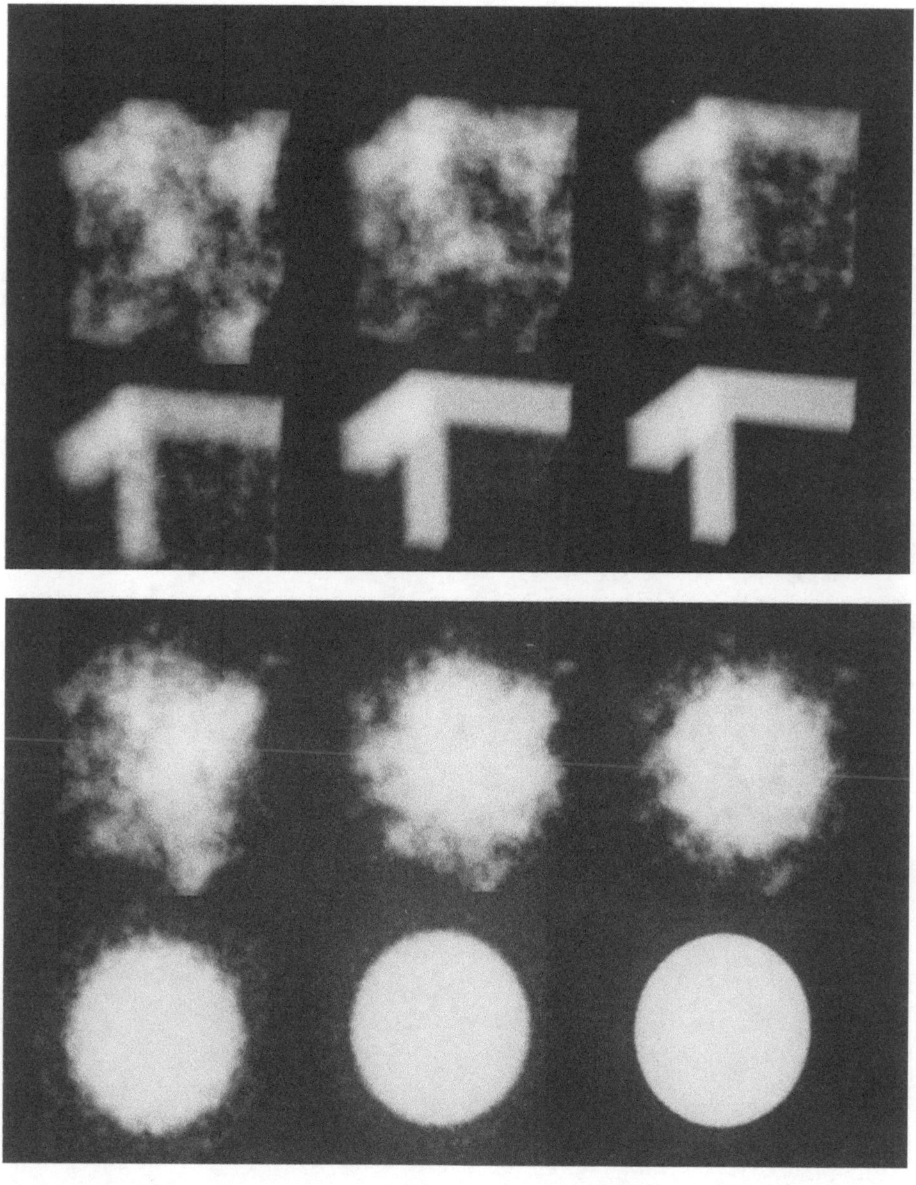

Fig 5. Varying the cut-off frequency for the phase and amplitude substitution method. The cut-off is form upper-left to lower-right 1, 2, 3, 5, 10, and 32; the texture resolution is 64^3; the energy equalization method has been used.

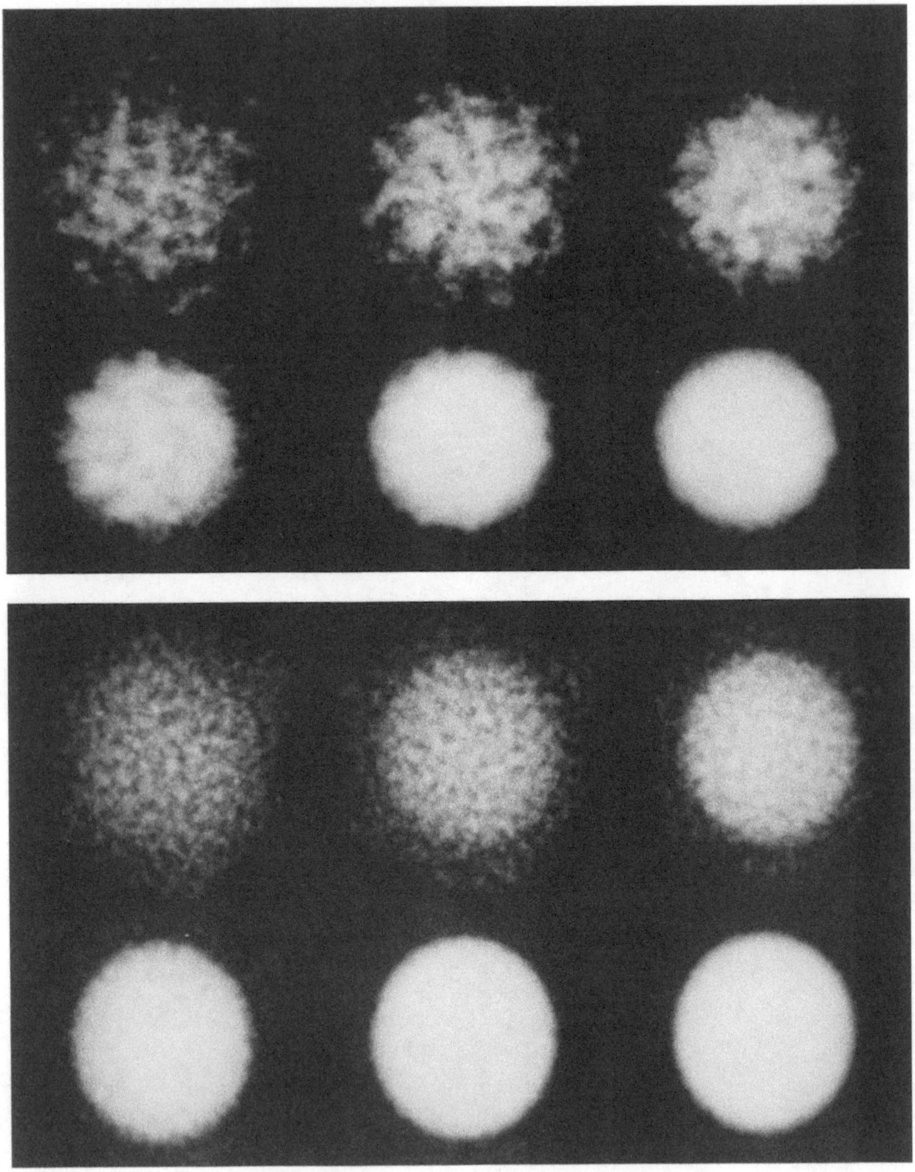

Fig 6. Varying the scaling factor from upper-left to lower-right. The texture resolution is 64^3. The cut-off frequency for the upper picture is 3.5, for the lower picture 7.

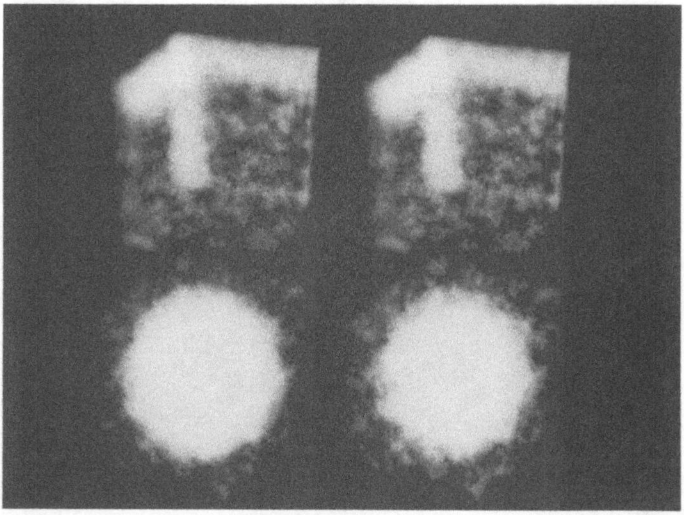

Fig 7. Demonstration of the amplitude equalization (upper) and the energy equalization method (lower) for a cut-off of 3.5 and a texture resolution of 64^3.

can be scaled by a particular factor prior to spectra unification. This is usually necessary since the two spectra generally have different scales. The choice of the scaling parameter is defined by the user. By means of the scaling factor the high-pass or low-pass character of the result can be amplified or attenuated without altering the texture database.

As a help for the user, two predefined methods are included in the system: amplitude equalization and energy equalization. In the case of amplitude equalization, the scaling factor is defined by the arithmetic averaging of the low-pass filtered amplitudes of the body and the texture. The arithmetic average of the body amplitudes is

$$\bar{B}_{amp} = \frac{1}{f_{cut}} \sum_{0}^{f_{cut}} B_{amp}$$

and of the texture

$$\bar{T}_{amp} = \frac{1}{f_{cut}} \sum_{0}^{f_{cut}} T_{amp}$$

Then the scaling factor sf is calculated as

$$sf = \frac{\bar{B}}{\bar{T}}$$

and the unified spectrum becomes

$$\begin{aligned} S_{amp} &= LowPass(f_{cut}, B_{amp}) \cup (sf \times HighPass(f_{cut}, T_{amp})) \\ S_{pha} &= LowPass(f_{cut}, B_{pha}) \cup HighPass(f_{cut}, T_{pha}) \end{aligned}$$

In the case of energy equalization, the factor is calculated in such a way as to equalize the energy filtered out. The calculation is the same as above, only in this case the squared amplitude is used and the geometrical average is calculated. The result for both methods is illustrated in fig.7. In our examples, for a given cut-off, amplitude equalization produces a denser and sharper boundary, while energy equalization produces a rather random or turbulent boundary. However, this empirical result has to be further verified.

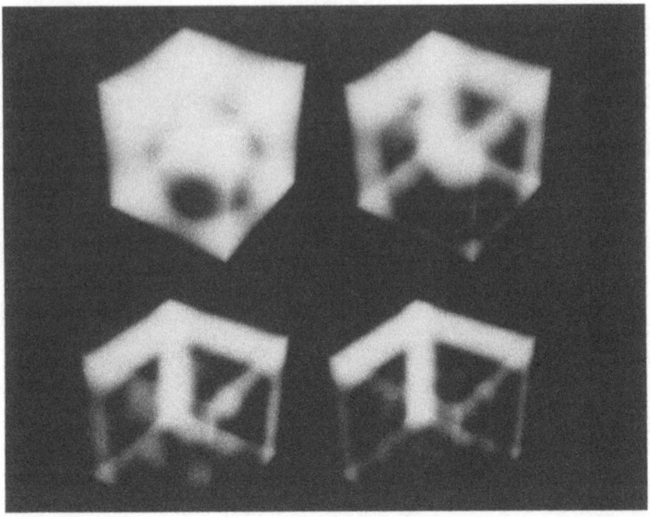

Fig 8. The 'axis' motive after low-pass filtering with a cut-off frequency of 1, 2, 3, and 4 (form upper left to lower right). With increasing cut-off the impulse width becomes smaller. The low-pass filtering errors can be seen.

5 Post-Filtering

A closer look at the 'axes' shape in fig.8 shows a different problem inherent to this method. In this picture, the boxes representing the axes are placed exactly at the boundary of the unity cube, low-pass filtered and back-transformed. After transformation, aliasing can be seen on the cube sizes opposite to those occupied by the axes: some kind of 'soup' or 'fog' can be seen where empty space is expected. A different problem can be seen in fig.5 (lower part): the space between the axes that should be empty shows some cloudy structure. A similar but much less detectable problem can be seen around the corners of the upper part of fig.5. These 'cloud residuals' in the empty space become smaller and finally invisible when the cut-off frequency is increased.

The reason for the first aliasing effect is well-known (see, e.g., [14]) and inherent to any signal-filtering method like the one used here. Low-pass filtering of a signal extracts

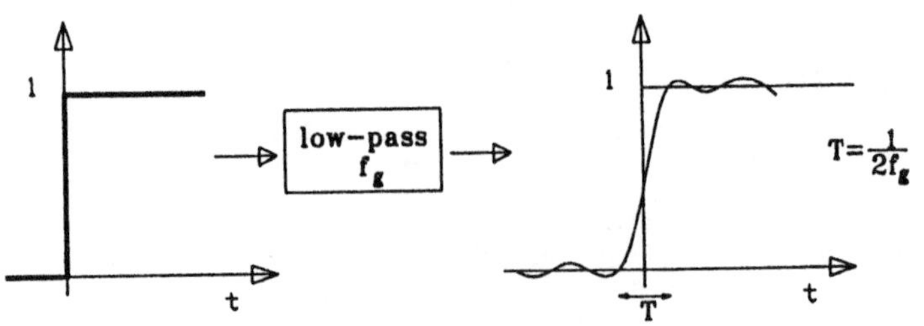

Fig 9. Low-pass filtering of a unit jump.

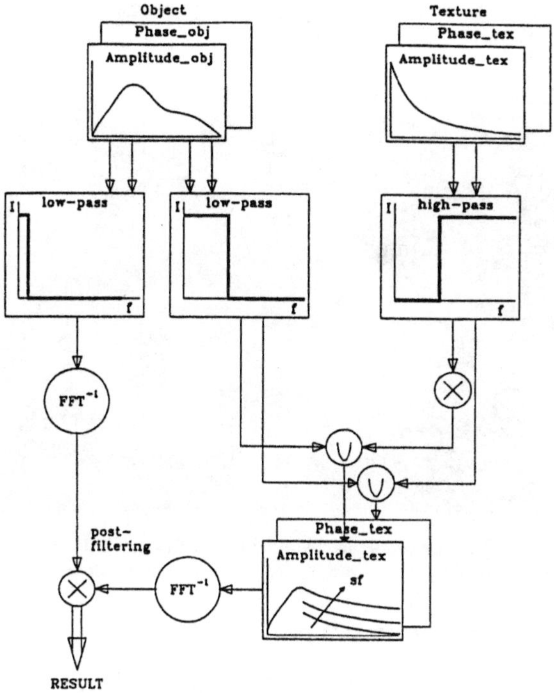

Fig 10. Post-filtering of a texture in Euclidean space.

information-carrying frequencies from the spectrum and, thus, alters the signal form. Fig. 9 (upper) shows the spectral representation of a unit jump. This spectrum is filtered by a low-pass filter with a cut-off frequency of f_g, and the result is back-transformed to Euclidean space. Two filtering results can be seen: 1) the signal slope becomes smaller with decreasing cut-off frequency, and 2) the newly gained impulse shows an oscillating character. The smaller impulse slope results in a wider impulse: the voxels representing each axis 'overflow' the cube boundary. Since FFT is a periodic transformation, the overflow over the left boundary appears on the opposite boundary as aliasing. This phenomenon is more detectable if a very sharp, non-symmetric shape (like the axis used here) is placed very close to, or even upon, a unity-cube boundary, and becomes less detectable if a symmetric shape is placed near the middle of the unity cube.

The second aliasing effect is also easy to explain. The cloud is defined everywhere within the unity cube, even in the space between the axes. High-pass filtering of the cloud and 'merging' with the main body do not eliminate these frequencies, so that they remain visible in the final picture.

In order to reduce the second effect, we decided to apply a post-filtering to the texture data after Euclidean-space transformation. By means of post-filtering we want to eliminate all texture values lying outside of the main body and its transitional zone. Since the shape of the body can be arbitrary, the formulation of an analytical 'smoothing function' will generally be not possible. For that reason, we gain our filter from the body itself. We first apply a low-pass filter with a very low frequency on the spectral representation of the body and then transform the result to the Euclidean space. After inverse transforming the texture data are multiplied by this filtering function (see fig.10). The result of such a post-filtering in Euclidean space for varying cut-off frequency can be seen in fig.11.

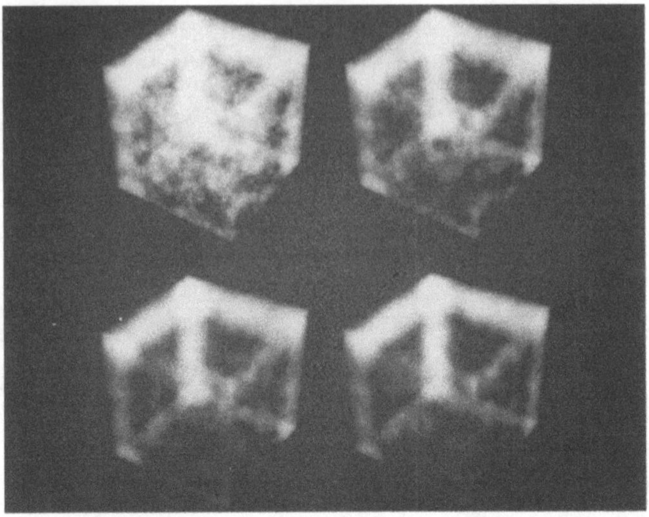

Fig 11. The axis-motive from 7 before (upper left) and after post-filtering in Euclidean space with cut-off of 1, 2 and 3. Aliasing (cloud residuals in the empty space) have been greatly reduced.

6 Discussion and Conclusion

Texture shaping has been introduced as a method for generating data describing density objects of arbitrary form. Thus, one has the possibility to 'press' his favorite texture into a predefined shape. In contrast to other existing techniques, our approach operates in the frequency domain and manipulates the spectra of the texture and the object. We examined two different methods for combining the spectra. The 'phase-and-amplitude' substitution converges to the desired shape for relatively low cut-off values, whereas the data produced by the 'phase substitution' method appear more random. The complete process is influenced by only two simple and intuitive parameters: the cut-off frequency and the scaling factor. For the second parameter, automatic defaults that give reasonable results are already built into the system; in any case, the user can use the calculated value as a starting point for further experimentation. Two different error types, which result from the filtering process and produce aliasing in the final picture, have been pointed out and discussed. For the most important case a satisfactory and practically available solution at a low computational cost has been proposed. Nevertheless, the development of different alias-reducing methods remains an important field of research for us.

The pictures generated here are, in several cases, similar to those given by Perlin (marble and solar corona in [17], fire ball in [15]) using his 'turbulence function'. This is not surprising since we also use a $1/f^{\beta}$ texture for our examples. The difference between the two approaches lies in the fact that Perlin operates exclusively in the Euclidean space, whereas we operate mainly in the frequency domain. In addition, our approach is more flexible since we can 'press' our texture into any arbitrary shape without major difficulties (in Perlin's hypertexture, points within the transition zone must be re-projected onto the surface of the objects, something which is easily done only for simple shapes like cubes, spheres, tori, etc.). We can easily influence the accuracy of the shape approximation and/or the random nature of the boundary within the transition zone, and thereby producing any intermediate shape between the two 'key frames'. We can even gradually change the parameters between the frames and, thus, generate a more or less continuous and smooth transition between the two shapes. Last but not least, using a texture resolution of 64^2 and applying the rendering techniques described in [20] and [19], we only need about 5-10 minutes on a SUN-4 and about one minute on a Silicon Graphics IRIS using the hardware support of the GL library for generating and rendering a frame in 512^2 resolution (Perlin's approach requires several hours for each frame).

On the other hand, our method also has some drawbacks: Since we apply the discrete FFT, we have to convert the continuous object representation into a discrete one. In general, this can be done without difficulty, but the result of the conversion remains discrete and, thus, resolution-dependent: in the case of a major camera movement (e.g., a zoom-in), a re-discretization according to the new object size as it appears on the screen will be necessary. In addition, a voxel representation of a body usually requires significantly more memory space than a functional one. Finally, in several cases aliasing effects can become visible. As a result, one can conclude that both methods have positive and negative features, so that they supplement each other. For a given problem one method may give a better solution than the other. For a more detailed discussion and comparison between spectral synthesis and the functional approach please refer to [21].

We want to examine the possibilities and the limits of the proposed method in the near future. Our first target is to reduce, or even eliminate, aliasing errors inherent to the involved signal filtering method. For that purpose we plan to test better filters than the used low-pass, windowing techniques, etc. Further, the idea of continuous shape changing through interpolation in Fourier space appears challenging and attractive for simulating effects like explosions and flames. At last, the precise definition of the form and thickness of the transition zone of complex objects is still under examination in addition to a more general form for preventing aliasing effects due to filtering.

Acknowledgements

The authors would like to express their thanks to C. Distasio and S. Wurster for proofreading the manuscript. This research has been carried out within the "Textur-Editor" project sponsored by the "Deutsche Forschungsgemeinschaft" (DFG; German Research Foundation), grant number En 123/10-2.

Literature

[1] Blinn, J. F.: *Light Reflection Functions for Simulation of Clouds and Dusty Surfaces*, ACM Computer Graphics, SIGGRAPH-82, Vol. 16, No. 3, pp. 21-29, 1982

[2] Bracewell, R.: *The Fourier Transform and Its Applications*, McGraw-Hill, 1965

[3] Ebert, D., Parent, R.: *Rendering and Animation of Gaseous Phenomena by Combining Fast Volume and Scanline A-buffer Techniques*, ACM Computer Graphics, SIGGRAPH-90, Vol. 24, No. 4, pp. 357-367, August 1990

[4] Englert, G., Schendel, M.: *Fractals and Formal Texture Specification*, in: Encarnacao et.al. (Eds.), 'Fractal Geometry and Computer Graphics', pp. 83-102, Springer Verlag, 1992

[5] Feibush, E., Levoy, M., Cook, R.: *Synthetic Texturing Using Digital Filter*, ACM, pp. 294-301, 1980

[6] Gardner, G. Y.: *Visual Simulation of Clouds*, ACM Computer Graphics, SIGGRAPH-85, Vol. 19, No. 3, pp. 297-303, 1985

[7] Gonzalez, R. C., Wintz, P.: *Digital Image Processing*, Second Edition, Addison Wesley Publishing Company, 1987

[8] Heckbert, P.: *Survey of Texture Mapping*, IEEE Computer Graphics and Applications, pp. 7-19, 1986

[9] Kajiya, J. T., von Herzen, B.: *Ray-Tracing Volume Densities*, ACM Computer Graphics, SIGGRAPH-84, Vol. 18, No. 3, pp. 165-174, July 1984

[10] Lewis, J. P.: *Algorithms for Solid Noise Synthesis*, ACM Computer Graphics, SIGGRAPH-89, Vol. 23, No. 3, pp. 263-270, July 1989

[11] Lovejoy, S., Mandelbrot, B.: *Fractal Properties of Rain, and a Fractal Model*, Tellus, Vol. 37 A, No. 3, pp. 209-232, May 1985

[12] Mandelbrot, B.B.: *The Fractal Geometry of Nature*, Freemann N.Y. 1983

[13] Max, N. L.: *Atmospheric Illumination and Shadows*, ACM Computer Graphics, SIGGRAPH-86, Vol. 20, No. 4, pp. 117-124, August 1986

[14] Oppenheim, A., Schafer, R.: *Digital Signal Processing*, Prentice-Hall International Editions, 1975

[15] Perlin, K., Hoffert, E.: *Hypertexture*, ACM Computer Graphics, SIGGRAPH-89, Vol. 23, No. 3, pp. 253-262, July 1989

[16] Peachy, D. R.: *Solid Texturing of Complex Surfaces*, ACM Computer Graphics, SIGGRAPH-85, Vol. 19, No. 3, pp. 279-286, 1985

[17] Perlin, K.: *An Image Synthesizer*, ACM Computer Graphics, SIGGRAPH-85, Vol. 19, No. 3, pp. 287-296, July 1985

[18] Rushmeier, H., Torrance, K.: *The Zonal Method for Calculating Light Intensities in the Presence of a Participating Medium*, ACM Computer Graphics, SIGGRAPH-87, Vol. 21, No. 4, pp. 293-302, July 1987

[19] Sakas, G., Gerth, M.: *Sampling and Anti-Aliasing Discrete 3-D Volume Density Textures*, EUROGRAPHICS'91 Award Paper, IEEE Computer and Graphics, Vol. 16, No. 1, pp. 121-134, Pergamon Press, 1992 *and* Proceedings EUROGRAPHICS'91, Vienna, Austria, pp. 87-102, North-Holland Publishers, September 1991

[20] Sakas, G.: *Fast Rendering of Arbitrarily Distributed Volume Densities*, Proceedings EUROGRAPHICS' 90, Montreux-Switzerland, pp. 519-530, North-Holland Publishers, September 1990

[21] Sakas, G.: *Modeling Turbulent Gaseous Motion Using Time-Varying Fractals*, in: Encarnação et.al. (Eds.), "Fractal Geometry and Computer Graphics", pp. 173-194, Springer Verlag, 1992

[22] Voss, R.: *Random Fractal Forgeries*, ACM Computer Graphics, SIGGRAPH-85 Tutorial Notes, 1985

Color Spaces and Human Color Perception

Ingeborg Tastl, Werner Purgathofer

This paper is concerned about the definition and features of color spaces in general, about the differences between geometrical and perceptual color spaces and about one possible way to create a colorimetric system, which is more based on human color perception than color spaces commonly used in computer graphics.
Keywords: color, perceptual color spaces, color perception, color difference.

1 Introduction

Color spaces are systems, which describe the connection between the radiant energy (electromagnetic wave), entering the eye of a person, and the color, which the person perceives. In order to understand this fact correctly one has to keep in mind that lights of different spectral compositions (with different physical properties) can be seen to be the same color. For example the color sensation white can be produced by light emitted from a color monitor or by light reflected from a sheet of paper. Although the spectral distributions of the reflected and the emitted light are certainly different, the color sensations or the perceived colors are the same. This means that a color spaces should assign all colors, which can't be distinguished by an observer, the same numerical value. On the other hand all colors with the same numerical value should look like the same (Thorell 1990).

With the help of color spaces one should be able to perform the following actions:
- to give an objective, more or less exact specification of a color,
- to select colors, which can easily be distinguished,
- to select colors with the same lightness,
- to select colors, which can be distinguished even by colordeficient persons.

Colorimetric systems can also be seen as orderly arrangements of colors in geometric solids or space. At the beginning of colorimetry color systems were usually represented by simple geometric objects, but later on human color perception was explored and the results were involved in the colorimetric systems, which as a consequence became more and more complex. Although the industry would need a color space completly based on human color perception to solve industrial color control problems and to decide whether two colors are different and how much they differ, the ideal colorimetric system hasn't yet been found.

When we talk about perceived color it is important to distinguish between nonluminous (perceived) color and luminous (perceived) color. The first one is color perceived to belong to an area that appears to be transmitting or diffusely reflecting light as a secondary light source. The other kind is color perceived to belong to an area that appears to be emitting light as a primary light source, or that appears to be specularly reflecting such light. Therefore the main difference is reflexion on the one hand and emission on the other hand. This paper is only concerned about luminous perceived color, and in detail about color perceived to belong to color monitors and about color spaces used in this context.

The exact aim of this paper is to examine the human sensitivity for luminous color and color differences and to include the results in an existing color space of computer graphics. Below, one way to construct a perceptual colorimetric system, where the spacing of the color points corresponds to the perceptual differences between them and where the metric includes information about the sensitivity for colors of human beings, is described.

In the first part of this paper the main differences between geometric and perceptual color spaces and the benefits of perceptual ones are discussed. After that the ideas and the implemented method to find a perceptual colorimetric system are described and in the end the results as well as ideas for improvements are presented.

2 Commonly Used Color Spaces Versus Perceptual Uniform Color Spaces

One of the most commonly color specification models used in connection with color monitors and especially in connection with computer graphics is the RGB-model. In this model each color is described as an additive mixture of the primary colors red, green and blue. On the monitor colors are physically produced by different levels of red-, green- and blue phosphor intensities and thus the RGB-color model is often announced to be a hardware oriented color space (Foly 1990; Durett 1987).

Unfortunately human beings aren't used to specify color compositions in terms of red, green and blue intensities. It is easier for us to describe a color with the attributes hue, lightness and saturation. Another disadvantage of the RGB-model is the vaguely relation to color appearance and to the color perception of human beings. The fact that a person is able to distinguish a lot of different colors in one part of the spectrum while he isn't so sensible in another part of the spectrum is not included in this color description system (Murch 1989).

Changing a color in one part of the color space by a certain amount will be visible while the same amount of change in another part won't be visible. As a consequence amounts of RGB-differences can't be used to express amounts of color differences perceived by human observers. Similar problems also exist in the HLS-color space and also in the CIE-color space.

In order to overcome these problems color spaces, which are more perceptual uniform have been developed. The most important features of such uniform color spaces are (Thorell 1990):

- *intuitive color specification:* a color description, which is oriented on the intuitive color description of human beings;
- *perceptual metric:* the spacing of the color points corresponds to the perceptual differences between them and the metric is a measure for the amount of perceptual color difference;
- *color experiments:* the results of color experiments concerning the color perception are included in the colorimetric systems.

Using perceptual color spaces makes it easier to define color scales for visualization. Other examples where the usage of uniform color spaces is beneficial are the great domain of medicine, the storage and transmission of images of computer graphics and computer graphics in general (Durrett 1987; Thorell 1990).

3 Existing Perceptual Uniform Color Spaces

As there are a lot of different perceptual color spaces, it is not possible to discuss them all in this paper. Therefore our idea how to develop a uniform color space is compared to five already existing methods without restricting the generality.

In the forties of our century MacAdam made a lot of experiments with a new constructed photometer to quantify the perceptual spacing of the chromaticity diagram of the 1931 CIE XYZ space (MacAdam 1942). During the experiments the subjects were shown various color samples with known chromaticities and were asked to match the samples by using adjustable color sources. In the end the standard deviations between the color samples and the new produced colors were calculated and as a consequence for each original sample an elliptical domain of colors, which couldn't be distinguished from the original one, could be found. These ellipses were of different sizes and demonstrated some facts concerning human color perception as well as the nonuniformity of the chromaticity diagram. But it should be mentioned that those examinations were only performed with colors of the same luminance.

Subsequently scientists have tried to find a linear transformation of the chromaticity diagram in order to gain a color representation with a uniform spacing. Until now there exist various proposals; the one which is recommended by the CIE is the *1976 UCS model.* (Durrett 1987). Unfortunately it wasn't possible to get a *real* perceptual color space, where the ellipses would have become circles, but only a *rather* uniform one. Besides, this color space can only be used to compare colors with the same luminance.

Another perceptual uniform color space is the *Munsell color space*, a series of color charts, originally proposed in 1905 by the artist A. Munsell (Durrett 1987; Schettini and Ventura 1990; Billmeyer 1987). Subsequently a lot of studies concerning the perceptual spacing were undertaken and resulted in a redesign of the system. Although it doesn't seem to be a bad model of human color perception and although a lot of other studies use this model for reasons of comparison, it is not suitable for computer graphics. The reasons are: this color space is only a series of color charts; it has an irregular shape and there are no mathematical expressions to transform Munsell color specifications into CIE and RGB color specifications. Moreover it has been designed for pigment colors and not for luminous colors. Nevertheless it shows another approach to define a perceptual uniform color space: At the beginning a lot of facts concerning the human color perception are used to find a global arrangement of the different groups of colors. In the next step experiments are performed in order to make local improvements. This way is contrary to definition of the UCS model, where an existing color space was the starting point.

Another quite interesting color specification system is the one of the Optical Society of America (*OSA color space*) (Billmeyer 1987; Durrett 1987; Wyszecki 1982). There the essential idea is that in the whole color space the distances between nearest color neighbours should be constant and that for each color the number of nearest neighbours should be maximized. In order to reach this aim a space lattice with a cubo-octahedron as basic element has been used. In order to find the right color for each point of the regular rhombohedral lattice a lot of judgments of relatibe magnitudes of color differences were performed. Like the Munsell color space, the OSA model is a collection of colored samples and it doesn't seem to be of great practical importance. The reasons are the small number of samples, the quite unfamiliar arrangement which hasn't an obvious connection to the perceptual attributes hue and saturation, and the fact that there is no simple relation between the OSA color specification and CIE XYZ color specification. Besides, like the Munsell color space this color space was developed in connection with nonluminous colors.

The last two color spaces, from which the main ideas shall be described in this chapter are the *CIELUV* and the *CIELAB* color models, which are both recommended by the CIE (Robertson 1990). The starting point of the LUV model was a rather simple projective transformation of the chromaticity diagram of the 1931 XYZ color space. In the following a lightness dimension based on experiments made by Wyszecki was added. The result was the CIE 1964 UVW space, which was improved by other scientists and finally renamed to CIE 1976 LUV space. The CIE 1976 LAB space is based in some way on formula gained with the help of the Munsell color space.

At the moment these two color spaces are probably the best approximations of visually uniform color spaces. But as the CIE points out that they are not exactly uniform and the color difference formula based on these color spaces are not entirely satisfactory (CIE 1986). One of the problems is that the experiments which lead to the definition of the CIELUV color models were performed under certain viewing conditions. Therefore the results of color difference calculations also depends on how well the actual conditions match the original ones. Another fact is that color difference equations will not give exact values if the luminace of the colors is relative low or the sizes of the color samples are rather low.

At the moment nobody knows if it is possible to find an exact perceptual uniform color space which includes all the features of human visual color perception (Wyszecki 1982). But nevertheless we think that's worth searching for.

The way how we exactly tried to find a perceptual uniform color space is discussed in the next chapters. But the main differences to the above described methods are: the fact that we tried to perform experiments concerning human color perception on a color monitor with the definitive aim to find a perceptual color space for color monitors. This is different to other approaches. Morover this color model shall be used used in the future under normal working conditions. Therefore we didn't perform the experiments under a special test environment. But naturally we tried to avoid circumstances, which would dramatically reduce the sensitivity for color differences. In the end we didn't try to find mathematical formula to include the results of our experiments in an existing color space but we iteratively changed the original arrangement of the colors with the help of the computer. More details are described in the following chapters. Besides, our work can in some way be seen as a continuation of the experiments performed by MacAdam.

4 The Method

The whole method can be subdivided into two parts. The first one is concerned about the color perception tests, while the second one is concerned about the inclusion of the results in a widly used color space of computer graphics.

4.1 Color Perception Tests

The aim of these tests was the experimental examination of the sensitivity of human beings for differences of colors produced on the screen of a monitor. Therfore it was interesting in which parts of a color space human beings can perceive very small amounts of color differences, and in which parts great changes are necessary in order to be perceived by human beings (MacAdam 1942; MacAdam 1985; Hurvich 1981; Wyszecki 1982).

In the actual implementation the HLS color space was used and a discrete amount of colors was selected. The arrangement of these so called standard-colors is shown in figure 1 (see color plates). The experiments should find out the domain of metameric colors, i.e. all those colors which can't be distinguished from the selected color, for each standard color.

Analysing the human color perception at discrete points in the color space by trying to find discrimination-thresholds between closely similar colors and using the results to find some facts concerning human color perception for the whole color space was the way, which we decided to realize. Other methods for the examination of the human color perception can be found in (Wyszecki 1982). In order to gain exact results it would have been necessary to examine as much standard-colors as possible. But on the other hand it was also necessary to think on the practical realization. Therefore the 29 standard-colors shown in figure 1 (see color plates) were chosen.

From the great variety of different methods, proposed to find discrimination thresholds concerning the perception of human beeings, the production-method was chosen (Sixtl 1982). This means that each standard-color was presented on the screen with colors of comparison. In general for each standard-color six different colors of comparison were used: two colors which had different hue-values and the same lightness- and saturation-values, two colors of comparison which had different lightness-values and two colors which had different saturation-values. In course of the tests the subject was asked to change the color of comparison until it couldn't perceive a difference between the changed color of comparison and the standard-color. This color, which the person couldn't distinguish from the standard-color any more, was called result-color. The numerical difference between the standard-color and the result-color was a measure for the amount of change of that standard-color, which was necessary before a different color was perceived. The numerical value was also a measure for the sensibility of human beings for changes of one component of the standard-color. Therefore this amount was called just-noticeable-difference and it was in some ways a unit for the perception of one color component, in one direction.

At the end of the test we had six distance values or just-noticeable-differences (i.e. the numerical difference between the component of the standard-color and the component of the respective result-color) for each standard color: two values which described how much the hue-component of the standard-color could be changed without getting perceivable different colors, two values, which described how much the lightness-component could be changed and two values, which described the possible changes of the saturation-component. These six distance values described for each standard-color approximatly the domain of metameric colors. As expected, the domains of metameric colors weren't spheres, which meant that a standard-color had to be changed by a certain amount in one direction to get a perceivable different color and the same standard-color had to be changed by a different amount in another direction. Furthermore the domains of metameric colors of different standard-colors were of different sizes. Thus it was necessary to change one component of a standard-color by a certain amount to get a perceivable difference and at the same time it was necessary to change the same component of another standard-color by a different amount to perceive the same result. Expressed the other way round the results showed that the spacing of the colors in the initial color space didn't correspond to the perceptual differences. Furthermore they showed the nonuniformity of the initial color space (Tastl 1989).

Besides, these examinations were performed on a color monitor and they were possible because a color monitor can produce colors, which are numerical different but can't be distinguished by persons (Conrac Division, Conrac Corporation 1985). Another quite important fact is that the perception of colors as well as the perception of differences is influenced by a lot of different features, like the color of the background, the size of the color fields, the light of the environment, and so on. Therefore it was on the one hand important to consider the various features when designing the test environment and on the other hand it was important to remember the concrete test environment for being able to reproduce the test situation.

In order to gain an rather objective impression of the human color perception the tests were performed with 20 different color-normal persons. After statistical calculations we got an impression of the average sensitivity of persons for color differences at certain points in the color space. How far these results correlated with the real color perception was difficult to decide. Nevertheless we were able to discover some tendencies and compare them with facts, which we got from literature (MacAdam 1942; Wyszecki 1982), especially from other studies concerning perceptual color spaces. One of the mentioned tendencies was the fact that human beings seemed to be more sensitive for lightness-differences than for saturation-differences and more sensitive for hue-differences than for saturation-differences.

4.2 Inclusion of the Results in the Initial Color Space

The experimental examinations demonstrated the nonuniformity of the initial colorimetric arrangement and at the same time they pointed out some features of the human color perception. The target of this work was to include these features in the intial color space in order to get a more perceptual uniform one. This meant that the original color arrangement had to be changed so that the domains of metameric colors would become domains of approximatly equal sizes and as a consequence the new metric would be a measure for the amount of perceptual color difference.

In order to understand how the transformation of the initial color space was performed, just imagine the standard-colors to be arranged in a regular three dimensional coordinate system

(Tastl 1989). Each standard-color has two neighbours with different hue-components, two neighbours with different lightness-components and two neighbours with different saturation-components. The geometrical difference between the standard-color and all its neighbours is equal. From the examinations described in chapter 4.1, we have got in general six distance values, which describe the amounts of changes possible without getting a perceivable different color.

As it is easier to demonstrate the algorithm in the one- dimensional space, we consider at first points, which are only connected with two neighbours. In the figure 2 three standard-colors A, B and C with different saturation-components are shown. The saturation-value of A is 0, the saturation-value of B is 10 and the saturation-value of C is 20. Therefore the "geometrical" difference between A and B as well as between B and C is 10; a2, b1, b2 and c1 are the above mentioned distance values of the saturation-component.

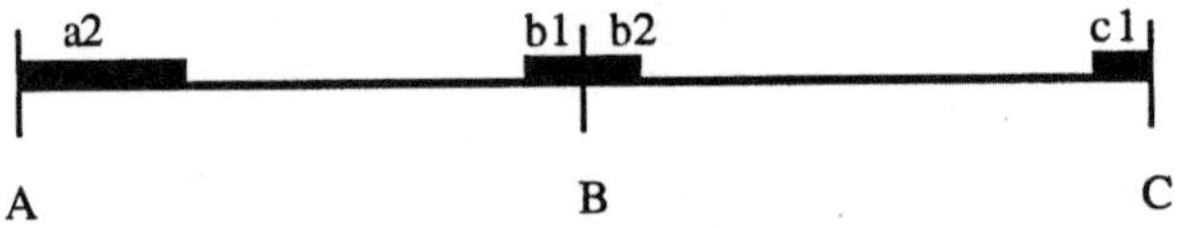

Fig. 2: The initial situation

When we calculate the average between the distance values a2 and b1 (b2 and c1) we get the value of one *perceptual unit* for the interval AB (BC). Now we can calculate the *amount of perceptual units* of the interval AB and BC. For example if we take the values 3 for a2, and 1 for b1, b2 and c1; then we get the perceptual unit 2 for the interval AB and the perceptual unit 1 for the interval BC. This also means that the amount of perceptual units in the interval AB is equal to 5 and in the second interval it is equal to 10. Going back to the original meaning, this example shows that a human being is twice as much sensible in the interval BC than in the interval AB.

Considering the above example the original geometrical differences between A and B and the difference between B and C are the same. In this example this also means that the differences of saturation are equal. What is different are the perceptual units of the two intervals: In the first interval the perceptual unit has a value of 2 and in the second interval the perceptual unit has a value of 1. This means that a unit for the discrimination of colors is combined in different domains with different values. Exactly this fact holds true in a nonuniform color space. On the contrary in a uniform color space a perceptual unit has approximatly the same numerical value in the whole color space. In order to gain a perceptual uniform situation we have to change the position of point B, until the perceptual units in the two intervals are approximatly the same. As a consequence the geometrical arrangement of the points A, B and C corresponds with the experimentally found amounts of units in the two intervals. For our example this means that the distance between B and C has to be twice as long as the distance between A and B. Having reached this situation is similar to the fact that the geometrical arrangement of the points corresponds approximatly with the human color perception. The quality of the correspondence naturally depends on the quality of the color perception tests. The final situation after the change of the position of point B is shown in figure 3.

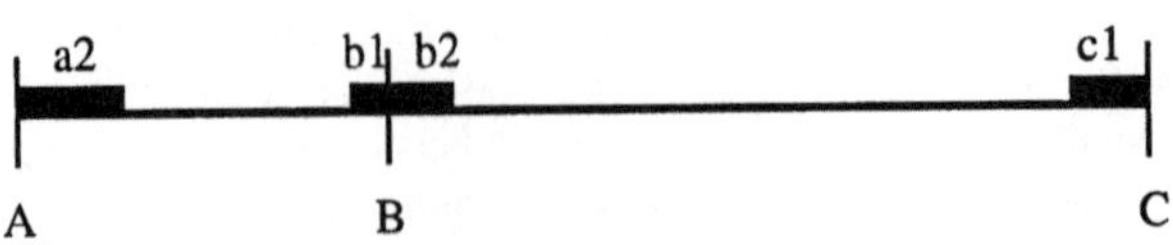

Fig. 3: The final situation

Changing the positions of the three points is equal to scaling and thus the distance values have got other numerical amounts: a2 has got the value 2, b1 has got the value 0.6, b2 and c1 have got the value 1.3. As the distance between point A and point B amounts to 6.6 and the distance between point B and point C amounts to 13.3 the perceptual unit has got the value 1.3 in both intervals. This has exactly been the situation, which we wanted to reach.

Until now we have only considered the one dimensional situation, but the same method can be used for the three dimensional situation. There a point can be moved not only in two direction, but in six directions. In the actual implementation a point in the 3D color space has in general 6 neighbours and we have 6 amounts of perceptual units. Then we have to change the position of the point until the geometrical situation corresponds as good as possible with the human color perception. When we consider not only a few points but the whole colorimetric system, we have to change the arrangement of the colors until the perceptual unit has got approximatly the same values in all intervals of the color space. At the same time this means that the proportions of the distances between the points are similar to the experimentally found amounts of perceptual units. The aim is an arrangement of colors, which corresponds with the human color perception.

Thinking over the starting situation and the situation, which we want to reach, it is obvious that there are different ways for realization.

In the actual implementation the described transformation was performed in an iterative way. The colorimetric arrangement was changed step by step. In detail this meant that for each pair of standard-colors the original distance between them and the experimentally found numerical value of the perceptual unit were used to calculate for each interval the amount of perceptual units. These amounts described how many different colors a color-normal test subject can discriminate between two respective standard-colors. It was a measure for the perceptual difference between two standard-colors. Therefore the amount of perceptual units was a suitable figure to define a perceptual metric, i.e. it described the ideal distances between the various standard-colors. Thus it was necessary to change in an iterative way the arrangement of the standard- colors until the distances between the color points were equal as good as possible to the ideal distances. This was similar to the fact that the numerical value of the perceptual unit got similar in the various intervals.

Changing the arrangement in an iterative way meant taking at first one color point and changing its position by a small amount, so that the differences between the actual distances from one color point to all its neighbouring color points and the respective ideal distances became smaller. After the same procedure was repeated for all the other standard-colors, the first point was changed again and so on. In this way it was possible to achieve a nearly simultaneous transformation. The change of the position of one point was repeated until a change couldn't improve the relation between actual distances and ideal distances. The whole transformation stopped when a change of none of the points could improve the relative positions of all standard-colors. This usually didn't mean that the distances between the standard-colors were exactly equal to the respective ideal distances. What we wanted and what was possible to reach was a an approximation of the ideal arrangement.

The whole situation can be compared with points, which are connected with one another by rubber bands. Originally these rubber bands are stretched in different intensities; and after the transformation the intensities are more or less equal.

5 Discussion of the Results

The results of the change of the colorimetric system corresponded approximately to the results of the experimental examinations (figure 4; see color plates). Those points of the initial color space, where great sensitivities for color changes had been discovered, were moved away from each other; points where only small sensitivities had been discovered, were moved towards each other.

Therefore at the one hand the transformation was successful but on the other hand some problems arose, which couldn't be solved in a satisfactory way. It was not enough to find a mathematical solution. Thinking over the result as an arrangement of colors it was not plausible in every detail. It would perhaps be necessary to include more problemspecific requirements in the definition of the ideal color arrangement, which should be reached after the transformation. The model behind this work can certainly be improved. Beside this fact it would also be necessary to study the human color perception in greater detail and to improve the color perception tests so that it is possible to get a more accurate impression of the human color perception and the human color dicrimination in particular. Even a higher amount of standard-colors would result in a more exact impression of the color perception. Examining thresholds for discriminating closely similar colors and extending the results to the whole color space is one possible way, but there are certainly alternatives.

This work should be understood as a study to the subject of uniform color spaces. We wanted to show one way to define a color arrangement which is more based on human color perception than those commonly used in computer graphics. We wanted to present an idea, but we still continue working on this topic and we try to consider several kinds of improvements in the next implementation. Thus we hope that in the end it will be possible to get a perceptual uniform color space, which is not only nice to look at but can be used in everydays work in computer graphics.

References

Billmeyer, F.W. (1987) Survey of Color Order Systems. Color Research and Application, Vol.12, Nr.4.

Commission Internationale de l'Eclairage (1986) Colorimetry. CIE Publication 15.2, Bureau Central of the CIE, Austria.

Conrac Division, Conrac Corporation (1985) Raster Graphics Handbook. Van Nostrand Reinhold Company, New York

Durett, H. (1987) Color and the Computer. Academic Press, London.

Foley, J.D. (1990) Computer Graphics. Addison-Wesley, New York.

Hurvich, L.M. (1981) Color Vision. Sunderland.

MacAdam, D.L. (1942) Visual Sensitivities to Color Differences in Daylight. In: Journal of the Optical Society of America, Vol. 32, Nr.5, pp. 247-274.

MacAdam, D.L. (1985) Color Measurement. Springer-Verlag, Berlin.

Murch, G.M., Taylor, J.M. (1989) Color in Computer Graphics: Manipulation and Matching Color. In: Advances in Computer Graphics V. Springer-Verlag, pp. 19-47.

Robertson, A.R. (1990) Historical Development of CIE Recommended Color Difference Equations. Color Research and Application, Vol.15, Nr.3, June.

Schettini, R., Ventura, A.D. (1990) Perceptual Color Scales for Data Display. Automatika 31.

Sixtl, F. (1982) Meßmethoden der Psychologie.

Tastl, I. (1989) Perceptual Color Spaces in der Computergraphik. Diplomarbeit. Technische Universität Wien.

Thorell, L.G.(1990) Using Computer Color Effectively. Prentice-Hall, New York.

Wyszecki, G., Stiles, W.S. (1982) Color Science. Wiley Interscience, New York.

Visulux: a Radiosity Based Lighting Design Tool

Andrew J. Anderson, Michael Grant

Abstract

This paper contains a description of VISULUX, a lighting design tool based on accepted computer graphics techniques. Development of the model is a collaborative project between the LiDEC division of Philips International Bv., and the University of Strathclyde. VISULUX accepts a description of a zone's geometry, surface finishes, contents and light sources. This data is used to generate a complete multi-chromatic view independent description of each surface's spectral luminance distribution. Outputs from the model include numerical luminance and illuminance values and coloured perspective images. This paper also describes the main requirements of a general purpose lighting design tool and the restrictions that this imposes on the implementation of such a tool.

1. Introduction

Predicting the performance of lighting systems and assessing lighting quantity and quality within luminous environments are essential components of the lighting design process. Over the years a variety of lighting design tools and models with different levels of sophistication and potential have been developed. These tools and models however, are subject to various geometrical limitations and simplifying assumptions relating to the propagation of light that restrict their accuracy and limit their application. Traditionally, these design tools produced metrics relating to quantitative assessment but were generally incapable of evaluating lighting quality.

Computer graphics owes its existence to the simulation of light and its interaction with materials, progress being measured by a more complete treatment of the optical phenomena involved, yet it is perhaps surprising that while most CAD applications are heavily reliant on computer graphics the usage of computer generated imagery in lighting design and the existence of purpose built lighting design tools are, in the main, conspicuous by their absence.

The pursuit of ever greater realism in synthetic images requires the use of a more accurate and physically correct treatment of light. The technique that emerged as the front runner in this area is acknowledged to be radiosity, where the solution of the global illumination equation is explicitly derived and the conservation of energy is guaranteed.

A situation now exists where software for the generation of computer graphics employs a more rigorous, and potentially more accurate, model of light than do many popular evaluation tools. Given that there is now the potential to equate accuracy and reality, to provide lighting designers with both qualitative and quantitative feedback, what is required, and what the future offers, is the luminous equivalent of desk top publishing's What-You-See-Is-What-You-Get. However, the gulf between the research and the commercial community is still too wide to allow a ready path for the uptake of the state of the art.

The classic design scenario allows for the description of a problem, the simulation of its solution and the evaluation of the result. This process is cycled until the solution iterates towards the desired result. To turn a research tool into a useable tool for design requires a delicate balance between the computational expense involved and the achievement of some desired level of accuracy. Given that this cost/performance ratio can be redressed there is currently a need and a market for the development of lighting design tools that combine quantitative and qualitative output and are tailored towards the working practice and available hardware within the design environment.

On the basis of the above it was decided to direct the research towards three main aims. Firstly to progress beyond the limitations of existing evaluation programs; secondly to move towards the ultimate aim of being able to evaluate visually the merits of a lighting installation; and finally to achieve these aims within a hardware environment that is accessible to the design profession.

2. Radiosity

In order to simulate the behaviour of light in an environment the light energy incident on each surface can be considered as originating from two sources, either directly from an emitter or indirectly from a secondary source such as a reflecting surface. This relationship can be formulated into an expression that describes the energy balance due to the hemispherical integral over a surface. This expression, known as the radiosity equation, relates the total rate of energy leaving a surface, to the rate of direct energy emission and to the amount of energy arriving at that surface that is reflected back into the hemispherical space above the surface.

$$B_j = E_j + \rho_j \sum_{i=1}^{n} B_i F_{ij} \text{ for } j = 1, n \tag{1}$$

where :
n = number of surfaces in environment
B_j = radiosity of surface j.
E_j = emissivity of surface j.
ρ_j = reflectivity of surface j.
F_{ij} = form factor between surfaces i and j.

The form factor in the above equation F_{ij} relates the fraction of energy leaving surface i that arrives directly at a second surface j, to the radiative energy leaving i through a hemispherical region above that surface. The form factor equation is given below [9].

$$F_{A_i - A_j} = \frac{1}{A_i} \int_{A_i}\int_{A_j} \frac{\cos\theta_i \cos\theta_j \, dA_j \, dA_i}{\pi r^2} \tag{2}$$

For a closed zone consisting of n surfaces conservation of energy requires that all the radiant energy leaving a surface must be accounted for. This implies that ;

$$\sum_{j=1}^{n} F_{ij} = 1 \text{ for } i = 1, n \tag{3}$$

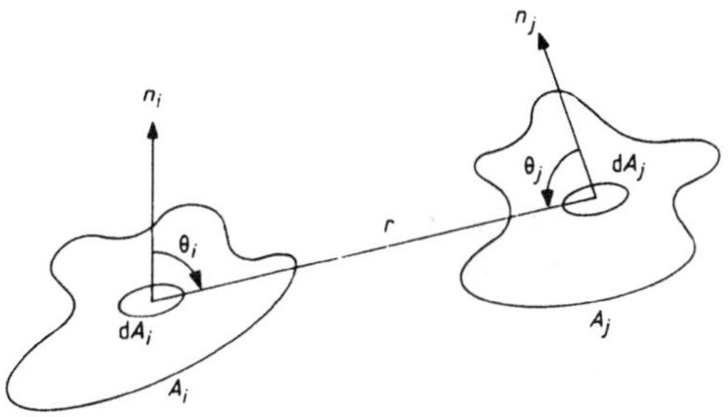

Fig. 1 Two communicating elemental areas. [After Clarke, 1].

The relationship given in (1) forms a system of linear equations which can be reformulated as the matrix given below.

$$\begin{bmatrix} 1-\rho_1 F_{1,1} & -\rho_1 F_{1,2} & \cdots & -\rho_1 F_{1,n} \\ -\rho_2 F_{2,1} & 1-\rho_2 F_{2,2} & \cdots & -\rho_2 F_{2,n} \\ \cdot & \cdot & \cdots & \cdot \\ \cdot & \cdot & \cdots & \cdot \\ -\rho_n F_{n,1} & -\rho_n F_{n,2} & \cdots & 1-\rho_n F_{n,n} \end{bmatrix} \begin{bmatrix} b_1 \\ b_2 \\ \cdot \\ \cdot \\ b_n \end{bmatrix} = \begin{bmatrix} e_1 \\ e_2 \\ \cdot \\ \cdot \\ e_n \end{bmatrix}$$

The energy balance over all participating surfaces is found through the simultaneous solution of the unknown radiosity terms in the above. This formulation presupposes that the emissivity is uniform over the surface. To ensure that this is the case it is common practice to reduce each of the surfaces in the model to its finite element equivalent. This has one advantage in that it reduces the double area integral in (2) to a more manageable form.

$$F_{dA_i - A_j} = \sum_j \frac{\cos\theta_i \cos\theta_j \, dA_j}{\pi r^2} \tag{4}$$

where j is number of subdivisions of A_j

The major disadvantage is realised in the computational expense required in the calculation. Applied in an occluded environment the rate at which the form factor can be evaluated precludes its use in general practice. A method to reduce this overhead was introduced by Cohen *et al* [2] in which a hemi-cube was imposed over each surface element and a sub set of form factors were calculated relating the centre of the cell to a number of pixels on each cube face. Each visible surface element was then projected onto the cube and associated with a delta form factor. This reduced the computational time but the storage requirements remained high at $O(n^2)$. A further method introduced to solve this problem is introduced by Cohen [3], where by utilising a novel solution scheme and the power of custom graphics hardware the required identities are generated "on the fly". This last scheme is patently efficient but relies on dedicated hardware with attendant portability and implementation difficulties.

Evidently the balance here is between speed and storage requirements, form factors are a function of geometry alone and hence need to be calculated only once. It is therefore logical to store the result of the initial calculation and reuse it as and when necessary. The gain in speed is however offset by the size of the resulting data set. In order to overcome these two limitations optimisation can only be realised by increasing processing power to overcome the first or by seeking a means of reducing the number of form factors to limit the second. A possible solution is to seek an approximation that will reduce the overheads of both calculation and storage that are a function of the combinatorial pairing of surface elements.

The research currently underway at ABACUS [6] is directed towards investigating these limitations and proposing a strategy that does not compromise the numerical integrity of the solution.

3. Current Work

The behaviour of light, under some set of conditions, is a function of interrelated variables concerning geometrical position and orientation and the directional, and spectral, distribution of light. These variables form a potentially infinite system of linear equations that must be solved simultaneously. To enable an economical solution the problem must be reduced to a realistic numerical equivalent by discretising surfaces, the visible spectrum and the range of incoming and outgoing directions.

In order to describe the illumination events above a surface it is common to employ the notational convenience of a hemisphere of unit radius situated above some reference point. This representation was employed by Nusselt and has been the basis of many experimental techniques.

Nusselt's theorum states that for some finite area the form factor is equivalent to the portion of the area of the base of the hemisphere covered when a finite area is projected first onto the hemisphere and then orthographically onto the base.

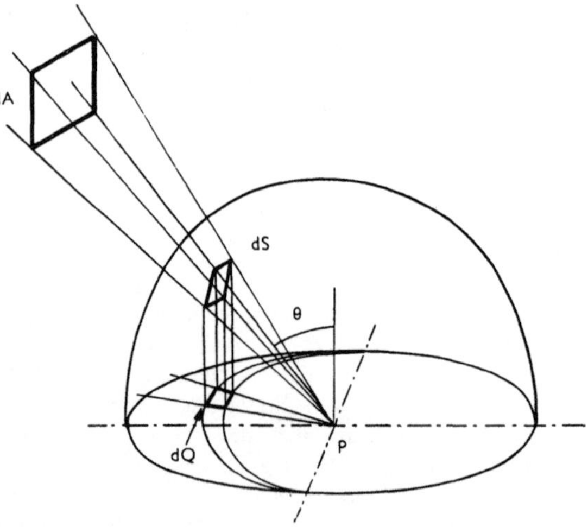

Fig. 2 Nusselt's Geometrical Analogue. Form factor at P due to area dA is given by area dQ of the projection of the surface element dA on the base of a hemisphere of unit radius and centre P.

In order to directly benefit from this analogue it is convenient to discretise the hemisphere into small, equal solid angles. This reduces the form factor calculation to a user controlled level of accuracy and provides a ready means of maintaining a record of the incoming or outgoing intensity within each solid angle.

A numerical method derived by the Martin centre [8] provides an efficient method of ensuring the discretisation of a hemisphere results in a number of equal solid angles. This algorithm can be explained as follows. Using the finite element form of the form factor equation (4) and since ;

$$d\omega = \frac{dA\cos\theta}{r^2}$$

we can rewrite it as ;

$$\sum_j \frac{\cos\theta_j d\omega_j}{\pi}$$

To satisfy conservation of energy this must sum to unity (3) therefore ;

$$\sum_j \cos\theta_j d\omega_j = \pi \qquad (5)$$

If the solid angles $d\omega_j$ are all identical we can write equation (5) as ;

$$d\omega_j \sum_j \cos\theta_j$$

To ensure equal solid angles they must all subtend equal areas on the surface of the unit hemisphere. The area of a horizontal strip on a sphere S is given by :

$$S = 2\pi rh$$

So to divide the hemisphere into strips of equal area requires equal values of h. Once the strips are of equal area, patches of equal area can be obtained by sub-dividing the length of the strips by the same denominator. These 'bubble' patches will then subtend equal solid angles at the centre of the hemisphere. In each case the width of the patch is defined by an angle θ in a plane parallel with the equator.

For a sphere of unit diameter (See Figure 3.) : $\cos(\theta) = \delta$

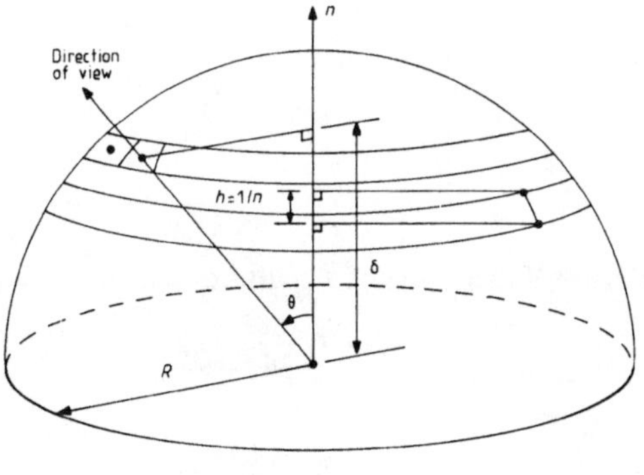

Fig. 3 Strip and Patch subdivision of a unit hemisphere.

where δ is the projection of the radius (in the direction of view) onto the normal. If the patches are identified by their axes, then the angles which these lines make with the normal to the viewing plane, will have cosines which are evenly spaced. If the hemisphere is divided into n strips these cosines will be:

$$\frac{1}{2n}, \frac{1}{2n} + \frac{1}{n}, \frac{1}{2n} + \frac{2}{n}, \ldots\ldots \frac{1}{2n} + (n - \frac{1}{n}).$$

Summing these, through the formula for arithmetic series, gives,

$$S_n = \frac{n}{2}(2a + (n-1)d)$$

where $a = \frac{1}{2n}$ and $d = \frac{1}{n}$ giving,

$$S_n = \frac{n}{2}(2\frac{1}{2n} + (n-1)\frac{1}{n})$$

i.e.:

$$S_n = \frac{n}{2}$$

If the strips are divided into m even patches, then there will be m of each of the terms in the series above, and so $\frac{mn}{2}$ for all patches. Since each patch will subtend a solid angle of $2\frac{\pi}{mn}$ steradians, this yields :

$$2\frac{\pi}{mn} \times m\frac{n}{2} = \pi$$

Which is the correct solution.

4. Illuminating Events

4.1 Direct Illumination

The excitance of the environment originates as the luminous intensity, for a given waveband (λ), in a given direction, from a light source as given by:

$$I_\lambda = \frac{d\phi}{d\omega}$$

where:

I_λ = *luminous intensity.*
$d\phi$ = *luminous flux.*
$d\omega$ = *differential solid angle.*

The direct component of the illumination of a cell of differential area dA is:

$$E_d = I_\lambda \frac{d\omega}{dA}$$

and since,

$$d\omega = \frac{dA\cos\theta}{r^2}$$

then,

$$E_d = I_\lambda \frac{\cos\theta}{r^2}$$

The total illuminance, over all wavebands in the spectrum is given by:

$$E_d = \sum I_\lambda \frac{\cos\theta}{r^2}$$

4.2 Reflection Function

The subsequent re-emission of the incident energy is a function of the reflective properties of the surface. If the surface exhibits diffuse characteristics then the reflected energy is distributed equally over all exit angles, given specular properties then a relationship exists between the direction of the incident light and the directional distribution of outgoing light. This relationship is a function of a material property commonly called the bidirectional reflectance. [12]

$$\rho_{(in,out)} = \frac{I_{out}}{E_{in}}$$

Given non-diffuse material characteristics the bidirectional reflectance function for a surface will have a unique value for each unique combination of incoming and outgoing directions. For any outgoing direction α the outgoing directional intensity I_α will be a function of the sum of the cell emission in direction α and the summation of the proportion of all light from a range of directions β that is reflected in direction α.

$$I_\alpha = E_\alpha + \sum_{\beta=1}^{\beta} \rho_{(\alpha,\beta)}\, I_\beta \cos\theta\, d\omega$$

The intensity of outgoing energy within each solid angle is retained thus preserving the history of each incoming ray and maintaining a discrete local radiosity field.

4.3 Indirect Illumination

Indirect illumination at a point is calculated by integrating over the hemispherical environment surrounding that point, as in Figure 2. The illumination produced at a point P on a horizontal plane i in a zone by a surface element j of area dA will be the same as that produced by the area dS on the surface of a hemisphere of unit radius given that dS subtends the same solid angle at P as does dA. If θ is the polar angle with which the direction of this projection makes to the normal at P and L is the luminance of surface element j this illumination will be $L_j\ dS \cos\theta$.

Therefore if the luminous intensity of dA is $L_j dA$

then

$$dE = L_j \frac{dA}{D^2}$$

where the differential illumination at P normal to $dA = dE$ and D is the distance of dA from P

It then follows that,

$$dE = L_j dS$$

since dS and dA subtend the same solid angle.

Therefore the integral over the entire hemisphere reduces to:

$$E = \sum_{j=1}^{ndir} L_j\ dS \cos\theta$$

and since the hemisphere is divided into $ndir$ equal solid angles;

$$E = \sum_{j=1}^{ndir} L_j \frac{2\pi \cos\theta}{ndir}$$

The intensity of the light travelling in the direction of projection is related to the radiant energy distribution on j which in turn is obtained by calculating which solid angle, on surface j, the projection enters. In order to reduce the computational overhead the simplifying assumption is made that the radiant intensity from the target in the direction of the projection is representative of the entire surface area subtended within that solid angle. From this it can be seen that the global solution reduces from $O(n^2)$ to $O(n \times ndir)$.

5. Implementation

A typical radiosity implementation falls into three functional sections, firstly a process whereby the essentially geometrical processing is performed, next a stage where the illuminating sources are introduced and the radiosity solution is computed and finally the imaging or result recovery post process. By separating the functionality of these stages into discrete program modules a number of benefits can be realised. Of these the most important is the ability to separate the geometrical from the photometric calculation. As these functions are independent it is possible to generate a single database that represents the geometrical relationship between surfaces, and then to use, and reuse, this precalculated data to explore a number of possible lighting schemes. This strategy has another benefit in that the memory of the processor is not shared between any of the functional tasks.

The first program module - VISUdis - accepts user data which describes a zone's topography and topology. The objective of VISUdis is to transform this continuous zone model to a discrete numerical equivalent by a 'surface and radiosity finite element' technique. The geometric input for the VISULUX system is a geometry data format that may be produced interactively using a geometry editor, or by another application program. The geometry data file consists of a set of body descriptions, which contain the coordinates of the vertices and the face specifications of each body. One data file may represent one object, or group of objects, or some part of an object. Bodies are by default solid polyhedral shapes, i.e., bounded by plane surfaces and may be either convex or concave. Bodies may contain holes where the holes are also properly bounded by planes.

Each surface must then be discretised into a finite element mesh by transformation into a local coordinate system over which a rectangular grid is then superimposed. (Figure. 4)

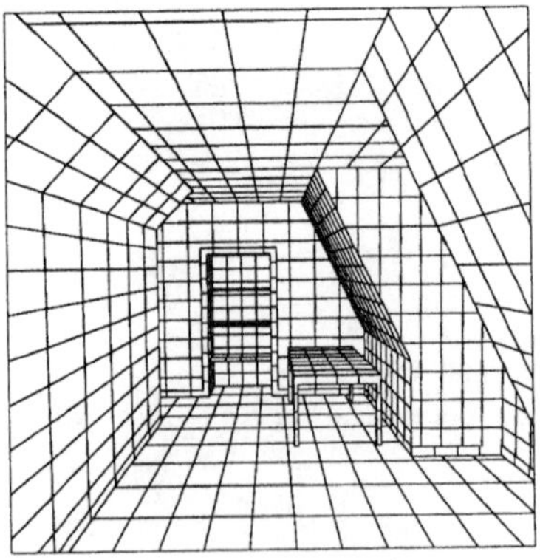

Fig. 4 Finite element equivalent of surface geometry.

The level of subdivision is controlled by a user directed dialogue that allows a number of options.

1 = user defined input
2 = fixed area cells
3 = fixed area with maximum
4 = examine surface
5 = list data
6 = continue

This system provides a great deal of flexibility in that a surface can be divided to a predetermined level, i.e., 10 × 10 cells, or into cells of a given dimension, i.e., 0.1 m^2 or, when applied to an area of the model, to a given size but with an upper limit imposed on the total.

Given a surface polygon that is complex in shape it is apparent that some of the cells created by the rectangular grid may not lie over the polygon. In this instance these cell are clipped to the polygon boundary or, if external, are discarded. Having reduced the surface model to a finite element equivalent the next step is to map the discrete radiosity field of each finite element cell. A unit hemisphere is generated and, by slicing this equally in the horizontal and vertical planes, its surface is represented as a set of finite patches, each subtending an equal solid angle at the base centre point.

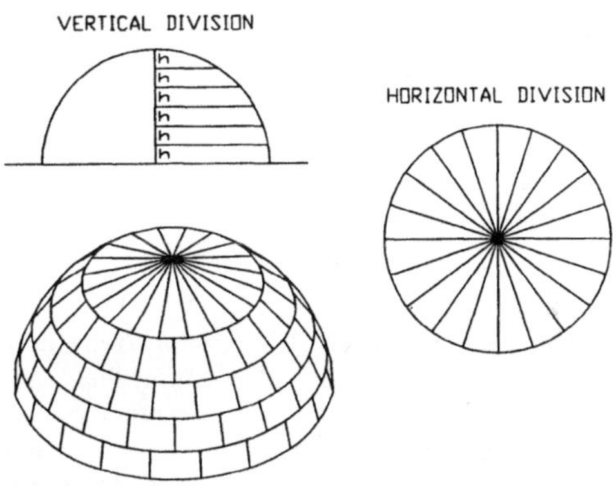

Fig. 5 Patched Unit Hemisphere.

When placed over a polygon cell, these surface patches then represent the cell's discrete radiosity field to a user controlled level of accuracy. The 'patched' unit hemisphere is then placed over each finite element in turn and a number of rays formed by connecting the cell's centre point with the centroids of each hemispherical patch. Each ray is then projected to locate the point of intersection with another polygon grid cell. Hidden surface elimination is optimised by culling surfaces against clipping planes derived from the hemispheres polar coordinates. In this way each projection gives rise to a ray, and each ray has a source and sink cell. The output from VISUdis is a data structure of rays representing the total number of possible light flow-paths.

The second program module is VISUcalc. This accepts the output from VISUdis and the user's specification and positioning of luminaires extracted from the fittings' database. In this database luminaire photometric data is represented as a number of vectors corresponding to a fitting's three dimensional candle-power distribution. These vector-sets are then assigned to several monochromatic wavebands representing the luminaire's spectral characteristics. The computational mission of VISUcalc is to process each vector through the list of rays representing the total inter-reflection possibilities.

With reference to the radiosity equation it is apparent that the emission of a cell is a function of the surface's own emissivity and also the rate at which incident energy arrives at that point. This requires the simultaneous solution of a very large matrix of linear equations. Since this matrix is diagonally dominant it is practical to seek a solution through iterative approximation, the solution converging after some three to five iterations depending of the average surface reflectivity [7]. The one disadvantage of using an iterative solution technique is the need to maintain three sets of variables to hold the initial values, the current estimate of reflected energy and a total value. This incurs a heavy penalty in the use of machine memory but can be minimised by the following strategy. In pseudo code the algorithm is structured as follows.

```
For every cell and direction DO
 set initial estimate of total to direct
END
For every iteration DO
  For every cell DO
    For every direction DO
    get incoming energy
    END
    For every direction DO
    distribute outgoing energy
    END
  END
    For every cell and direction DO
    improve value of total
    END
END
```

The direct contribution from the light source(s) is written to the database, then the initial estimate of the total light energy is set to this value. Using this strategy allows the initial value to remain in a disk file, the running total is kept in memory and the current reflected estimate is only held for the cell under consideration. When the improved total value is calculated the initial value is read from the disk, added to the new estimate of the reflected contribution and used to overwrite the old total value. For each iteration, each cell is taken in turn and effectively samples the environment by integrating over the hemisphere and calculating the incoming indirect contribution from the target in each direction. This incoming energy then becomes a component of the cell's emissivity by invoking the bidirectional reflectance function and distributing the energy over the hemisphere. The output from VISUcalc is a database giving, for each surface finite element, the spectral luminance distribution.

VISUpost is the analysis module which allows the recovery and presentation of this performance data. It has two capabilities. The first allows the performance data to be recovered for each surface. This yields, for any set of design parameters, surface illuminance or luminance. This data can be tabulated by a report generator giving a full break down of the user input data, the luminaire types, position and orientation and the resulting light levels as a function of wavelength and surface characteristics. Alternatively the numerical output can be formatted for post processing to display histograms or isolux contour diagrams.

The second capability concerns the transformation to terminal RGB to enable, via a depth sorting algorithm, the production of an enhanced, coloured image. This is configured to allow display on a wide range of hardware platforms and hardcopy devices varying from one bit deep images for monochrome workstations through eight bit pseudo colour displays to twenty four bit true colour devices.

With the advent of the new generation of lighting evaluation models it is obvious that the traditional result display formats are incapable of dealing with either the range or the wealth of information now produced. Using the VISULUX system a designer might typically be calculating light levels at around 10,000 locations in 3-D space. To tabulate these values in an easily assimilated format is a non trivial task especially when the surfaces are no longer constrained to be either orthogonal or rectangular. The ability to condense this data into an image is certainly one of the major goals in current computer graphics research, however while it is relatively straightforward to represent the information graphically and assess the quantity and distribution of light within an enclosure it is still unresolved as to whether the user can make safe, qualitative judgements on the basis of a synthetic image.

An image is constructed by taking each grid cell and restructuring it as a polygon. The intensity at the centroid is exported to each vertex of this polygon by interpolating between it's neighbours. This process recreates the 3-D model and allows the generation of a view from any given eye point. The cell intensities are found by projecting a vector from the eye position towards the centre of each visible cell. The solid angle through which this vector enters can then be calculated and the surface luminance for that direction read from the database. Given a suitable sample range of the visible spectrum, these luminous intensity values can be transformed into CIE tristimulous chromaticity coordinates [5] or a one to one mapping can be assumed if the wavelength discretisation matches the Red, Green and Blue primaries.

6. Validation and Testing

The model has been validated to the extent that all partners have an agreed degree of confidence in the predictions. These activities have concentrated on an analytical approach rather than empirical comparisons, the results of these tests have proven that the model behaves as the theory and experience dictates.

6.1 Methodology Tests

These activities have concentrated on two areas, firstly the effects of discretisation on the surfaces and the hemispheres. A number of different discretisation strategies for the hemisphere were evaluated in an attempt to improve the distribution of sample points. These included;

(1) Dividing the hemisphere on the basis of equal subdivision of the vertical axis and generating equal area patches as previously described.

(2) Creating the patches as described above except that every second ring was offset by an angle equal to half of the azimuthal angle between adjacent patches.

(3) The current implementation, the hemisphere subdivision was generated on the basis of equal spacing along the horizontal axis as opposed to the vertical axis. The latter provides a better distribution of sample points which improves the quality of the radiosity map and hence leads to a more accurate solution. Results of parametric test runs indicate that for a very crudely discretised environment (less than eight cells per surface) predictions were unreliable. As the degree of discretisation increases the predictions

quickly converge and thereafter further subdivision makes little impact. The effect of the level of discretisation on the hemispheres is harder to define, given a uniformly lit enclosure a modest subdivision (say, sixteen directions) produces very quick and reasonably accurate results. Increasing the number of hemisphere patches produces more reliable results for a less uniformly lit space but it proved to be difficult to lay hard and fast rules for any optimal degree of subdivision. The best blend of surface discretisation and hemisphere discretisation was shown to vary according to the characteristics of individual models and lighting schemes.

6.2 Inter-model Validation

The second area of activity regarding testing was related to comparisons between existing models, individual algorithms and existing theoretical knowledge. While the model produced the expected predictions and trends for the more easily defined cases it was more difficult to acquire suitable data against which to compare the reliability of the treatment of specular surface properties and wavelength dependency. Philips are currently capturing measured data for a number of test cases, as this data becomes available, these tests shall continue.

6.3 Design Trials

Over the research period VISULUX has been used in a number of design trials both to test the program and to increase it's profile within Philip's design department [10,11]. These trials have ranged from street lighting studies, assessing the efficacy of the installation with regard to illumination levels and uniformity and also with regard to security issues and light pollution, through the testing of new products for office lighting and to specialised applications in tunnel lighting, airfield lighting and sports field lighting.*

7. Future Directions

Future work will continue to seek advances in performance and to the attainment of greater accuracy and relevance to design requirements.

One major enhancement in the functionality of the model will be to integrate existing natural lighting with the current artificial lighting capabilities. This aspect of the work is currently being funded by the CEC Joule initiative [4]. The objective being to develop control strategies for artificial lighting in the presence of daylight.

Over the period of this research it has become obvious that there is a growing interest in the use of computer graphics to communicate quality as well as quantity in lighting design. Although this is a complex subject, relying on psychological perception as well as physical representation, the methods and the targets can now be clearly defined.

Acknowledgements

This project was a collaborative venture between ABACUS, the LiDEC division of Philips and CALIBRE at the Technical University of Eindhoven. Thanks are due to all who contributed.

---

* For examples, see Fig. 6-9 on color plates

References

1. Clarke, Joseph A,."Energy Simulation in Building Design," Adam Hilger Ltd., Bristol and Boston, 1985.

2. Cohen, Michael F., Greenberg, Donald P., "The Hemi-cube a Radiosity Solution for Complex Environments," Proceedings of SIGGRAPH '85, Computer Graphics Volume 19, Number 3, July 1985.

3. Cohen, Michael F., Shenchang, Eric C., Wallace, John R., Greenberg, Donald P., "A Progressive Refinement Approach to Fast Radiosity Image Generation," Proceedings of SIGGRAPH '88, Computer Graphics Volume 22, Number 4, August 1988.

4. Grant, M., Trommelin T., "Investigation into Comfort and Energy Related Aspects of Daylighting," Case for Support to the Commission of the European Communities, 1989.

5. Judd, D B., Wyszecki G., "Color in Business, Science, and Industry," Wiley and Sons, New York, 1975.

6. Maver, Thomas W., Grant M., Anderson Andrew J., "Collaborative Development of a Design Aid Based on an Inter-Reflective Multi-Chromatic Lighting Model," Report to SERC, ABACUS 1990.

7. Modest, Michael., "A General Model for the Calculation of Daylighting in Interior Spaces," Energy and Buildings, Volume 5, 1982.

8. Numan, M Y., Moore, Greg R.,"Form Factors: The problem of Partial Obstruction," Working Paper for the Martin Centre for Architectural and Urban Studies, University of Cambridge, Department of Architecture, July 1982.

9. Sparrow, Ephraim M., Cess R. D., "Radiation Heat Transfer," Hemisphere Publishing Corporation, Washington, 1978.

10. Tenner, Ariadne D., Tan, B G P., "The Digital Illumination Model, a Computer Program for Calculating and Visualising Artificial Lighting," Proceedings IES Annual Conference 1989.

11. Tenner, Ariadne D., Jacobs, Michel., "Computer Graphics; From Toy to Tool," International Lighting Review, Volume 3, 1989.

12. Torrance, Kenneth E., Sparrow, Ephraim M., "Biangular Reflectance of an Electric Nonconductor as a Function of Wavelength and Surface Roughness," Journal of Heat Transfer, 1965.

Synthesis of Multispectral Images of Natural Landscape

Pietro A. Brivio, Paolo Furini, Massimo Righetti and Daniele Marini

Abstract

Simulation of natural phenomena by computer is one of the most interesting applications of computer science. This paper presents a methodology for synthetic generation, visualisation and rendering of Earth natural landscapes. A software package TISS (Territorial Image Synthesis System) has been developed in the framework of a research project concerning the study of satellite remote sensing images and their integration with informations derived from DEMs. TISS is able to simulate multispectral images of a natural landscape in mountainous areas starting from spectral responses data-base of materials spread out in the area, digital elevation model (DEM), computer-graphics rendering techniques, color typology attribution and energy equilibrium illumination model. First experiment has been conducted over a typical Alpine high-relief terrain, where elevation ranges from 1800 to 3500 meters above sea level, belonging to Eastern Italian Alps. Multispectral synthetic images generated by TISS are presented and compared with multispectral images acquired by Landsat-TM on the same area. The synthetic images and the underlying model can be used for many applications, e.g. flight simulator, land planning and management, and others related fields.

1 Introduction

Visual simulation of natural phenomena by computer is one of the most exciting and challenging applications of computer graphics. Visual simulation consists of those processes computable with mathematical methods whose result can be visually represented: the resulting image supports the model and the image becomes a knowledge tool.

Recently, with the advent of remote sensing satellites, Earth landscapes representation has proved of a continually increasing importance both in land planning and resources management and in traditional computer graphics applications. Computer graphics with the rapid development of hardware and software has given in turn significative contributions to accurate illumination models and rendering techniques.

A landscape is usually perceived as a continuously varying surface that cannot be modelled appropriately by a simple model. A digital representation for the continuous variation of relief over space is known as a digital elevation model (DEM) or digital terrain model (DTM). In high mountain terrains the availability of a numerical topographic surface is a necessary condition for further improvements in studying and simulating related natural phenomena.

The Remote Sensing Department of Institute of Geophysics of the Lithosphere - CNR, Milan in collaboration with Laboratory of Eidomatics, Department of Information Sciences, University of Milan, developed a software system (TISS - Territorial Image Synthesis System) in the framework of research activities concerning the study of satellite remote sensing images and the integration of these data with all the informations derived from DEMs, for a better knowledge of natural processes taking place in a mountain region landscape. TISS is able to generate synthetic images of a natural landscape in mountainous areas, at different wavelength from visible to near infrared. Visual simulation is based on spectral responses data-base of Earth surface materials, different tools for developing digital elevation model (DEM), computer-graphics rendering techniques, color typology attribution and energy equilibrium illumination model.

Our research aims to develop a complete methodology for generation and visualisation of natural landscapes by integrating the different approaches, computer graphics and remote sensing, in order to create reliable models of a territorial area, both from the elevation or geometric viewpoint, and from its radiometric behaviour.

2 Overview of the TISS System

Figure 1 represents the whole software system TISS. Input data are a materials spectral responses data-base, topographic maps, remote sensing images and thematic maps.

From a data base of Earth features spectral responses, using a color attribution procedure and an illumination model, we can obtain the three R,G,B colours to represent a selected material, as snow, soil, vegetation etc. Two different approaches for reconstructing the DEM of an area are available in the system. The first one is based on classical interpolation techniques, and the other one on fractal geometry principles. Inputs to these modules come from digitized contour lines or from regular grids of known elevation values. The obtained terrain altitude matrix (DEM) is used to derive other characteristic attributes such as surface slope, exposure and insolation maps. The information concerning the spatial distribution of materials over an area can be acquired from existing thematic maps or from remote sensing satellite images, using classification procedures that are well established in the image processing field.

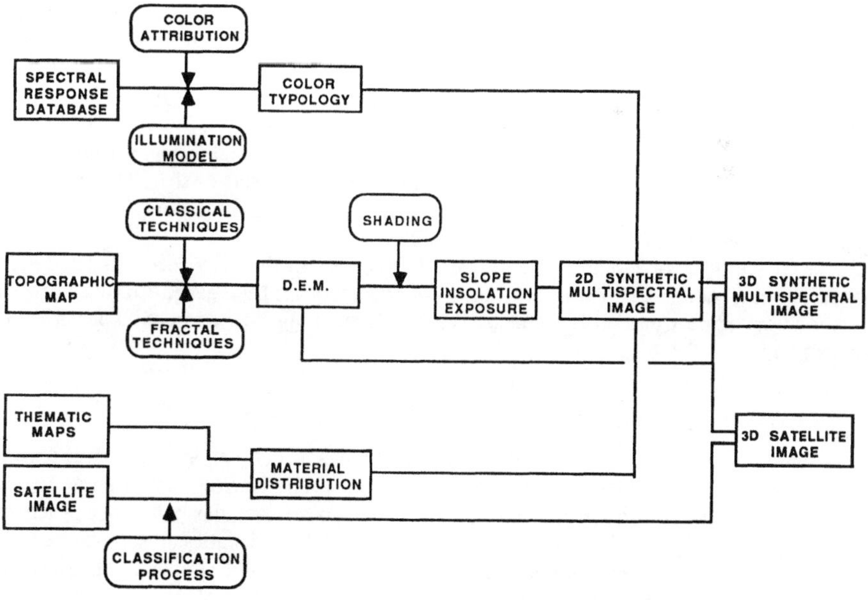

Fig. 1 : Diagram of TISS (Territorial Image Synthesis System)

The final part of TISS is devoted to simulation and visualization, and is based on computer graphics techniques. It allows us to generate, from all information previously elaborated, multispectral synthetic images of the terrain, that can be compared to satellite acquired ones, and to produce any perspective display of both satellite and synthetic images.

A synthetic image generated using TISS is a function

Im(x,y) = F(f1, f2, f3, f4, f5)

where:

f1 = El(x,y)	Elevation
f2 = Cl(x,y)	Classification of satellite image
f3 = Sh(x,y)	Shading
f4 = Co(x,y)	Color attribution
f5 = Il(x,y)	Illumination

Some combination or partial selection of above functions gives any intermediate result. The TISS software system is a modular package, completely developed in C language and running on VAX/VMS and UNIX machines, with a window-based user interface.

3 Digital Elevation Model

The most common data structure for DEM is a bidimensional array, whose elements are interpreted as elevation data. The approaches to the estimation of above data from few measured samples are many, but all of them present limits and drawbacks. Some methods use photometric techniques on stereoscopic photographic images; this solution provides good results, but stereo couple images are often difficult to acquire and very expensive too. Other families of numerical approaches use input data coming from contour lines of topographic maps; these methods are less expensive, but present the problem of the acquisition of the isolines, a boring and error prone operation that is usually done by hand. All these algorithms for DEM construction interpolate between two adjacent contour lines or scattered data, producing elevation value grids. The interpolation or approximation methods range from: weighted distance method, maximum slope/minimum path between two contour lines, polynomial interpolation, splines, Fourier series and kriging method for scattered data (Bohm et al, 1984; Frederiksen et al, 1985; Burrough, 1986).

Figure 2 shows a DEM of a 7.5 x 7.5 km mountainous area in the Eastern Alps covering parts of the Aurina and Isel Valleys, on the border between Italy and Austria. This region is a typical high-relief terrain, where the landscape has rapidly changing slopes and elevation ranging from 1800 to 3500 meters above sea level.

From two IGM topographic maps, scale 1:25000, contour lines have been digitized every 100 meters elevation; a DEM with a resolution of 15 meters has been generated by an original methodology: a linear interpolation method along the maximum slope direction perturbed by local kriging.

3.1 Fractal Topographic Relief

The research of new techniques for DEM construction has recently adopted methods derived from fractal geometry, which provides a new approach where the irregularity, erraticity and self-similarity of fractal structures mimics at best the typical irregular behaviour of natural terrains/landscapes. The basic idea in creating a fractal model is to identify an appropriate value for the fundamental parameter that characterizes the erratic behaviour of natural scenes: fractal dimension D (Mandelbrot, 1983; Barnsley, 1988). Once the fractal dimension has been estimated, it is possible to reconstruct a surface using different fractal methods which interpolate the original set of elevation data (Raffy et al., 1986; Cadisco et al., 1984; Brivio et al., 1989).

Assuming that the topographic relief surface Z(x,y) with $Z \in \Re$ is a realization of a fractional Brownian Motion (fBM) which is characterized by zero mean and variance defined as

$$E [| Z_H (p) - Z_H (p) |^2] = C . || \Delta ||^{2H} \quad (1)$$

being C a constant and $|| \ ||$ the Euclidean norm where $0 \leq H < 1$ is a real number, the determination of fractal dimension D can be achieved applying some special mathematical method on finite sets of samples. The term H is associated to the fractal dimension D of relief by the relation

$$D = E + 1 - H \quad (2)$$

where E is the usual topological dimension (E=2 in the case of relief surface).

From the altitudes matrix previously generated we extracted a finite set of data regularly sampling every fourth pixel: this is the data set taken as input for the estimation procedure of fractal dimension and for the fractal interpolation. Let Z(x,y) = Z(p), the algorithm implemented computes a variance analysis of the vertical displacements according to the simplified relationship

$$\mathrm{Var}\, Z(p+K) = K^{2H} \qquad (K = 1, 2, ..., N) \quad (3)$$

where K is the lag or distance over which variance is measured and N is the maximum distance, that have to be limited between one fifth and one third of the total distance over the data (Cohen et al. 1990). The coefficient H was estimated using least square approximation of logarithm of (3) and a fractal dimension D = 2.243 was obtained.

Interpolation process of a sampled elevation data set consists in finding a rule which relates the elevation Z of a generic point p to the known elevation data of some p_i neighbours:

$$Z(p) = f(Z(p_i)) \qquad (i = 1, ..., M) \quad (4)$$

Let us consider a point p and its neighbour containing M points of known altitude, and the elevation as a fBM function, we can state that

$$E [Z_H(p)] = \mathbf{L}^T \mathbf{Z}_H \quad (5a)$$

$$\mathrm{Var} [Z_H(p)] = || p - 0 ||^{2H} - \mathbf{L}^T \mathbf{V} \quad (5b)$$

V being vector of distances $|| p - p_i ||^{2H}$ with i = 1, ..., M. From statistical properties of fractional Brownian Motion, **L** coefficients can determined as the solution of the system of equations:

$$\mathrm{Cov} [\mathbf{Z}_H (p)] \mathbf{L} = \mathbf{V} \quad (6)$$

where Cov is covariance matrix of M neighbour points of p. Elevation data Z(p) are then generated as random numbers belonging to a gaussian distribution defined by relationship (5a, 5b).

Accuracy evaluation of fractal DEM was assessed by comparison with height values of classical DEM: a maximum difference of 104.000 m. and an R.M.S. error of 10.933 m. resulted.

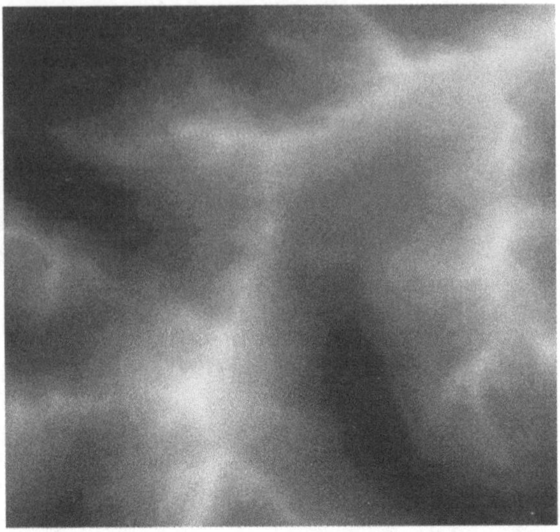

Fig. 2 : The Aurina and Isel Valleys DEM: in Light Gray the High Areas, in Dark the Bottom Valleys

4 Rendering of the DEM

One of the most popular techniques for the visualization of digital topographic surface uses 3D fish-net or wire-frame representation. To improve visual quality in the representation of hilly and mountainous landscapes, cartographers have developed many techniques; one of the most successful is the shading relief, an "artistic" method realized by hand shading and using airbrush techniques to produce the desired effects.

With the availability of numerical maps, it became possible to automatically compute shaded relief maps of terrain (Horn, 1981). These projections of three-dimensional landforms not only have a high degree of realism, but are finding growing applications in quantitative geomorphology. A further advance was reached when this technique was used in combination with information derived from digitized thematic maps and from remote sensing satellite images (Dubayah et al, 1986; Jones et al, 1988). Moreover, various works in computer graphics represented wonderful landscapes, built by fractal methods, with a high degree of realism, but they showed imaginary worlds (e.g. Voss, 1988).

In this research efforts have been made to conjugate the remote sensing physical approach to the real world with the high capabilities of computer graphics in simulating natural phenomena. Techniques and algorithms of the image synthesis world (Foley et al., 1984; Rogers, 1985) have been analyzed in order to render as likely as possible the natural reality with colours and perspective effects of a photographic kind. On the other hand, multispectral simulation of Earth terrain requires us to take care of all interaction mechanisms of electromagnetic energy with surface features.

In the following paragraphs various steps of the proposed methodology and related problems will be examined and described.

4.1 Material Distribution

Two main factors determine the color (RGB) of an image pixel: a natural feature represented by the pixel and its illumination conditions. In order to assign a reliable color typology to a landscape surface, a color is given to each pixel; then a spatial distribution model of the material present in that area is required.

A first possible solution is to subdivide the DEM in different elevation zones and to assign a material to each zone, for example coniferous trees from 1000 to 1500 meters, pasture from 1500 to 2000, rocks and snow beyond. This solution doesn't give particularly exciting results; natural terrains have a more erratic land cover types distribution and the limits between two different materials are not so sharply defined.

Another approach consists in digitizing an existing thematic map of the area or drawing by hand a material distribution, that is in dividing the DEM in thematic areas, each having a code, and in associating a material to every area. In this case the result is better: you can easily map glaciers, woods or arable lands, etc. and modify their distribution. The problem of the border between two materials still remains.

Multispectral remote sensing satellite imagery carries more information on natural features present in an area. Using a classification procedure we can build a stratified image in which pixels are assigned to different categories, representing materials spatial distribution.

In this study a Landsat-5 Thematic Mapper (TM) image, acquired on September 7, 1985 and covering a mountainous region of the Eastern Alps which includes the Aurina and Isel valleys, has been used. The image was registered on the DEM with an RMS error of 1.4 pixel, and by means of the parallelepiped classifier, five classes, i.e. coniferous, pasture, rocks, snow and ice, have been mapped. This classifier works as a thresholding process based on 3 spectral bands in the visible range. A class is defined as a 3D interval (a parallelepiped), and each pixel is classified according to its three spectral components.

4.2 The Spectral Color Code

The equation $\alpha(\lambda) + \rho(\lambda) + \tau(\lambda) = 1$ represents electromagnetic radiation interaction with physical surfaces, where $\alpha(\lambda)$ is the spectral absorptivity, $\rho(\lambda)$ is the reflectivity and $\tau(\lambda)$ is the transmissivity of a given material. Reflectance characteristics of surface features may be quantified measuring the reflected portion of incident energy; these measurements are a function of wavelength and are termed the spectral reflectance curve. In Remote Sensing applications the Earth features are normally considered as opaque bodies, therefore $\tau(\lambda)$ vanishes and $\alpha(\lambda) = 1 - \rho(\lambda)$. Within the visible part of the spectrum, these parameters give the "color effect" and allow to discriminate between different objects.

In recent years, research on photorealistic visual simulation, has led to the development of new techniques, based on spectral responses of objects measured in the laboratory (Hall, 1989). This approach studied simple objects in a closed environment in order to simulate, for example, the color of a teapot on a wood table in a poorly lighted room. A natural environment, such as a mountainous region, is more complex, thus more difficult to simulate. Nevertheless, in the remote sensing literature spectral reflectance curves for many natural objects are reported (Colwell, 1983; Swain et al., 1978; Lillesand et al., 1979; Wolfe et al., 1978).

The TISS system includes a database of spectral reflectance curves, from 380 to 2500 nm, with a resolution of 1 nm, for different types of material, such as soils, rocks, vegetation, water.

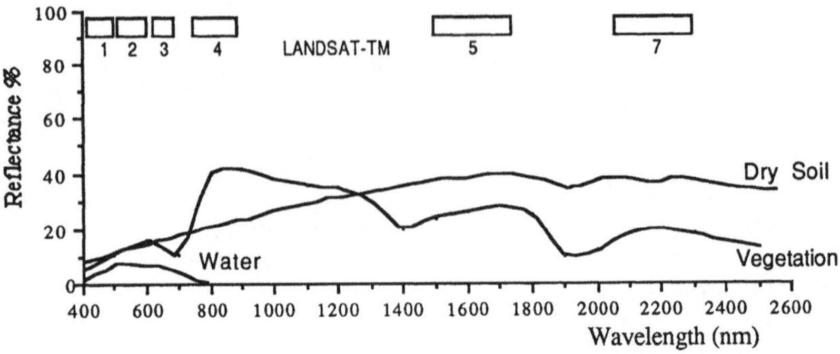

Fig. 3 : Typical Spectral Reflectance Curves and Landsat-Thematic Mapper Reflective Bands

Figure 3 shows a typical spectral reflectance curve for healthy green vegetation, representing average reflectance values obtained by a large sample of data. In the same figure are also reported the reflective spectral bands of Landsat-TM sensors: band 1, 452-518 nm; band 2, 528-609 nm; band 3, 625-693 nm; band 4, 776-904 nm; band 5, 1568-1784 nm and band 7, 2097-2348.

The color code attribution to each material is obtained, in the TISS system, by multiple convolutions of spectral irradiance of the energy source, the Sun, materials spectral reflectance curves, spectral interference of the atmosphere and MTF (modulation transfer function) of three out of six reflective TM bands.

4.3 The Illumination Model

In the synthesis process the pixel color will depend not only on the spectral characteristics of the material, but also on the illumination conditions. In mountainous regions these are determined by geometric relationships of the Sun, the elementary surface orientation and the satellite sensor, variable from one pixel to another. From the DEM of the area we derive an illumination factor for each image pixel, which depends on the microfacets orientation with respect to the Sun (slope and exposure maps) and on the cast shadows (insolation map) (fig. 4).

The irradiance of an elementary surface illuminated by a source with an illumination angle α (where α is the angle between the normal to the terrain and the illumination vector), is given by equation:

$$E_{S1}(\lambda) = E_S(\lambda) \cos(\alpha) \, d\omega \tag{7}$$

where $E_S(\lambda)$ is the solar irradiance (watt m^{-2} mm^{-1}), i.e. the amount of energy incident on an elementary area perpendicular to the illumination vector, measured on the surface after attenuation and scattering through the atmosphere (Wolfe, 1978), and $d\omega$ is the infinitesimal solid angle from the light source to the illuminated surface area (fig. 5).

The term $E_{S1}(\lambda)$ becomes zero for $\cos(\alpha)<0$ and for cast-shadowed pixels too. After the interaction with the material, the radiant energy $E(\lambda)$ leaving the terrain surface can be expressed by the following equation:

$$E(\lambda) = E_a(\lambda) + \rho_{bd} \, E_{S1}(\lambda) \tag{8}$$

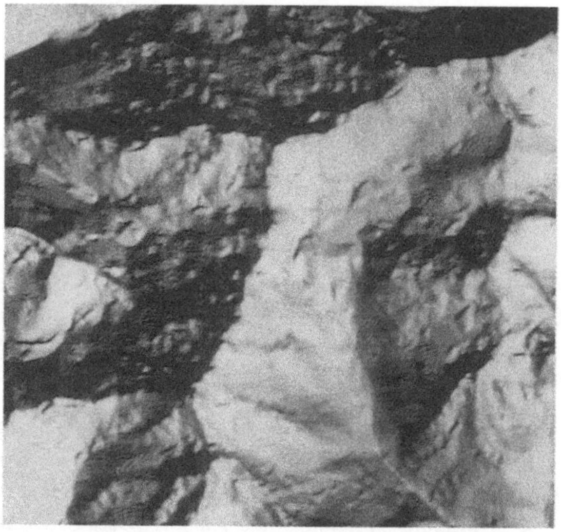

Fig. 4 : Insolation Map of the DEM, Simulating the Sun Position when the Satellite Image has been Acquired

where $E_a(\lambda)$ is the atmosphere or skylight radiance and ρ_{bd} is the bidirectional reflectance, which gives the ratio of incident energy reflected in a given direction. This ratio can be considered as composed of two terms: $\rho_{bd} = (dR_d + sR_s)$, the first is a diffusive and the second is a specular reflective component. This last component, which is not considered in the TISS model, is a combination of multiple factors such as the Fresnel equations, a roughness function, etc., parameters that are costly to simulate; their influence is not so important outdoors as in a closed environment and, moreover, they are not well known for natural features of Earth landscape.

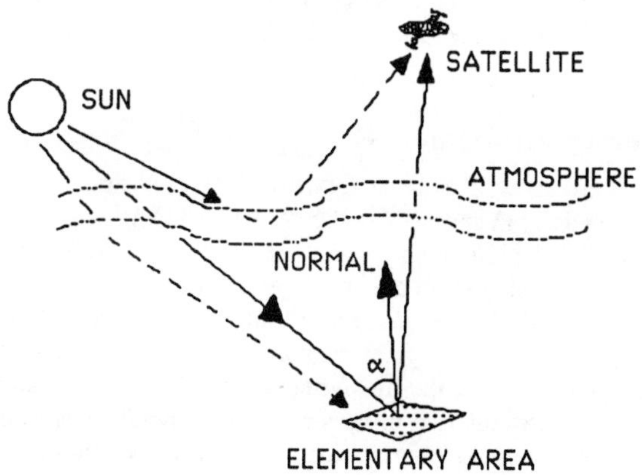

Fig. 5 : Energy Interaction between Sun, Target and Satellite

Therefore we approximate the last term of equation (8) with a simple diffusive component $E_d(\lambda)$, that can be obtained by:

$$E_d(\lambda) = E_{s1}(\lambda) * S_m(\lambda) \tag{9}$$

where $S_m(\lambda)$ is the material spectral reflectance curve. The first term of right hand side of eq. (8) is approximated by:

$$E_a(\lambda) = E_h(\lambda) * S_m(\lambda) \tag{10}$$

where $E_h(\lambda)$ is the skylight contribution. Its computation can be very costly, but it is customary (Hall 89) to approximate this term multiplying $E_s(\lambda)$ by a constant $0<c<0{,}3$ that represents the part of solar energy diffused in the atmosphere; low values of c simulates night conditions, while medium values simulate day light.

Then the term $E(\lambda)$ in eq. (8) represents the total energy leaving the elementary surface. Not all this energy arrives to the satellite: it is attenuated by atmospheric absorption and scattering and by distance. Absorption and scattering effects are simulated in TISS system by the simple function λ^{-4}, which approximates the Raleigh scatter (Swain, 1978); distance influence is ruled by r^{-2}. The energy $E_R(\lambda)$, reflected by an elementary area and received by the remote sensor, can therefore be approximated:

$$E_R(\lambda) = E(\lambda)\, \lambda^{-4}\, r^{-2} \tag{11}$$

The TM sensors have different MTF_i (modulation transfer function) which characterize their relative spectral response to the incident energy. Moreover acquired energy level are sampled in an interval [0,255]. In order to approximate the image acquisition process we convolve $E_R(\lambda)$ with each sensor's MTF_i and normalize the results in the interval [0,255]:

$$Im\,(x,y,\Delta\lambda) = k \int_{\lambda_0}^{\lambda_0+\Delta\lambda} E_R(\lambda) * MTF_i\,(\lambda)\, d\lambda \tag{12}$$

These values can be directly compared with the Multispectral image effectively acquired. If the integration intervals are chosen in the visible range, a realistic simulated image can be obtained.

5 Synthetic Versus Real Images

Figure 6 (see color plates) shows the classical false color composition using the infrared band TM4, and the visible bands TM3 and TM2 of the Aurina and Isel valleys, and the multispectral image simulated by the TISS system; down, another composition of TM and synthetic bands. A visual comparison can give an idea of the quality of the results obtained.

A quantitative analysis of all six couple of images is illustrated in the Table 1, where statistical parameters are computed on pixels value. The presence of the snow covered areas is responsible for the coincidence of the histogram peak and the max value for both sets of multispectral images. The most relevant difference between synthetic and acquired images can be recognized in band 1, where a histogram displacement is evident due to the atmospheric haze contribution, not corrected on the satellite images.

The evocative potentialities of this approach are highlighted in Figure 7 (see color plates), where we show two perspective 3D representations of the area as observed from the valley. These images have been embellished with some rendering effects, as texturing in the pine areas, smoothing filter over the snow covered regions; the 3D display is obtained mapping the image function over the DEM and perspective projecting it.

Table 1: Statistics of Remote Sensing (TM) and Synthetic (SY) Images for 6 Spectral Bands

Spectral	MinValues		MaxValues		MeanValues		St. Dev.s		Correlation
Bands (nm)	TM	SY	TM	SY	TM	SY	TM	SY	(TM,SY)
452-518	41	6	255	255	128.04	68.69	78.56	66.83	0.794
528-609	12	6	255	255	76.44	69.58	72.41	65.72	0.888
625-693	9	6	255	255	81.93	69.60	77.37	66.04	0.881
776-904	4	6	255	255	78.47	82.11	62.86	61.28	0.844
1568-1784	1	7	188	183	50.21	46.90	31.57	31.42	0.758
2097-2348	1	5	97	115	25.67	32.77	16.00	21.12	0.723

6 Conclusions and Future Works

Future work is needed in order to improve the illumination model, considering scattering and absorption effects of the atmosphere and albedo contribution. Moreover the simulation of spectral sampling of present remote sensing satellite (e.g. Landsat-TM have 50-250 nm bandwidth) drastically reduces the information detail/richness of materials spectral responses. The proposed model of multispectral synthetic image generation, which computes spectral radiance with 1 nm bandwidth, could give a better term of comparison and tuning, in new remote sensing data acquired by imaging spectrometers, which will have spectral sampling interval of 5-10 nm, today available only from some experimental aerial surveys.

The Territorial Image Synthesis System is a software package able to produce a wide variety of multispectral synthetic images, and as such can help in better understanding natural processes taking place in mountain regions landscapes. Some data here described as intermediate results, such as slope, exposure and insolation maps, can be regarded as important tools in many applications.

Interesting applications can be found also in land planning and environmental resources management, environmental impact of human activities (infrastructures, dams, roads), study and mitigation projects of natural hazards (landslides, floods, volcanic eruptions, etc.), scenarios for flight or driving simulators.

Acknowledgements

This research project was partially supported by ASI (Italian Space Agency) and MURST 60%.

References

Barnsley, M. (1988) *Fractals everywhere.* Academic Press Inc., Boston, p. 394.

Bohm, W., Farin, G., Kahman, J. (1984) *A survey of curve and surface methods in CAGD.* Computer Aided Geometric Design 1.

Brivio, P.A., Righetti, M. (1989) *Metodi frattali per la costruzione di modelli digitali del terreno.* Workshop CNR-GNIM "Conoscenza per immagini", Roma, 27-28 Novembre 1989, pp. 277-280.

Burrough, P.A. (1986) *Principles of geographical information systems for land resources assessment.* Clarendon Press, Oxford, p. 193.

Cadisco, F., Crosta, G., Marini, D. (1984) *Landscape synthesis by fractal approximation.* Proceedings Premier Colloque Image, Biarritz 1984, pp. 195-199.

Cohen, W.B., Spies, T.A., Bradshaw, G.A. (1990) *Semivariograms of digital imagery for analysis of conifer canopy structure.* Remote Sensing of Environment, vol. 34, n. 3: 167-178.

Colwell, R.N. (1983) *Manual of Remote Sensing.* American Society of Photogrammetry and Remote Sensing (Second Edition), Falls Church, Virginia, USA.

Dubayah, R.O., Dozier, J. (1986) *Orthographic terrain views using data derived from digital elevation models.* Photogrammetric Engineering and Remote Sensing, vol. 52, n. 4, pp. 509-518.

Foley, J.D., Van Dam A. (1984) *Fundamentals of interactive computer graphics.* Addison Wesley, Reading, Massachussets, p. 660.

Frederiksen, P., Jacobi, O., Kubik K. (1985) *A review of current trends in terrain modelling.* ITC Journal 85 n.2, Enschede, pp. 101-106.

Furini, P., Righetti, M., Brivio, P.A., Marini, D. (1991) *Una metodologia per la simulazione di scene ambientali.* I.Co.Graphics '91, 5-8 marzo 1991, Milano, Mondadori , Milano, pp. 405-412.

Hall, R. (1989) *Illumination and color in computer generated imagery.* Springer-Verlag, Berlin, p. 282.

Horn, B.K.P. (1981) *Hill shading and the reflectance map.* Proc. IEEE vol.69, n.1, pp.14-47.

Jones, A.R., Settle, J.J., Wyatt, B.K. (1988) *Use of digital terrain data in the interpretation of SPOT-1 HRV multispectral imagery.* Int. Journal of Remote Sensing, vol.9, n.4, pp. 669-682.

Lillesand, T.M., Kiefer, R.W. (1979) *Remote sensing and image interpretation.* John Wiley & Sons, New York, p. 612.

Mandelbrot, B.B. (1983) *The fractal geometry of nature.* W. H. Freeman and Company, New York, p. 468.

Raffy, M., Ramstein, G., Becker, F. (1986) *Fractal analysis of a relief and its remote sensed image.* Second Image Symposium, International Electronic Image Week, Nice, April 1986, pp. 79-82.

Rogers, D.F. (1985) *Procedural elements of computer graphics.* McGraw-Hill, New York, p. 462.

Swain, P.H., Davis, S. M. (1978) *Remote sensing: the quantitative approach.* Mc Graw-Hill, New York, p. 396.

Voss, R. (1988) *Fractals in nature: from characterisation to simulation.* In: The Science of Fractal Images, eds. Peitgen, H.O., Saupe D., Springer Verlag, New York, pp. 21-69.

Wolfe, W.L., Zissis, G.J. (1978) *The infrared handbook,* ERIM, Ann Arbor, Michigan.

A Rendering Pipeline for Street Lighting Simulation

Salvador Clavé, Markus Gross

Abstract

This paper presents an application in the field of illumination and traffic engineering with the definition of its rendering pipeline. The system incorporates a new solution for the calculation of the energy that arrives at each polygon of the scenery from point light sources in the direct illumination step. Also, we introduce an extension of the radiosity method to incorporate physical surface light sources with intensity distribution curves and a set of image transformation for the correct display of the results according with the behavior of the human visual system. The simulation methods are based on spectral sampling allowing photorealistic rendering of planned lighting hardware as well as safety checks of dark areas in the street.

1 Introduction

In Germany the setting of street lighting follows the standard DIN 5044 (DIN 81) (DIN 82). Illumination engineers dispose, for that purpose of calculation, methods based on physical and geometric attributes of the light sources they use (Hentschel 87). The standard methods for visualizing their results are isocandela lines or simple dithered grey level presentations. These techniques, however, deliver only a limited information and do not provide any relationship to the visual perception of the street scenery during night time.

On the other hand, Computer Graphics offers a lot of sophisticated rendering methods for the photorealistic presentation of artificial and real sceneries (Foley et al. 90). Respectively the presence of powerful graphics hardware opens the way to high quality rendering in a wide field of applications. The advantages of these simulation methods, often using ray–tracing and radiosity techniques, are presented with great success in interior decoration (Sillion and Puech 1989) and driving simulation (Nakamae et al. 1990) (Groß et al. 1991b). For this reason, the goal of our research was to find an approach for the realistic presentation of street lighting distribution.

Especially sensitive problems, like e.g. the perceptibility of pedestrians or traffic signs, the lighting's exemption from glare, the influence of the road surface, wetness or buildings can be ascertained and documented. Furthermore, a scrutiny of the rules laid down in the DIN 5044 as well as an identification of incorrectly installed systems are made possible by the analysis of the simulation results with the street geometry.

Examining the requirements of this application in detail and contrasting the available methods of lighting calculation, we saw, that in particular the recently developed combinations of the ray–tracing method (for the first time presented by Whitted (Whitted 1980) and Kay (Kay and Greenberg 1979) with the radiosity method (derived from thermal engineering by Goral (Goral et al. 1984)) describe an appropriate way of solving the problems. However, the inclusion of realistic point and surface sources defining their spectral energy and luminous intensity distribution is here of particular interest for street lighting.

Apart from a correct simulation of the illumination, there is also the question of the correct presentation of the results on a display due to its limited physical capabilities. Thus, our approach takes also into account the behavior of the human visual system during night time using psychophysical transformations.

The remainder of this paper is as follows: First, we describe the rendering pipeline and its requirements derived from illumination engineering. Then the light source model is introduced. After that, the simulation concepts and methods are presented using direct illumination, an extension of the traditional radiosity method and ray tracing. Finally, we discuss the results from our process and give an outlook for further work.

2 The Rendering Pipeline

In order to obtain a realistic image on the display three different aspects must be defined: a real scenery with physical properties and a realistic light source model; a model for the light propagation in the environment (simulation methods) and third, a way for displaying the results of the simulation. The definition of these aspects and the concatenation of their computer solutions give the rendering pipeline as a result.

According to this definition we divide the rendering pipeline in three modules (figure 1): preprocessing, spectral sampling renderer and imaging transformations.

A correct physically based rendering has to provide spectral sampling methods. Thus in the preprocessing module the polygons of the scenery (street, buildings, street furniture...) are defined with the spectral distribution of their reflection or transmission coefficients. Moreover, light sources are given by their spectral energy distribution and their luminous intensity surface, which is supposed to be independent of the wavelength. In addition, it is also necessary to define textures, especially the street textures taking influence on the light distribution. One possibility to simulate its roughness is using bump mapping techniques, fractals or a combination of both (Nakamae et al. 1990).

The spectral sampling renderer module of the application incorporates sophisticated simulation models. Respectively in the case of safety aspects indirect light and interreflection often play an important role. Dry streets usually have high diffuse reflection coefficients while their specularity is very low. In this case, a radiosity method (Goral et al. 1984) is an elegant technique for the simulation. However, the inclusion of the luminous intensity distribution is required.

On the other hand, wet streets' specularity cannot be neglected any longer. For this reason, we use a two-pass general shading approach combining radiosity and ray-tracing as introduced in (Sillion and Puech 1989).

The renderer module uses spectral sampling which is a very expensive method compared with the common rendering using the three primaries R,G and B. Based on a spectral distribution of the energy of a light source and the spectral distribution of the reflection coefficient which is an attribute of the surface element, the corresponding illumination equation must be solved for each sample. The result is a function which describes the spectral distribution of light to be reflected or transmitted from the surface.

The aim of the imaging transformation module is to present the spectral energy distribution $P(\lambda)$ of each pixel scanned by the ray-tracer according to psychophysical human behavior and display possibilities.

3 Modelling the Light Sources

For most of the specific applications of engineering illumination the assumption that light sources emit light equally in all directions cannot be accepted any longer. A Lambertian source (perfect diffuse emitter) is a big simplification for most actual light sources. Most published illumination engineering papers include this effect (Verbeck and Greenberg 1984) (Nishita et al. 1985) but often assume rotational symmetric intensity distribution that is not applicable for physical based street light simulation.

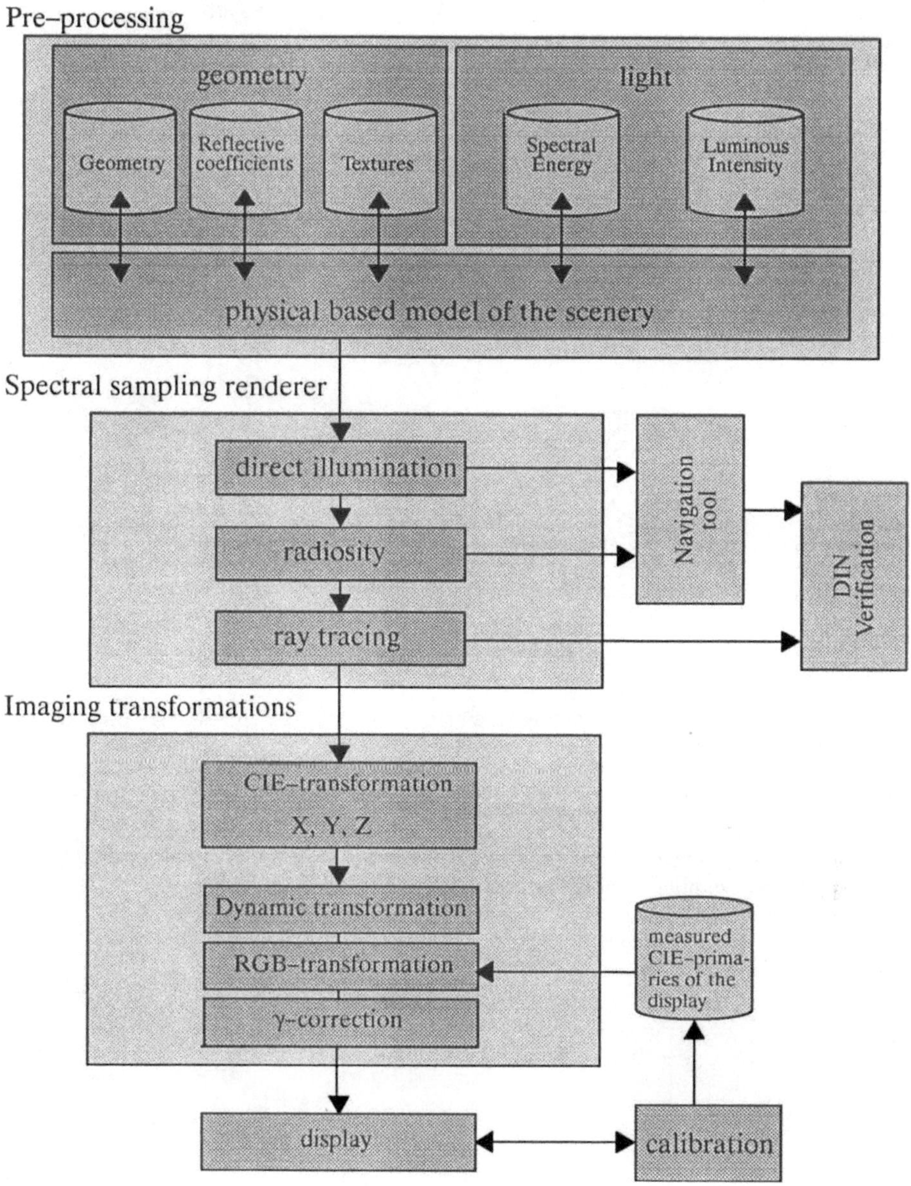

Fig. 1: Physically based rendering pipeline

The luminous intensity distribution of the light is expressed by different curves (figure 2a) which are defined in several planes called C–planes. The plane including the light centre and the illumination axis is called C_0–C_{180} (figure 2b) and all other planes are defined from this one just with the angle between them (θ_i = C_{20}–C_{200}, C_{40}–C_{220} ... C_{90}–C_{270}). The luminous intensity data for each plane are given by samples at different values of β (β_i = 0,10,...180), β describing the angle between the illumination axis and the lighting ray. Section 4.1 explains how to manage these values.

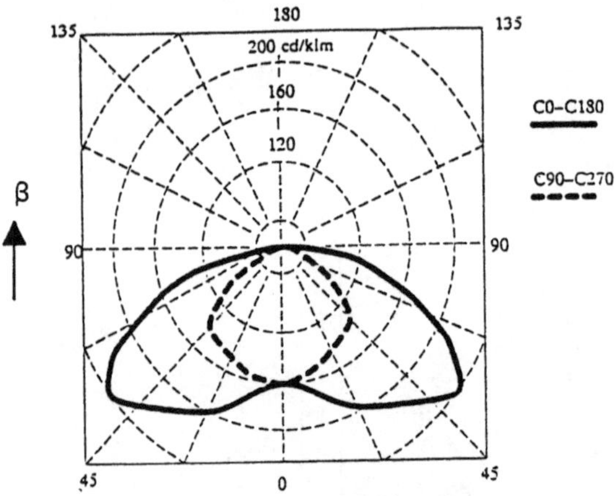

Fig. 2a: Luminous intensity distribution

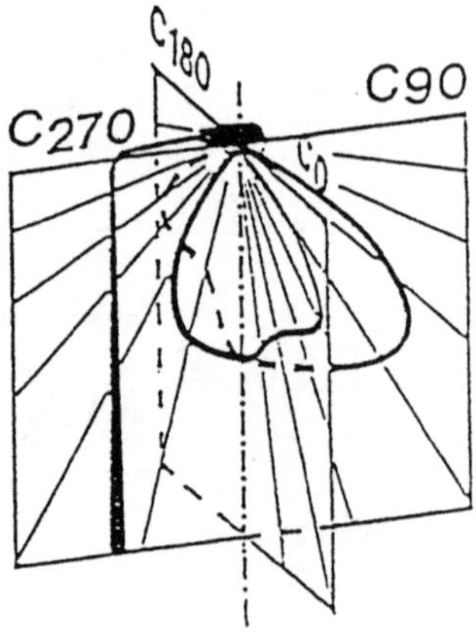

2b: C-planes

The complexity of the luminous intensity distribution depends not only on the geometry of the lamp but also on the lighting material (halogen, high pressure–sodium vapor, etc.) and the material of the refractors and reflectors of the lamp. These curves are obtained by measurements taken from special hardware equipment. The interested reader may find it more detailed in (Hentschel 1987).

The color of physical light sources is specified with their spectral energy distribution curve $\overline{E}(\lambda)$.One example is shown in figure 3 (Richter 1981).

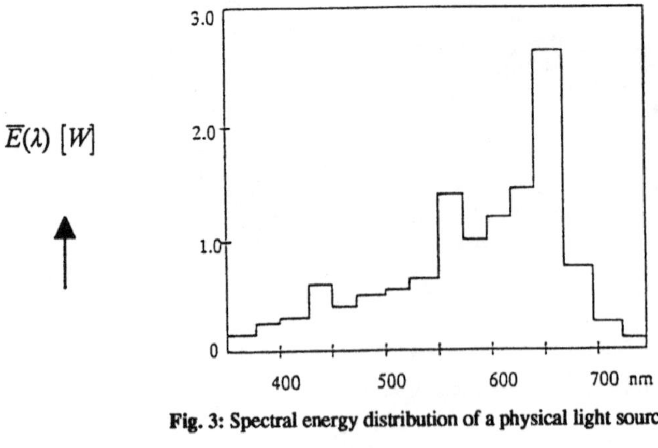

Fig. 3: Spectral energy distribution of a physical light source

4 Simulation Methods

4.1 Direct Illumination

In order to calculate the spectral energy arriving at a patch from a point light source the luminous intensity distribution and the spectral distribution of the source must be combined.

P is the representative point of the patch that defines the plane C_{θ_P} (at θ_P degrees from the plane C_0, see section 3) with the illumination axis. This plane is between the planes C_{θ_i} and $C_{\theta_{i+1}}$ where the curves of luminous intensity are defined (figure 4).

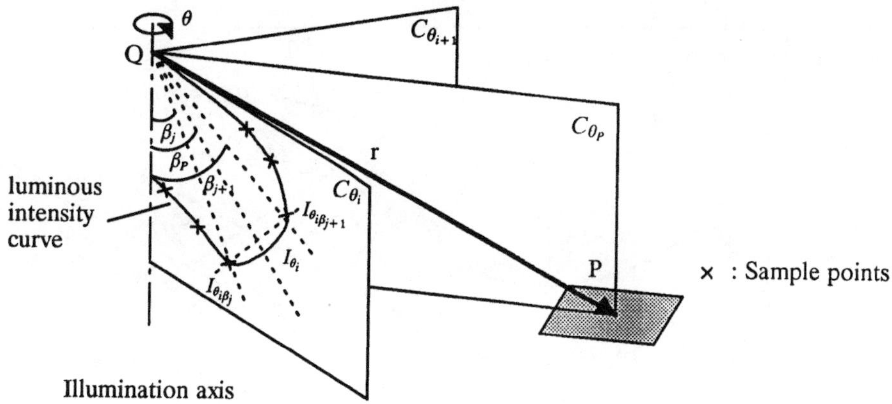

Fig. 4: Calculation of the luminous intensity of the point P

P also defines a lighting ray with the light point Q. This line is defined β_P degrees from the illumination axis and is located between the line β_j and β_{j+1} where the samples of the luminous distribution at each plane C_{θ_i} are given.

First, a linear interpolation is accomplished in the planes C_{θ_i} and $C_{\theta_{i+1}}$ between the given values at β_j and β_{j+1} degrees to obtain the I_{θ_i} and $I_{\theta_{i+1}}$ values:

(1)
$$I_{\theta_i} = I_{\theta_{i\beta_j}} + (I_{\theta_{i\beta_{j+1}}} - I_{\theta_{i\beta_j}}) \cdot w$$

where

$$w = (\beta_P - \beta_j)/(\beta_{j+1} - \beta_j)$$

Second, a linear interpolation is accomplished between these two values to obtain the I (β_P, θ_P) assigned to the point P in the plane C_{θ_P} :

(2)
$$I(\beta_P, \theta_P) = I_{\theta_i} + (I_{\theta_{i+1}} - I_{\theta_i}) \cdot t$$

where

$$t = (\theta_P - \theta_i)/(\theta_{i+1} - \theta_i)$$

The result of this interpolation process is the luminous intensity which is a representative for the patch.

The luminous energy of a light source E_L is defined by the integral over the spectral energy distribution $\overline{E}(\lambda)$ with the spectral luminous efficiency curve $V(\lambda)$:

(3)
$$E_L = \int_{\lambda} \overline{E}(\lambda) \cdot V(\lambda) d\lambda$$

But, the luminous intensity distribution I (β, θ) also describes the luminous energy of the light source E_L by an integral over the solid angle Ω of the hemisphere (figure 5).

(4)
$$E_L = \int_{\Omega(\beta,\theta)} I(\beta,\theta) d\Omega$$

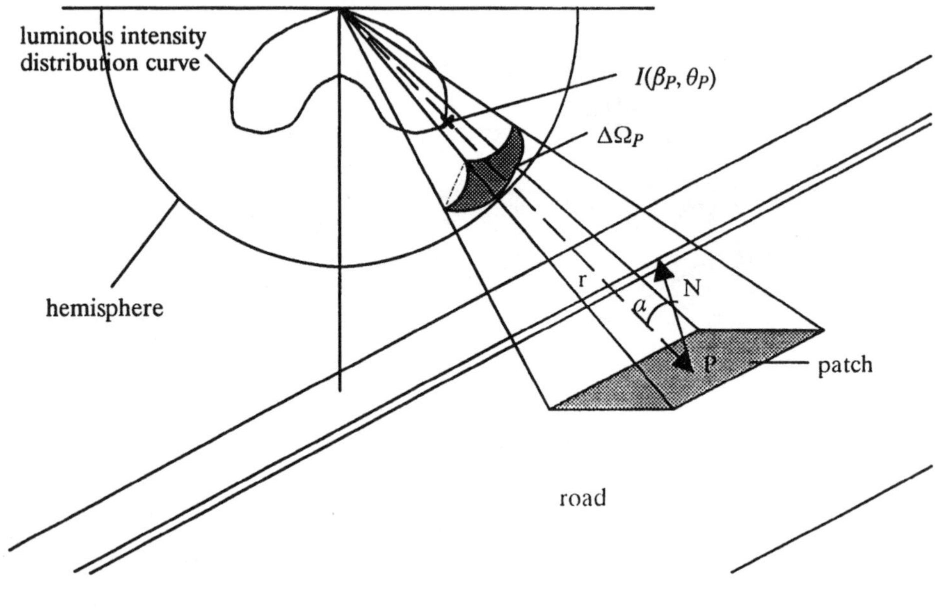

Fig. 5: Energy arriving at the patch

Thus, the amount of luminance energy E_P from the light source, arriving at the patch is obtained by projecting the patch onto the hemisphere and multiplying the solid angle $\Delta\Omega_P$ covered with the luminous intensity I (β_P, θ_P):

(5)
$$E_P = \Delta\Omega_P \cdot I(\beta_P, \theta_P)$$

E_P expresses the illuminance at the patch. Considering spectral sampling methods the part of energy $E_P(\lambda_K)$ for each sample λ_K must be calculated with the following equation:

(6)
$$E_P(\lambda_K) = \overline{E}(\lambda_K) \cdot \Delta\lambda_K \cdot \frac{E_P}{E_L}$$

where $\Delta\lambda_K$ represents the distance between the samples. Thus, the values of $E_P(\lambda_K)$ describe spectral samples of the energy arriving at the patch.

4.4.1 Solid Angle Computation

It is well known that the solid angle from a surface patch $\Delta\Omega$ is the area of the projection of this surface onto the hemisphere centered in the light source point divided by the square of the radius of the hemisphere.

Most of the times this projection is not easy to calculate but can be approximated as the projected area in the direction of the lighting ray divided by the distance square.

$$\Delta\Omega = \frac{A_{projected}}{r^2} = \frac{A \cdot \cos a}{r^2} \tag{7}$$

where A is the area of the surface patch.

But, for engineering applications it is necessary to know how accurate this approximation is. It is easy to find the real solid angle for a disk surface with $\alpha = 0$ and to calculate the relative error of the approximation. R is the radius of the test disk surface and r the distance from the point source. Thus, we have found the following table:

%error	r/R
< 10%	2.7
< 5%	4.0
< 3%	5.0
< 1%	9.0

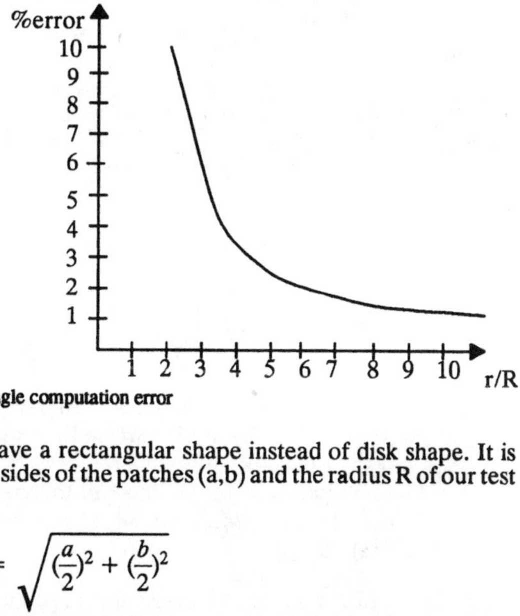

Fig. 6: Solid angle computation error

However, normally the patches have a rectangular shape instead of disk shape. It is possible to find a relation between the sides of the patches (a,b) and the radius R of our test disk:

$$R = \sqrt{\left(\frac{a}{2}\right)^2 + \left(\frac{b}{2}\right)^2} \tag{8}$$

In our application, the minimum distance between light sources and road is between 7 and 9 meters. So, the size of the patches, or better the size of the elements after the geometric subdivision process, can be 1 square meter to obtain an error of about 3% which is less than the error of the intensity distribution data of the light sources.

In all the cases with α different from 0 we also use the values of figure 6 to obtain the size of the elements patches. The value of the real error in these cases is always smaller than the one calculated with the table.

4.2 Extended Radiosity Method

As described below we use a combination of radiosity and ray–tracing techniques to render the scene. Before starting the radiosity method for each sample λ_K we distribute the energy from the point light sources to each patch in the scene combining the luminous intensity distribution curve of the light with its spectral distribution (see section 4.1).

The light arriving from M light sources at a patch i at one sample λ_K is given by if the patch is visible from the source.

$$\sum_{l=1}^{M} E_{il}(\lambda_K)$$

Now, we introduce physical surface light sources, that means the emitting patches of the scenery are also evaluated with measured luminous intensity curves. The advantage of this method is pointed out in Meyer (Meyer et al. 1986) who proved that the calculations made by assuming directional light sources are significantly more accurate than those made by assuming a perfectly diffuse light sources. For these reasons the extension of the radiosity method consists in the computation of the light reaching the patches from the surface light sources in the direction of the lighting ray.

For emitting surfaces the amount of energy arriving from reflection from the other patches of the scene or directly from other sources can be neglected in most applications. Thus, figure 7 expresses the energy model for the non emitting surfaces and the mathematical expression is shown in equation (9).

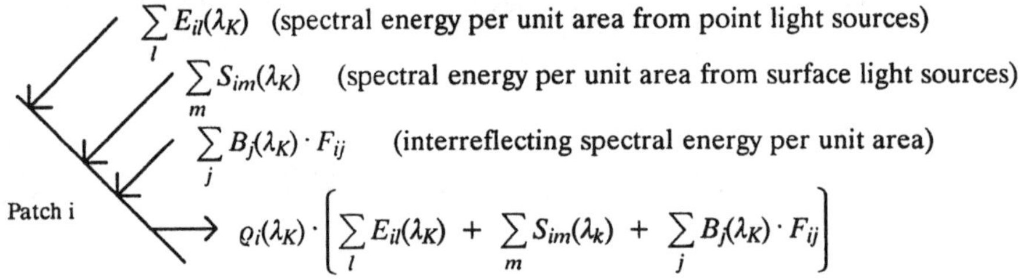

Fig. 7: Energy model for radiosity

$$B_i(\lambda_K) = \varrho_i(\lambda_K) \cdot \left[\sum_l E_{il}(\lambda_K) + \sum_m S_{im}(\lambda_k) + \sum_j B_j(\lambda_K) \cdot F_{ij} \right] \tag{9}$$

with $B_i(\lambda_K)$ Radiosity of the patch i (energy per unit area)
$S_{im}(\lambda_K)$ Energy from the surface light source m arriving at i
$\varrho_i(\lambda_K)$ Reflectivity of patch i
$B_j(\lambda_K)$ Radiosity of the non emitting patch j
F_{ij} Form–factor from patch i to patch j
$E_{il}(\lambda_K)$ Energy from the point light source l arriving at i

An equation for it exists for every surface patch in the scene which can be supposed as the result of a subdivision process. The system of simultaneous equations is solved to determine the radiosity of each patch. The equation is solved at different samples, as E_{il} (λ_K), S_{im} (λ_K) and ρ_i (λ_K) are given as spectral distribution curves. The computation of S_{im} (λ_K) is explained in the section 4.2.1.

The form–factor F_{ij} gives the fraction of energy leaving the patch i that arrives at the patch j. All form–factors depend only on the geometric relationship between patches, but not on viewer positions or attributes of the surface and they are constant over all spectral samples. To calculate them the Hemi–cube technique (Cohen and Greenberg 1985) is often used.

So far we explained the evaluation of the energy in one patch which arrives from point and surface lights and from the other patches. In order to simulate the wetness of the streets (increment of specular coefficient) the amount of reflected light arriving at each diffuse patch from the specular patches must be calculated (Wallace et al. 1986). In our application the street furniture (houses, traffic signs, etc.) interchanges diffuse energy with the road and when we simulate a wet street it is necessary to add the energy from the light sources that are specular reflected in the street and transmitted, for example, to the houses as diffuse. Both mechanisms of transport of energy are solved in the radiosity solution (figure 8) .

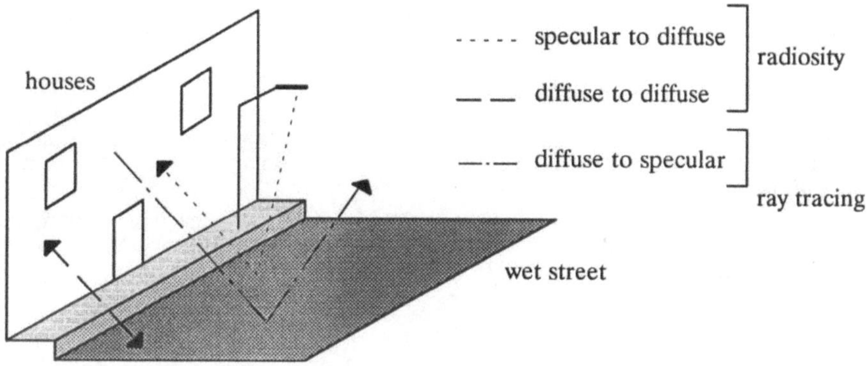

Fig. 8: Mechanisms of transport of energy

4.2.1 Surface Light Sources in Radiosity

For a perfect Lambertian emitter the luminous intensity I_d is related with the luminous energy of the light per unit area E_L^{rel} by the factor π. Thus, The luminous energy S_{im} arriving at the patch i from the surface source m can be expressed as:

$$(10) \qquad S_{im} = E_L^{rel} \cdot F_{im} = \pi \cdot I_d \cdot F_{im}$$

where F_{im} is the form factor of the patch i to the surface light source m and I_d is the diffuse luminous intensity of the source which is constant in all directions.

For a non perfect diffuse sources, we introduce a light source directional factor called α. This factor relates the real luminous intensity value in one direction (known by measurements, section 4.1) with the diffuse luminous intensity I_d :

$$(11) \qquad a = \frac{I(\beta,\theta)}{I_d} = \frac{I(\beta,\theta) \cdot \pi}{E_L^{rel}}$$

Thus, we can write again the equation 10 for non perfect diffuse sources as follows:

$$(12) \qquad S_{im} = E_L^{rel} \cdot F_{im} \cdot a = \pi \cdot I(\beta,\theta) \cdot F_{im}$$

This is the same equation that we receive when the surface source m is considered a perfect diffuse emissor of π I(β,θ) lumens per square meter.

Considering spectral sampling calculations the part of luminous energy $S_{im}(\lambda_K)$ for each sample λ_K must be calculated by the following equation:

$$(13) \qquad S_{im}(\lambda_K) = \overline{E}(\lambda_K) \cdot \Delta\lambda_K \cdot \frac{S_{im}}{E_L^{rel}}$$

4.3 Ray Tracing

The following pass in the rendering pipeline is done by a ray tracer. The ray tracing can calculate the amount of specular reflected light from specular patches arriving at the view point. For instance, the light arriving from the houses is reflected in the wet street (figure 8).

In the images without any high specular surface the ray tracer is also applied to get a smooth image. An adaptive antialiasing algorithm introduced by Whitted (Whitted 1980) is used to trace the ray for each spectral sample λ_K from the eye to the scenery.

5 Imaging Transformations

The result of the spectral sample render process is a spectral energy distribution for each pixel P(λ). In order to present it on the screen a set of image transformations has to be performed

In this way, first, we obtain the CIE–primaries X_P, Y_P, Z_P of each pixel combining the color matching functions with its energy distribution. The luminance Y_P for example is calculated according to equation (14):

$$Y_p = \int_{380nm}^{780nm} y(\lambda)P(\lambda)d\lambda \qquad (14)$$

where y(λ) : CIE–color matching function

Since the dynamics of the visual perception enclose a luminance extent of about 130 dB – it is no possible for the eye to modulate this extent simultaneously – the current brightness sensation depends on the level of adaptation, which matters for a street illumination under exempt from glare condition at about 1 – 2 cd/m^2. A modern graphic screen offers a modulatable extent of about 1 – 100 cd/m^2 at a dynamics of about 20 dB (Groß 1991a). For the description of the relationship between the physiological measure brightness (B) and the physical measure luminance (Y) there are competing arrangements, like the psychophysical basic rule of Weber–Fechner. Figure 9 shows an example for a brightness – luminance curve.

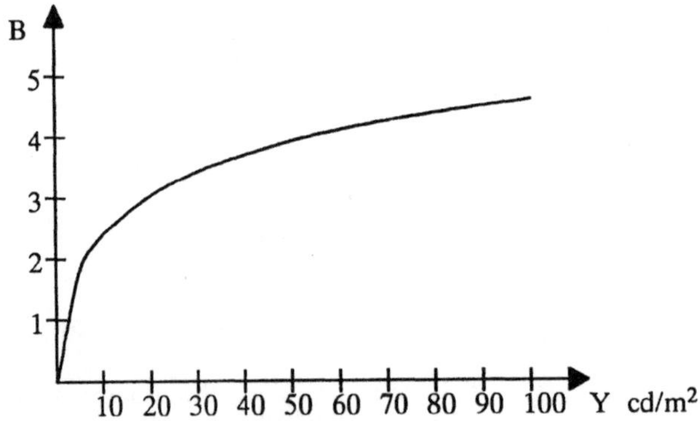

Fig. 9: Luminance-Brightness relation

Based on the Weber–Fechner rule, the following transformation according to equation (15) may be used to reduce dynamics:

$$B = \ln(1 + Y) \qquad (15)$$

However, this transformation often applied in the area of image processing is limited in its description of psychophysical aspects. Therefore, a lot of more complex descriptions were found in the past (Hentschel 1987), but none of them describes the B (Y)-function exactly.

After the calibration of the display the RGB triplet is found applying the transformation matrix [X_{RGB}] based on the chromaticity coordinates of the monitor phosphors and the monitor white point following equation (16):

$$(R_P, G_P, B_P) = [X_{RGB}]^{-1} (X_P, Y_P, Z_P) \quad (16)$$

Since not every CIE color may be presented on a display there may be negative R,G,B values. In this case, a next neighbor has to be found. Finally, a gamma correction of the calculated R,G,B values according to (Cowan 1983) provides a linear behavior of the display and may be calculated according to equation (17):

$$(R_P^{\gamma}, G_P^{\gamma}, B_P^{\gamma}) = (R_P^{1/\gamma_R}, G_P^{1/\gamma_G}, B_P^{1/\gamma_B}) \quad (17)$$

with γ_R, γ_G, γ_B : γ-values for the primitives of the tube.

6 Images

The upcoming images* show first results from our methods. Figure 10 shows one of the luminous intensity surfaces used to obtain the images. It belongs to one high pressure sodium vapor light source of 320 watts of power called "OURW-250". Figure 11 shows the lighting distribution on the street surface according to a bird sight 7 m above the road. It is possible to observe the characteristic intensity distribution on the street. Figure 12 shows the same view as Figure 11 but with the side street illuminated. Now, it is possible to perceive the traffic sign and the cylinder (a dummy which represents a pedestrian), placed on the right side walk. So, although the planned system follows the DIN rules at the regulations of the DIN the dark areas in the street may give dangerous traffic situations. The Figure 13 shows the street scenery from the view point defined in the DIN standard. It is located in the middle of the right lane, 1.50 m above the road surface..

7 Conclusion and Further Work

The application introduced has shown how Computer Graphics Techniques may help finding solutions towards serious problems of engineering disciplines. Respectively safety aspects of street lighting may be simulated and examined in advance when using physically based methods and display transformations which perceive psychophysical knowledge of human vision. Moreover, we use a new method to compute the energy from point light sources and real surface light sources in the radiosity step.

Our further work aims on simulating real street surfaces by bump mapping and glare effects by a postprocess as Nakamae introduced (Nakamae et al. 1990). Also the influence of different headlights and different weather conditions are of interest for this application

In the future we will also face improved virtual reality hardware e.g. high resolution head mounted displays which should help in solving the problems in the field of a correct presentation.

Acknowledgements

This work was supported by the Technische Werke der Stadt Stuttgart. The authors would like to thank to Dr. H. Freund and Mr. Roth for their kind cooperation and their deep interest in our research. Also a special reference to Mr. Zhang and Mr. Huth for the basic implementation of the software.

* See color plates

References

Cohen M.F., Greenberg D.P. (1985): The Hemi-Cube. A Radiosity Solution for Complex Environments. Computer Graphics, Vol. 19, No. 3, July 1985, 31 - 40 (Proc. SIGGRAPH 85)

Cohen M.F., Greenberg D.P., Immel D.S., Brock P.J. (1986): An Efficient Radiosity Approach for Realistic Image Synthesis. IEEE Computer Graphics & Applications, Vol. 6, No. 2, 26 - 35

Cowan W.B. (1983): An Inexpensive Scheme for Calibration of a Colour Monitor in Terms of CIE Standard Coordinates. Computer Graphics, Vol. 17, No. 3, July 1983, 315 - 321.

DIN 5044 Teil 1 (1981): Ortsfeste Verkehrsbeleuchtung; Allgemeine Gütemerkmale und Richtwerte, September 1981

DIN 5044 Teil 2 (1982): Ortsfeste Verkehrsbeleuchtung; Berechnung und Messung, August 1982

J.Foley, A. van Dam, S. Feiner, J. Hughes (1990): Computer Graphics. Principles and practice. Second Edition.

Goral C.M., Torrance K.E., Greenberg D.P., Battaille B. (1984): Modelling the Interaction of Light Between Diffuse Surfaces. Computer Graphics, Vol. 18, No. 3, July 1984, 213 - 222 .

Groß M. (1991a): Physiological aspecs of human vision and computer graphics. Eurographics 91 Tutorial. Wien. To be published.

Groß M., Berg A., Clave S. (1991b): Simulation von Straßenbeleuchtungsanlagen Einsatz fortschrittlicher Methoden der Graphiscen Datenverarbeitung. Licht 5-6, 1991. To be plublised.

Hentschel H.J. (1987): Licht und Beleuchtung, 3.Auflage, Hüthig Verlag Heidelberg 1987

Immel D.S., Cohen M.F., Greenberg D.P. (1986): A Radiosity Method for Non-Diffuse Environments, Computer Graphics, Vol. 20, No. 4, August 1986, 133 -142 (Proc. SIGGRAPH 86)

Kay D.S., Greenberg D.P. (1979): Transparency for Computer Synthesized Images. Computer Graphics, Vol. 13, No. 3, August 1979, 158 - 164 (SIGGRAPH 79)

Nakamae E., Kaneda K., Okamoto T., Nishita T. (1990): A Lightning Model Aiming at Drive Simulators. Computer Graphics, Vol. 24, No. 4, August 1990, 395 - 404 (Proc. SIGGRAPH 90)

Nishita T., Okamura I., Nakamae E. (1985): Shading Models for Point and Linear Sources. ACM Transactions on Graphics, Vol. 4, No. 2, April 1985, 124 - 146

Richter M. (1981): Einführung in die Farbmetrik, 2. Auflage, de Gruyter, Berlin 1981

Sillion F., Puech C. (1989): A General Two-Pass Method Integrating Specular and Diffuse Reflection. Computer Graphics, Vol. 23, No. 3, July 1989, 335 - 344 (Proc. SIGGRAPH 89)

Verbeck C.P., Greenberg D.P. (1984): A Comprehensive Light-Source Description for Computer Graphics. IEEE Computer Graphics & Applications, Vol. 4, No. 7, July 1984, 66 - 75

Wallace J.R., John, Cohen, Greenberg (1986): A two pass solution to the rendering Equation: A synthesis of Ray Tracing and Radiosity Methods.Computer Graphics, Vol. 21, No. 4, July 1986, (Proc. SIGGRAPH 87)

Whitted T. (1980): An Improved Illumination Model for Shaded Display. Communications of the ACM, Vol.23, No. 6, June 1980, 343 - 349

Color Plates

The State of the Art in Physically-based Rendering and its Impact on Future Applications

François Sillion

pp. 1–10

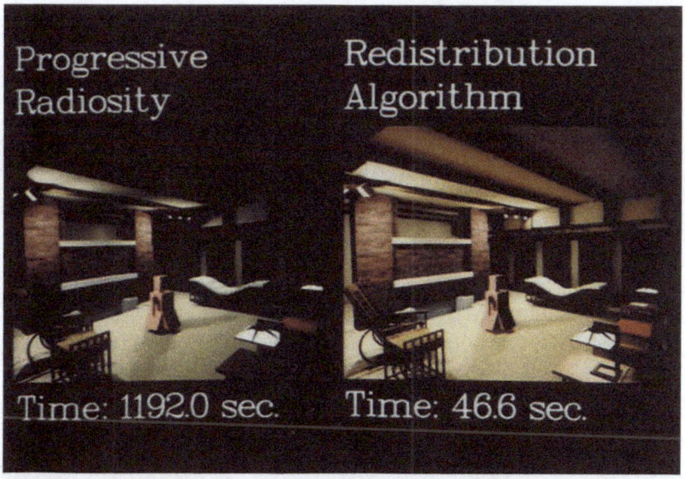

Fig. 8
Redistribution of radiosity after an object has been added to the scene: by redistributing only the most important contributions (right), a usable image is obtained much faster than by restarting the computation from the beginning (left). Both techniques ultimately converge to the same solution. Image courtesy of David W.George, Cornell University

Fig. 9
Simulation of global illumination with arbitrary reflectance functions. Note the highlights on the cupboards

An Adaptive Sampling Technique for Multidimensional Integration by Ray-Tracing

Christophe Schlick

pp. 21–29

Fig. 7
CHECK, BALLS, GLOSS and FOCUS pictures

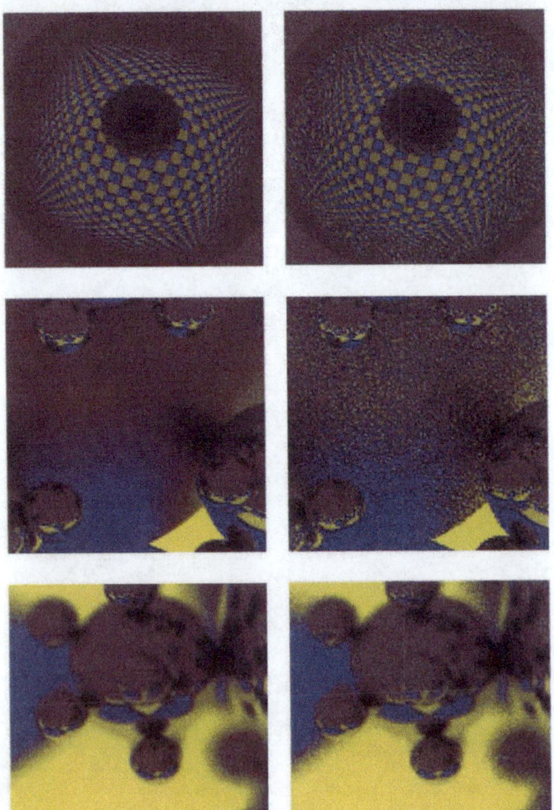

Fig. 8
Improvement in quality for the same rendering time

The Constant Radiosity Assumption Syndrome

Filippo Tampieri
Dani Lischinski

pp. 83–92

Fig. 4
Radiosity distribution of the receiver due to the shooting source as computed by the ray-traced form-factor method (left) and the new method (right).

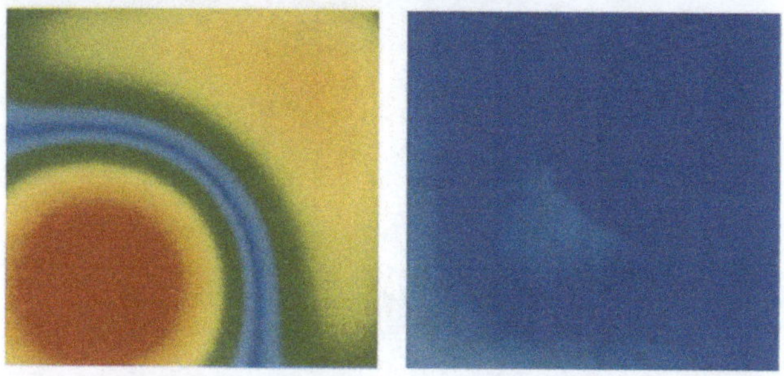

Fig. 5
Pseudo-color image of the absolute error in the radiosity distribution across the receiver as computed by the ray-traced form-factor method (left) and the new method (right). The error increases exponentially as the color changes from blue (0), through cyan (0.03), green (0.09), and yellow (0.21) to red (>0.45)

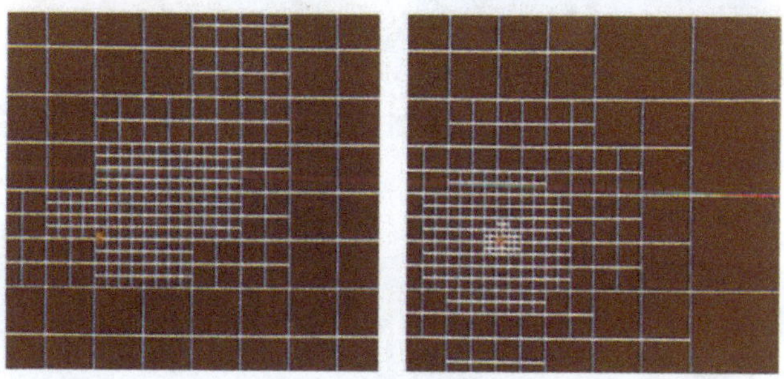

Fig. 6
Subdivision of the for a receiving point placed at (0.74, 0.62) as computed by the ray-traced form-factor method (left) and the new method (right). The location of the target point is marked by a red dot

Physics-Based Lighting Models: Implementation Issues

Pierre Tellier
Kadi Bouatouch

pp. 112–121

Fig. 6
Result of the first pass; perfectly diffuse materials

Fig. 7
Result of the first pass; diffuse and specular materials

Fig. 8
Result of the two passes

On the Division of Environments by Virtual Walls for Radiosity Computation

Bruno Arnaldi
Xavier Pueyo
Josep Vilaplana
pp. 198–205

Fig. 5
Complete environment solution

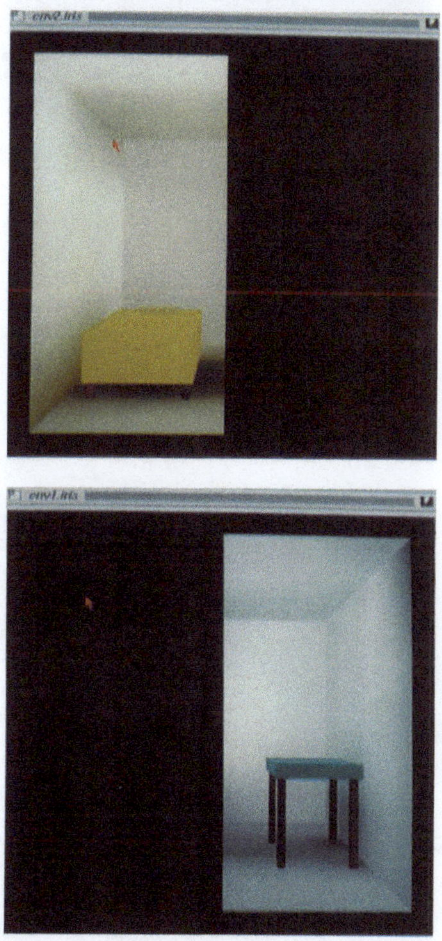

Fig. 6
Separate local environments with energy transfer through the virtual wall

Color Spaces and Human Color Perception

Ingeborg Tastl
Werner Purgathofer
pp. 219–226

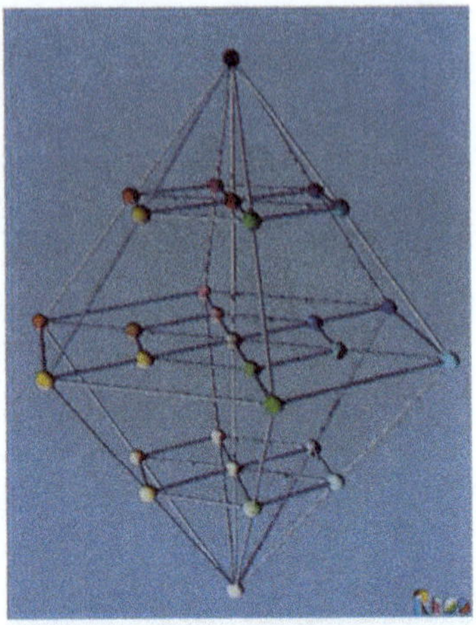

Fig. 1
Initial arrangement of the standard-colors in the HLS-color space

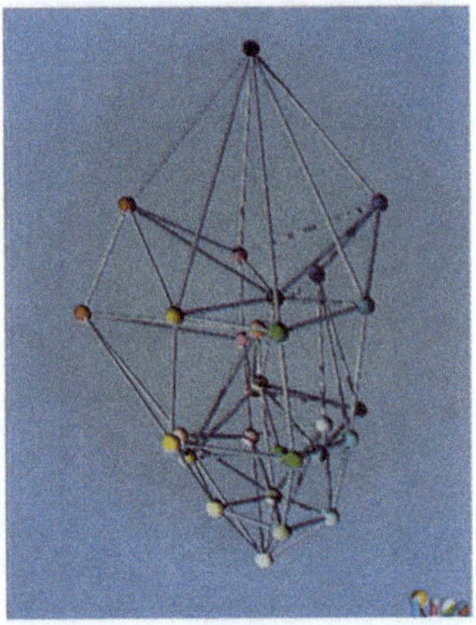

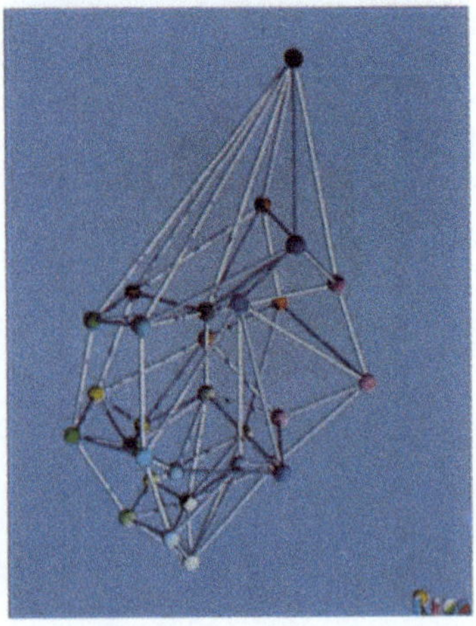

Fig. 4
Final arrangement of the standard-colors in the HLS-color space

Visulux:
A Radiosity Based Lighting Design Tool

Andrew J. Anderson
Michael Grant
pp. 227–239

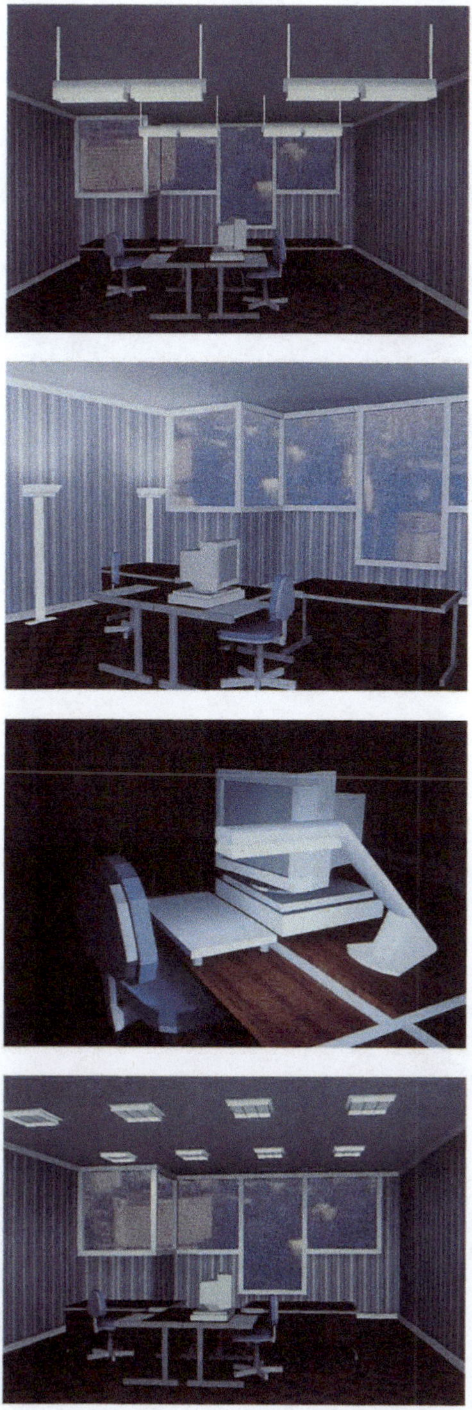

Figs. 6–9
Four design scenarios for a typical office layout. All created from same geometry description with different luminaire specifications

Synthesis of Multispectral Images of Natural Landscape

Pietro A. Brivio
Paolo Furini
Massimo Righetti
Daniele Marini

pp. 240–250

Fig. 6
Satellite Image – RGB: TM4, TM3, TM2 – and the Correspondent Synthetic Image – RGB: Synth4, Synth3, Synth2 – Simulated for Same Spectral Bands (upper). Satellite Image – RGB: TM5, TM4, TM2 – and Synthetic Image – RGB: Synth5, Synth4, Synth2 (down)

Fig. 7
3D visions of the Synthetic Image, adding sky, clouds and textures

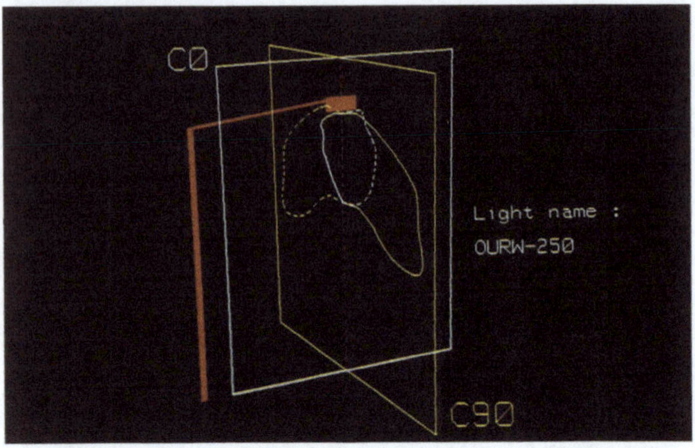

Fig. 10
Luminous intensity distribution curves from the "OURW-250" light source

Fig. 11
Bird-view, dry street,
side-street
not illuminated

Fig. 12
Bird-view, dry street,
side-street
illuminated

Fig. 13
DIN-position, dry street,
side-street
illuminated

List of Contributors

Andrew J. Anderson
ABACUS, Department of Architecture and Building Science, University of Strathclyde, 131 Rottenrow, Glasgow G4 0NG, Scotland, email: abacus@strath.ac.uk

Bruno Arnaldi
IRISA/INRIA, Campus de Beaulieu, 35042 Rennes Cedex, France, email: arnaldi@irisa.fr

James Arvo
Program of Computer Graphics, Cornell University, Ithaca, NY 14853, USA

Cevdet Aykanat
Bilkent University, Computer Engr. and Inf. Science Dept., 06533 Bilkent, Ankara, Turkey

Kadi Bouatouch
IRISA/INRIA, Campus de Beaulieu, 35042 Rennes Cedex, France

Pietro A. Brivio
Remote Sensing Department, Institute of Geophysics of the Lithosphere - CNR, 56, Via Ampère, 20131 Milan, Italy, email: pixel@imisiam.bitnet

Alan G. Chalmers
Department of Computer Science, University of Bristol, Bristol BS8 1TR, United Kingdom, email: alan@compsci.bristol.ac.uk

Salvador Clavé
Darmstadt Computer Graphic Center, Wilhelminenstrasse 7, D-6100 Darmstadt, Germany, email: zdiscm01@ebcesca1.bitnet

René Caubet
IRIT, Université Paul Sabatier, 118, Route de Narbonne, 31062 Toulouse Cedex, France, email: paulin@irit.fr

George Drettakis
Department of Computer Science, University of Toronto, 10, Kings College Road, Toronto, Ontario, Canada M5S 1A4, email: dret@dgp.toronto.edu

Martin Feda
Technische Universität Wien, Institut für Computergraphik, Karlsplatz 13/186, 1040 Wien, Austria, email: feda@eigvs4.una.ac.at

Eugene Fiume
Department of Computer Science, University of Toronto, 10, Kings College Road, Toronto, Ontario, Canada M5S 1A4, email: elf@dgp.toronto.edu

Paolo Furini
Remote Sensing Department, Institute of Geophysics of the Lithosphere - CNR 56, Via Ampère, 20131 Milan, Italy

Neil Gatenby
Computer Graphics Unit, University of Manchester, Oxford Road, Manchester M13 9PL, United Kingdom, email: gatenby@v2.cgu.mcc.ac.uk

Michael Grant
ABACUS, Department of Architecture and Building Science, University of Strathclyde, Glasgow G4 0NG, Scotland, email: abacus@strath.ac.uk

Markus Gross
Darmstadt Computer Graphic Center, Wilhelminenstrasse 7, D-6100 Darmstadt, Germany, email: gross@zgdvda.uucp

Pascal Guitton
Université Bordeaux 1 LaBRI, 351 Cours dela Liberation, 33405 Talence, France, email: guitton@geocub.greco-prog.fr

Eric A. Haines
3D/Eye Inc., 2359 North Triphammer Road, Ithaca, NY 14850, USA, email: erich@eye.com

Terry Hewitt
Computer Graphics Unit, University of Manchester, Oxford Road, Manchester M13 9PL, United Kingdom, email: hewitt@cgu.mcc.ac.uk

Veysi İşler
Bilkent University, Computer Engr. and Inf. Science Dept., 06533 Bilkent, Ankara, Turkey, email: isler@trbilun.bitnet

Frederik W. Jansen
Fac. Technical Math. and Informatics, Delft University of Technology, Julianalaan 132, 2628 BL Delft, The Netherlands, email: fwj@twi.tudelft.nl

Jean-Pierre Jessel
IRIT, 118, Route de Narbonne, 31062 Toulouse, France, email: jessel@irit.fr

Bertram Kernke
Technische Hochschule Darmstadt, GRIS, Wilhelminenstrasse 7, D-6100 Darmstadt, Germany

David Kirk
California Institute of Technology, Computer Graphics 350-74, Pasadena, CA 91125, USA, email: dk@egg.gg.caltech.edu

Arjan J.F. Kok
Fac. Technical Math. and Informatics, Delft University of Technology, Julianalaan 132, 2628 BL Delft, The Netherlands, email: arjan@duticg.twi.tudelft.nl

Brigitta Lange
Fraunhofer Institut for Computer Graphics, Wilhelminenstrasse 7, D-6100 Darmstadt, Germany, email: lange@igd.fhg.de

Robert van Liere
Department of Interactive Systems, Center for Mathematics and Computer Science (CWI), Kruislaan 413, 1098 SJ Amsterdam, The Netherlands, email: robertl@cwi.nl

Dani Lischinski
Program of Computer Graphics, 580 Eng. & Theory Center Bld., Cornell University, Ithaca, NY 14853, USA, email: danix@graphics.cornell.edu

Daniele Marini
Laboratory of Eidomatics, Department of Information Sciences, University of Milan, 39, Via Comelico, 20135 Milan, Italy

Bülent Özgüç
Bilkent University, Computer Engr. and Inf. Science Dept., 06533 Bilkent, Ankara, Turkey

Derek J. Paddon
Department of Computer Science, University of Bristol, Bristol BS8 1TR, United Kingdom

Mathias Paulin
IRIT, Université Paul Sabatier, 118, Route de Narbonne, 31062 Toulouse Cedex, France, email: paulin@irit.fr

Claude Puech
Laboratoire d'Informatique, Ecole Normale Superieure, 45, rue d'Ulm, F-75230 Paris Cedex 05, France, email: puech@dmi.ens.fr

Xavier Pueyo
Dept. LiSI, Universitat Politècnica de Catalunya, Avda. Diagonal, 647 Planta 8, 08028 Barcelona, Spain, email: xavier@lsi.upc.es

Werner Purgathofer
Technische Universität Wien, Institut für Computergraphik, Karlsplatz 13/186, 1040 Wien, Austria, email: purgathofer@eigvs4.una.ac.at

Massimo Righetti
Remote Sensing Department, Institute of Geophysics of the Lithosphere - CNR, 56, Via Ampère, 20131 Milan, Italy

Jean Roman
ENSERB LaBRI, 351, Cours de la Libération, 33405 Talence, France, email: roman@geocub.greco-prog.fr

Georgios Sakas
Technische Hochschule Darmstadt, GRIS, Wilhelminenstrasse 7, D-6100 Darmstadt, Germany, email: gsakas@igd.fhg.de

Christophe Schlick
Université de Bordeaux I - LaBRI, 351, Cours de la Libération, 33405 Talence, France, email: schlick@geocub.greco-prog.fr

Peter Shirley
Dept. of Computer Science, Indiana University, Lindley Hall, Bloomington IN 47401, USA, email: shirley@cs.indiana.edu

François Sillion
Laboratoire d'Informatique, Ecole Normale Superieure, 45, rue d'Ulm, F-75230 Paris Cedex 05, France, email: sillion@dmi.ens.fr

Filippo Tampieri
Program of Computer Graphics, 580 Eng. & Theory Center Bld., Cornell University, Ithaca, NY 14853, USA, email: fxt@graphics.cornell.edu

Ingeborg Tastl
Technische Universität Wien, Institut für Computergraphik, Karlsplatz 13/186, 1040 Wien, Austria, email: tastl@eigvs4.una.ac.at

Pierre Tellier
IRISA/INRIA, Campus de Beaulieu, 35042 Rennes Cedex, France, email: tellier@irisa.fr

Christophe Vedel
Laboratoire d'Informatique, Ecole Normale Superieure, 45, Rue d'Ulm, 75230 Paris Cedex 05, France, email: vedel@dmi.ens.fr

Josep Vilaplana
Dept. LiSI, Universitat Politècnica de Catalunya, Avda. Diagonal, 647 Planta 8, 08028 Barcelona, Spain, email: vilaplana@lsi.upc.es

John R. Wallace
3D/Eye Inc., 2359 North Triphammer Road, Ithaca, NY 14850, USA, email: johnw@eye.com

Changyaw Wang
Dept. of Computer Sci., Indiana University, Lindley Hall, Bloomington IN 47401, USA

Gregory J. Ward
Lighting Systems Research, Lawrence Berkeley Laboratory, 1 Cyclotron Rd. 90-3088, Berkeley, CA 94720, USA, email: greg@hobbes.lbl.gov

Authors' Index

Focus on Computer Graphics

(Formerly EurographicSeminars)

Eurographics Tutorials '83. Edited by P. J. W. ten Hagen.
XI, 425 pages, 164 figs., 1984. Out of print

User Interface Management Systems. Edited by G. E. Pfaff.
XII, 224 pages, 65 figs., 1985. Out of print (see below, Duce et al. 1991)

Methodology of Window Management. Edited by F. R. A. Hopgood, D. A. Duce, E. V. C. Fielding, K. Robinson, A. S. Williams. XV, 250 pages, 41 figs., 1985. Out of print

Data Structures for Raster Graphics. Edited by L. R. A. Kessener, F. J. Peters, M. L. P. van Lierop. VII, 201 pages, 80 figs., 1986. Out of print

Advances in Computer Graphics I. Edited by G. Enderle, M. Grave, F. Lillehagen. XII, 512 pages, 168 figs., 1986

Advances in Computer Graphics II. Edited by F. R. A. Hopgood, R. J. Hubbold, D. A. Duce. X, 186 pages, 96 figs., 1986

Advances in Computer Graphics Hardware I. Edited by W. Straßer.
X, 147 pages, 76 figs., 1987

GKS Theory and Practice. Edited by P. R. Bono, I. Herman.
X, 316 pages, 92 figs., 1987. Out of print

Intelligent CAD Systems I. Theoretical and Methodological Aspects.
Edited by P. J. W. ten Hagen, T. Tomiyama. XIV, 360 pages, 119 figs., 1987

Advances in Computer Graphics III. Edited by M. M. de Ruiter.
IX, 323 pages, 247 figs., 1988

Advances in Computer Graphics Hardware II. Edited by A. A. M. Kuijk, W. Straßer. VIII, 258 pages, 99 figs., 1988

CGM in the Real World. Edited by A. M. Mumford, M. W. Skall.
VIII, 288 pages, 23 figs., 1988. Out of print

Intelligent CAD Systems II. Implementational Issues. Edited by V. Akman, P. J. W. ten Hagen, P. J. Veerkamp. X, 324 pages, 114 figs., 1989

Advances in Computer Graphics IV. Edited by W. T. Hewitt, M. Grave, M. Roch. XVI, 248 pages, 138 figs., 1991

Advances in Computer Graphics V. Edited by W. Purgathofer, J. Schönhut.
VIII, 223 pages, 101 figs., 1989

User Interface Management and Design. Edited by D. A. Duce, M. R. Gomes, F. R. A. Hopgood, J. R. Lee. VIII, 324 pages, 117 figs., 1991

Advances in Computer Graphics Hardware III. Edited by A. A. M. Kuijk. VIII, 214 pages, 88 figs., 1991

Advances in Object-Oriented Graphics I. Edited by E. H. Blake, P. Wisskirchen. X, 218 pages, 74 figs., 1991

Advances in Computer Graphics Hardware IV. Edited by R. L. Grimsdale, W. Straßer. VIII, 276 pages, 124 figs., 1991

Advances in Computer Graphics VI. Images: Synthesis, Analysis, and Interaction. Edited by G. Garcia, I. Herman. IX, 449 pages, 186 figs., 1991

Intelligent CAD Systems III. Practical Experience and Evaluation. Edited by P. J. W. ten Hagen, P. J. Veerkamp. X, 270 pages, 116 figs., 1991

Graphics and Communications. Edited by D. B. Arnold, R. A. Day, D. A. Duce, C. Fuhrhop, J. R. Gallop, R. Maybury, D. C. Sutcliffe. VIII, 274 pages, 84 figs., 1991

Photorealism in Computer Graphics. Edited by K. Bouatouch, C. Bouville. XVI, 230 pages, 118 figs., 1992

Advances in Computer Graphics Hardware V. Rendering, Ray Tracing and Visualization Systems. Edited by R. L. Grimsdale, A. Kaufman. VIII, 174 pages, 97 figs., 1992

Multimedia. Systems, Interaction and Applications. Edited by L. Kjelldahl. VIII, 355 pages, 129 figs., 1992. Out of print

Advances in Scientific Visualization. Edited by F. H. Post, A. J. S. Hin. X, 212 pages, 141 figs., 1992

Computer Graphics and Mathematics. Edited by B. Falcidieno, I. Herman, C. Pienovi. VII, 318 pages, 159 figs., 1992

Rendering, Visualization and Rasterization Hardware. Edited by A. Kaufman. VIII, 196 pages, 100 figs., 1993

Visualization in Scientific Computing. Edited by M. Grave, Y. Le Lous, W. T. Hewitt. XI, 218 pages, 120 figs., 1994

Photorealistic Rendering in Computer Graphics. Edited by P. Brunet, F. W. Jansen. X, 292 pages, 175 figs., 1994

From Object Modelling to Advanced Visual Communication. Edited by S. Coquillart, W. Straßer, P. Stucki. VII, 305 pages, 128 figs., 1994

SPRINGER NATURE

GPSR Compliance

The European Union's (EU) General Product Safety Regulation (GPSR) is a set of rules that requires consumer products to be safe and our obligations to ensure this.

If you have any concerns about our products, you can contact us on ProductSafety@springernature.com

In case Publisher is established outside the EU, the EU authorized representative is:

Springer Nature Customer Service Center GmbH
Europaplatz 3
69115 Heidelberg, Germany

Papier | Fördert gute Waldnutzung
FSC® C083411

Kolibri 360 GmbH
Ferdinand-Jühlke-Straße 7
99095 Erfurt, Deutschland
produktsicherheit@kolibri360.de